To Mary Ochs
and the staff at
Mann Library.

For their extraordinary
assistance in producing
this book.

   Kraig Adler

Ithaca, 5 Sept. 1989.

JOHN EDWARDS HOLBROOK

Contributions to the History of Herpetology

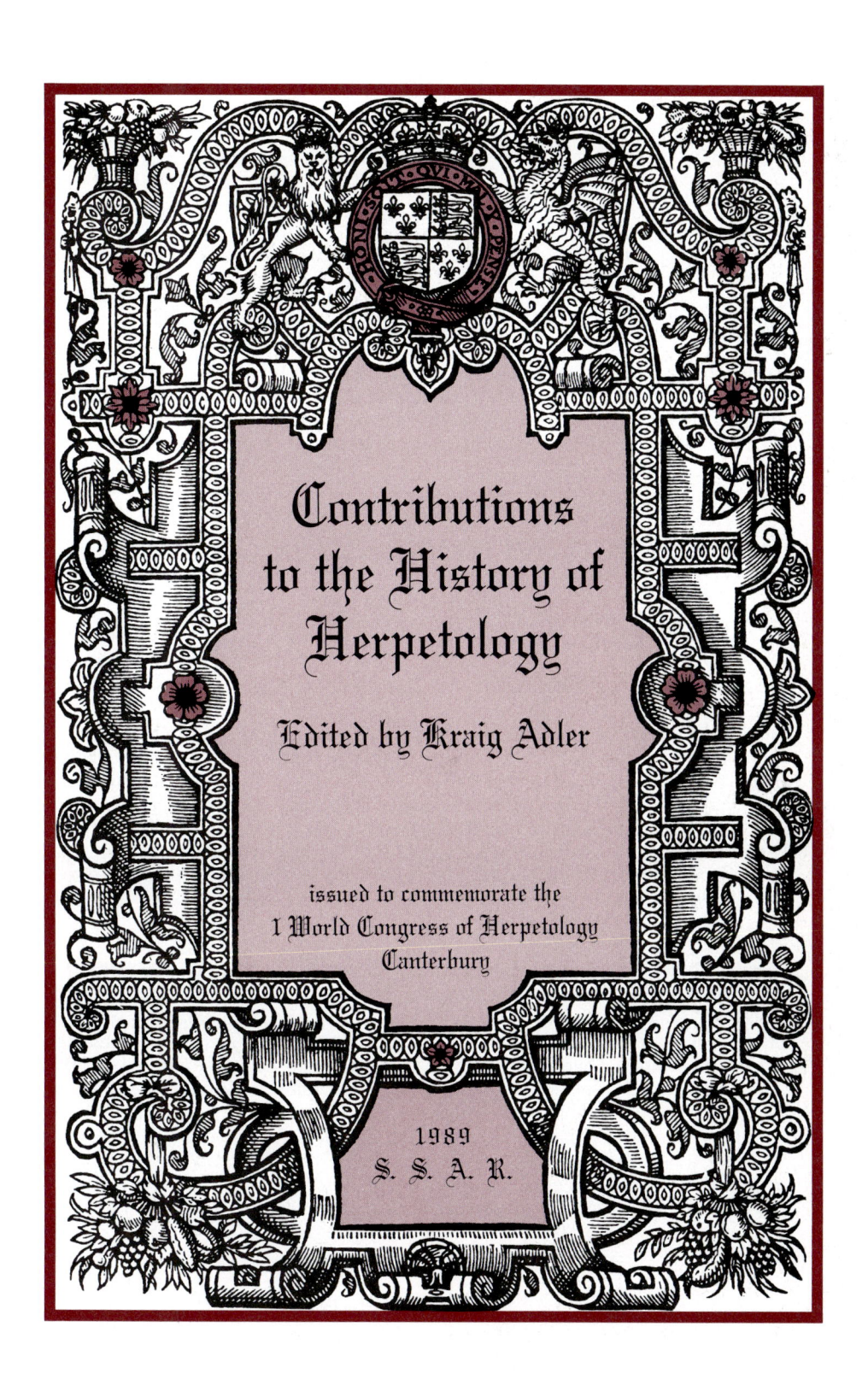

Edited by Kraig Adler

issued to commemorate the
1 World Congress of Herpetology
Canterbury

1989
S. S. A. R.

EDITOR'S NOTE

The frontispiece is a reproduction of the portrait of John Edwards Holbrook painted by Daniel Huntington in 1856 or 1857. The original is in oils on canvas, measuring 0.718 x 0.587 m (28 1/4 x 23 1/8 in.). The painting is part of the Andrew W. Mellon Collection at the National Gallery of Art, Washington, D.C., U.S.A. The portrait is reproduced with the kind permission of Ira Bartfield, Coordinator of Photography at the National Gallery.

The title page border is reproduced from "Title-Page Borders Used in England and Scotland, 1485-1640," by R. B. McKerrow and E. S. Ferguson, Printed for the Bibliographical Society by Oxford University Press, London, 1932. The compartmented border is a typical style for 16th Century title pages; this example is dated 1584.

This book serves to commemorate the I World Congress of Herpetology, held in Canterbury, United Kingdom, 11-19 September 1989. The Society for the Study of Amphibians and Reptiles is meeting jointly with the World Congress, which therefore also serves as the Society's 32nd Annual Meeting and its second outside the United States.

CONTRIBUTIONS TO HERPETOLOGY, NUMBER 5

KRAIG ADLER, *Editor*
Section of Neurobiology and Behavior
Cornell University
Ithaca, New York 14853, U.S.A.

JOHN S. APPLEGARTH
Associate Editor
3293 West 14th Avenue
Eugene, Oregon 97402, U.S.A.

Numbers in the *Contributions to Herpetology* series can be purchased from the Publications Secretary, Robert D. Aldridge, Department of Biology, St. Louis University, 3507 Laclede, St. Louis, Missouri 63103, U.S.A. A list of all Society publications, including those of The Ohio Herpetological Society and the *Catalogue of American Amphibians and Reptiles*, is printed at the end of this book; additional copies are available from Dr. Aldridge. *Contributions* are published irregularly and ordered by separate subscription, although Society members receive a special pre-publication discount. Authors who wish to have manuscripts considered for publication in the *Contributions* series should contact the Editor.

Members of the Society receive the quarterly technical journal (*Journal of Herpetology*) and a quarterly newsletter (*Herpetological Review*). Currently, dues are $25.00 for students, $35.00 for all others, world-wide; institutional subscriptions are $60.00. All memberships and subscriptions outside the U.S.A. require either a $5.00 surcharge to defray the *additional* surface mailing costs in excess of domestic rates or $25.00 for airmail delivery. Society members receive special discounts on *Herpetological Circulars*, *Facsimile Reprints*, and on books in the *Contributions* series. The *Catalogue* is available by separate subscription. Apply to the Society's Treasurer, Douglas H. Taylor, Department of Zoology, Miami University, Oxford, Ohio 45056, U.S.A. Payments from overseas should be made in U.S.A. funds, by International Money Order, or charged to MasterCard or VISA (include account number and expiration date of credit card).

© 1989 SOCIETY FOR THE STUDY OF AMPHIBIANS AND REPTILES

Library of Congress Catalog Number: 89-50341. ISBN: 0-916984-19-2.
Production specifications are given on the last page of this book.

CONTRIBUTIONS TO THE HISTORY OF HERPETOLOGY
edited by Kraig Adler

TABLE OF CONTENTS

HERPETOLOGISTS OF THE PAST, by Kraig Adler

Introduction 5

Biographies:
1500-1600: • Gessner 7
1600-1700: • Redi 8
1700-1760: • Seba 9 • Klein 9 • Roesel von Rosenhof 10 • Linnaeus 11
1760-1800: • Laurenti 12 • Schneider 13 • Lacepède 14 • Schoepff 14 • Merrem 15 • Russell 16 • Shaw 17
1800-1820: • Cuvier 17 • Latreille 19 • Daudin 20 • Bojanus 20
1820-1830: • Boie 21 • Wied-Neuwied 22 • Spix 23 • Wagler 23 • Fitzinger 24 • Rafinesque 25 • Say 26 • Harlan 27
1830-1840: • Gravenhorst 28 • Wiegmann 28 • Bonaparte 29 • Schlegel 30 • C. Duméril 31 • Bibron 32 • Holbrook 33 • Gray 34
1840-1850: • Bell 35 • A. Smith 36 • Tschudi 36 • W. Peters 37 • Blyth 38
1850-1860: • Agassiz 39 • Baird 40 • Girard 42 • A. Duméril 43 • Reinhardt 44 • Burmeister 45
1860-1870: • Günther 45 • Cope 46 • Strauch 49 • Jan 49 • Sordelli 50 • Krefft 51
1870-1880: • Bocage 51 • Jiménez de la Espada 52 • Gundlach 53 • Steindachner 54 • Anderson 54
1880-1890: • Boulenger 55 • Boettger 56 • F. Müller 57 • Bedriaga 58 • Vaillant 58 • Dugès 59 • Baur 60 • Garman 61
1890-1900: • Stejneger 62 • Philippi 63 • Berg 64 • Goeldi 64 • Méhelÿ 65 • Peracca 66 • Rollinat 66 • Lönnberg 67
1900-1910: • Andersson 68 • Nikolsky 68 • Siebenrock 69 • Werner 70 • Waite 71 • Wall 71 • Van Denburgh 72 • Ditmars 73 • Ruthven 74
1910-1920: • Wright 75 • Roux 76 • De Rooij 76 • L. Müller 77 • Annandale 78 • Ōshima 79 • Hewitt 80 • Camp 80 • Gaige 81 • Taylor 82 • Barbour 84
1920-1930: • Fejérváry 85 • M. Smith 86 • A. Lutz 87 • Miranda Ribeiro 87 • Amaral 88 • Nicéforo Maria 89 • Noble 90 • Schmidt 91 • Dunn 92 • Pope 94 • Bishop 94 • Blanchard 95 • Klauber 97 • Mertens 98 • Kopstein 99 • Kinghorn 100 • Glauert 101
1930-1940: • Haas 101 • Klingelhöffer 102 • Chernov 103 • Terentjev 103 • Scortecci 104 • Pitman 105 • Parker 106 • Angel 106 • Bourret 107 • Boring 107 • Okada 108 • Maki 109 • Logier 109 • Cochran 110 • Loveridge 111 • Gloyd 112 • Myers 113
1940-1950: • Carr 114 • Goin 115 • Cowles 116 • Kauffeld 117 • Vogel 118 • Martín del Campo 119 • B. Lutz 119 • Deraniyagala 120 • FitzSimons 121 • De Witte 122 • Sato 122 • Liu 123
1950-1960: • Bannikov 124 • Villiers 124 • McCann 125 • Mitchell 126 • Donoso-Barros 126 • Hoge 127 • Fuhn 127 • Romer 128 • J. Peters 129
1960-1970: • Blair 130 • Tinkle 131 • Başoğlu 132 • Medem 132 • Barrio 133

Literature Cited and Further Reading 134

Index to Biographies 136

INDEX OF AUTHORS IN TAXONOMIC HERPETOLOGY, by John S. Applegarth

Introduction 143

Index of Taxonomic Authors 144

ACADEMIC LINEAGES OF DOCTORAL DEGREES IN HERPETOLOGY, by Ronald Altig

Introduction 179

Lineages, by country: • Argentina 179 • Australia 179 • Austria 180 • Brazil 180 • Canada 180 • Denmark 180 • France 180 • Germany 180 • Hungary 181 • Italy 181 • Japan 181 • Mexico 181 • The Netherlands 181 • Poland 181

(Lineages, continued): • South Africa 182 • Spain 182 • Sweden 182 • Switzerland 182 • Turkey 182 • United Kingdom 182 • Union of Soviet Socialist Republics 182 • United States of America 183 • Zimbabwe 196

Index to Lineages 196

JOHN EDWARDS HOLBROOK (1794-1871), the Father of American Herpetology, a physician living in Charleston, South Carolina, was the first person to publish a comprehensive summary of the North American herpetofauna. To do so, he assembled an extensive collection of amphibians and reptiles for his own study and for use by his several artists in the production of his *magnum opus*, "North American Herpetology" (first edition, 1836-1840; second edition, 1842). Wherever possible, Holbrook preferred living specimens for his artists, in order that the animals could be rendered more accurately, and he went to great effort and expense to acquire them. Specimens were obtained from Holbrook's own collecting in all of the Atlantic coastal states and from his numerous correspondents throughout eastern North America. In this letter (transcribed at the right), Holbrook requests two kinds of turtles from his distant relative, Elias Nason, then residing in Augusta, Georgia. He also gives instructions on where to collect, how to ship them alive, and payment. For further details, see page 33. (Original letter in Adler collection.)

Charleston July 1 - 1837

Dear Nason

I received your letter some time since but have neglected until this moment to answer it because this is the first time since that I have had occasion to use you. As it is I want a certain flat species of terrapin with a yellow belly [probably Pseudemys scripta] *- found in the river where the bed is rocky - These are often exposed for sale in the market - but should none be there you must have some people to collect - half a dozzen or so and send them to me at once - They may be put in a small box with a few air holes & sent by the Rail Cars - giving a line by mail the same day - also I want a soft shell turtle* [Trionyx] *- he may be sent the same way taking care only to place a little moist Earth in the bottom - and giving directions not to let them be placed in the sun - any expense I will cheerfully pay - or you may present it to my friend Dr Coutsiren of the Bank at Hamberg* [Hamburg, South Carolina] *& he will pay it for me - yours*
Truly
answer this by return } J E Holbrook

HERPETOLOGISTS OF THE PAST
by Kraig Adler

THE OCCASION OF herpetology's first world congress is a proper time to evaluate the present state of the discipline and to consider its future. But it should also be a time, however brief, to reflect on herpetology's rich past and to recognize the men and women whose contributions have developed the field to its present state. Unlike several other biological disciplines, herpetology has never had its singular history written. To be sure, this book is not that history—which is well beyond the scope of this project—but this volume does represent an attempt to bring together some of the basic materials necessary for that untold story.

Rather than a history, the first part of this book focuses instead on 152 individuals, all deceased, who have advanced the study of amphibians and reptiles. One is immediately struck by the divergent careers of these persons. Some were gifted amateurs, often having no formal education at all, and others were among the most distinguished personages of their day. Several of them devoted their life's fortune to zoology, and to herpetology in particular; a few went bankrupt in doing so. Others accomplished their work despite severe physical handicap or prolonged illness, and some even lost their lives in carrying out their research. There is much here to admire and even to provide inspiration.

The main part of this book had its genesis in an audiovisual program developed by David M. Dennis and myself for the 25th Anniversary Meeting of the Society for the Study of Amphibians and Reptiles in 1982. The Society commissioned a brief sound-and-slide presentation on the history of herpetology—entitled "Herpetologists Then and Now"—which utilized six computer-controlled projectors. To our amazement, this show, which was scheduled to be shown but once, was in fact screened numerous times, by popular demand. With such interest, we decided to enlarge our production and, eventually, the program was expanded to a 40-minute show during which several thousand 35mm slides of people, institutions, and herpetological illustrations were displayed. This program subsequently has been shown dozens of times, both at national and regional herpetological society meetings in the United States over the last six years, and for the last time at Ann Arbor in 1988.

A common complaint at those showings was that no permanent record existed of the hundreds of portraits of herpetologists that moved past the viewer so quickly. This book was originally intended partly to satisfy that need but it has been expanded to include brief biographical sketches of some of the most prominent people. To this has been added two useful compilations of information that were started as independent projects by my colleagues, John S. Applegarth and Ronald Altig. I had known of their work and, because of the opportunity of the First World Congress of Herpetology this year, it seemed appropriate to pool our efforts into a book of greater reference value. I thank them for their willingness to do so and for their cooperation.

Content and Organization.—While Applegarth's and Altig's sections are intended to be comprehensive, mine has had to be more selective. My most difficult task was to choose the persons to feature in these biographies, representing only a small proportion of the many individuals worthy of this recognition. No one will agree with all of my choices and, indeed, I am painfully aware of the many omissions dictated by the severe space constraints.

The biographies are arranged in an approximately chronological order, so that the development of the discipline of herpetology can be visualized and also that each person can be viewed in a broader context. In addition, this arrangement permits the reader to see the influence of some key individuals not featured in these biographies (for example, the renowned early teachers, J. F. Blumenbach and Johannes Müller). Since some biographees had short careers of less than a decade whereas others were active for more than half a century, each person is positioned in sequence according to the period in which he or she first rose to prominence in herpetology. This organization is reflected in the Table of Contents and in the titles at the top of each page, in which each person has been categorized according to some time period. Admittedly this procedure has been somewhat arbitrary. Due to this arrangement, some persons at first may seem to be oddly placed. For example, Giorgio Jan, born in 1791, appears in the section for 1860-1870 because he only began his herpetological work at the age of 62. In another instance, the pupil (Johann J. von Tschudi) precedes his teacher (Louis Agassiz) because the former's contributions to herpetology occurred first. Because of this sequencing and because numerous other persons are mentioned in these biographies, I have included a comprehensive index (pages 136-141). Names abbreviated in the text are given in full in the index.

Format of Biographies.—Each person's name is given in the title of their biography in its usual form, that is, the style most frequently encountered in their own publications. Where alternative spellings of a name have been used, these are listed, but preference is given to the form used by the person, as indicated by the style of his or her signature. All names are spelled in their romanised form.

In reviewing a person's contributions, emphasis has been given to their herpetological work. Thus, non-herpetological aspects of a person's life, which in several instances were more prominent than their work with amphibians and reptiles, are given little discussion here. Since herpetology as a biological discipline is a 20th Century phenomenon, the earlier contributors were general naturalists or zoologists for whom herpetology was only one of several interests. Certainly they were not herpetologists in the modern sense.

References to obituaries or other biographical information are provided for each person, although for some no such articles have been published. In other instances, the cited references provide little information and I have had to depend upon primary sources. Thus, much information in this book is published for the first time. References abbreviated to author and date are given in full in the Literature Cited section (pages 134-135) or in that person's biography.

Dates.—Virtually all dates given in these biographies are reckoned according to the modern calendar now in general use throughout the world. This calendar—the Gregorian calendar, or sometimes termed the reformed Julian calendar—first came into use in 1582, but was not then adopted universally. By that time, the accumulated discrepancy between the old Julian and Gregorian calendars was ten days; thus, ten days of 1582 were omitted to bring the calendar into correspondence again. However, Great Britain and her colonies did not adopt the new calendar until 1752, and the Soviet Union and some neighboring countries did not do so until 1918, by which times the discrepancies were 11 and 12 days, respectively. Sweden, on the other hand, switched over gradually during the period 1696 to 1753. For events in these countries before these dates, therefore, it is common in the literature to see double dates (for example, the date "1/13 April 1839" in Russia, with the second being that according to the modern calendar).

In this book, I have given dates in their original form. Effectively, this means that most dates are in modern Gregorian format, except for the countries listed above that switched at various times after 1582. However, in the biographical information available to me, it was not always clear whether the old or modern form was used and in some cases I have not been able to determine the correct situation. Thus, readers must use such dates for these countries with caution if they occur before the calendar was changed to the Gregorian format.

Literature Cited.—I provide a brief list of literature, beginning on page 134, which is intended as an introduction to the historical literature of herpetology. A number of the references are ostensibly to other disciplines, but include persons who also published on amphibians and reptiles.

Acknowledgments.—The compilation of this sort of book requires the active cooperation of many professional colleagues and library personnel, who have generously acceded to my numerous requests. I am indebted to several colleagues who have critiqued all or part of my manuscript. Ronald Altig, John S. Applegarth, Ellin Beltz, Charles M. Bogert, Carl Gans, Marinus S. Hoogmoed, Konrad Klemmer, Alan E. Leviton, M. Graham Netting, Jay M. Savage, Hobart M. Smith, and Heinz Wermuth read the entire manuscript, whereas the following persons read specific sections: Natalia B. Ananjeva, Angus d'A. Bellairs, Donald G. Broadley, Clifford J. Choquette, Harold G. Cogger, Hajime Fukada, Alice G. C. Grandison, Toshijiro Kawamura, Peter K. Knoefel, Janis A. Roze, Andrew F. Stimson, and George R. Zug.

For assistance in locating biographical materials, I thank Michele L. Aldrich, Natalia B. Ananjeva, Charles P. Blanc, Wolfgang Böhme, Charles M. Bogert, William R. Branch, Donald G. Broadley, Leo D. Brongersma, Patricia R. Brown, Gustavo Casas Andreu, Ronald I. Crombie, Ilya S. Darevsky, Alain Dubois, Josef Eiselt, Wolfgang Engelmann, Oscar Flores Villela, Ramón Formas, Carl Gans, Arnold B. Grobman, Rainer Günther, W. D. Haacke, Marinus S. Hoogmoed, the Countess of Huntingdon, Benedetto Lanza, Raymond F. Laurent, C. J. McCoy, Göran Nilson, G. Pilleri, Jaime E. Péfaur, Olivier Rieppel, Zbyněk Roček, José P. Rosado, Petr Roth, Margaret S. Shaw, Anselm de Silva, Zdeněk V. Špinar, Christine Stocker, Franz Tiedemann, P. E. Vanzolini, Richard Wahlgren, David B. Wake, Yehudah L. Werner, and Er-mi Zhao.

Individual photographs are credited where appropriate, but I do wish to specially acknowledge the following persons for sending important collections of portraits: Natalia B. Ananjeva, Harold G. Cogger, Oliver G. Dely, Alain Dubois, the late Howard K. Gloyd (via Kathryn J. Gloyd and Roger Conant), Zoltán Korsós, Alan E. Leviton, Toshijiro Kawamura, the late Walter L. Necker (via Kaethe Necker), M. Graham Netting, Andrew F. Stimson, and Heinz Wermuth.

Library personnel at numerous institutions cooperated by sending relevant materials, and I cite the following for their aid: Mary A. Ochs and Caroline T. Spicer (Cornell University Libraries), Roxane Coombs (Museum of Comparative Zoology, Harvard University), V. Van de Ponseele (Museum National d'Histoire Naturelle), Marsha Gross (Academy of Natural Sciences), K. Moore (Royal Society), William Cox (Smithsonian Institution Archives), Ann Datta (British Museum [Nat. Hist.]), Anita L. Karg (Hunt Institute for Botanical Documentation, Carnegie Mellon University), and Christine Pankewycz (Royal Ontario Museum).

Several natural history booksellers assisted by sending useful references or other information. I thank Erich Bauer, Joseph R. Dinardo, Jr., Donald E. Hahn, and Dieter Schierenberg, and especially Howard K. Swann, John and Betty Johnson, and Rudolph Wm. Sabbot.

For assistance with translations, I thank Satoshi Amagai, Natalia B. Ananjeva, Douglas G. Brust, Frank Dodd, Davina Eisenberg, Hajime Fukada, T. Ulmar Grafe, Show-ling Shyng, Zbigniew Szyndlar, and Er-mi Zhao.

Much of the biographical material that I have assembled was accumulated as part of my duties as Editor of the *Facsimile Reprints in Herpetology* series of the Society for the Study of Amphibians and Reptiles, beginning in 1961. Over that period many persons, acting incidental to their duties as research technician or student assistant, have located valuable materials, but there have been too many persons involved to acknowledge individually. I will single out, however, two student assistants, Yu-shi Lin and Wilson C. Cheng, who have helped during this past year when a special drive was mounted to finish the task. To all of these assistants I proffer my thanks. I also acknowledge my debt to Dawn Potter and Carol Abarbanell, my secretaries, for their careful attention to detail, so very important in a document of this sort.

Corrections and Missing Information.—I am cognizant of the possibility that some information presented here is incorrect, although every effort has been made to insure accuracy. Since further editions or volumes of this book might be undertaken, I earnestly solicit comments and corrections, with supporting documentation wherever possible. Although I was successful in obtaining signatures for all biographees, I did not obtain three portraits (those for Laurenti, Daudin, and Bourret), despite extensive efforts. These portraits thus represent prime desiderata.

Correspondence may be addressed to: Kraig Adler, Cornell University, Section of Neurobiology and Behavior, Seeley G. Mudd Hall, Ithaca, New York 14853, USA.

BIOGRAPHIES

GESSNER, Conrad (1516-1565).

Conrad (sometimes Cuonrat) Gessner—or in its Latin form, Conradus Gesnerus (see *Note*)—was one of the great scholars of the Renaissance. Among his many seminal contributions, Gessner began the transformation of botany and zoology from the writings of Aristotle some eighteen centuries before. Gessner was born in Zürich, Switzerland, on 16 March 1516, the son of a furrier of German ancestry. He had an early interest in plants and animals, but because he also had a gift for languages most of his schooling was in Latin and Greek grammar. He was a language teacher for a while in Zürich, but in 1536 he went to Basel to study medicine.

Later, Gessner studied at the University of Montpellier (1540-1541), then the foremost European medical school, where he learned zoology from Guillaume Rondelet and also conducted field studies along the Mediterranean Sea. He returned to Basel and finished his M.D. degree in 1541 by defending a thesis on the question of whether the heart or the brain was the seat of emotion and sensory impressions; Gessner decided in favor of the brain, a bold assertion at the time since it was contrary to Aristotle's writings. This break with the tradition of the Middle Ages, which had been to follow slavishly the classical Greek scholars, was immortalized by Copernicus and Vesalius, both of whom were contemporaries of Gessner.

Gessner then returned to Zürich where, for the rest of his life, he taught at his former elementary school, the Carolinum, a theological seminary, and also conducted a medical practice. In the evenings he performed his great work, a series of several dozen books on virtually all aspects of human knowledge—botany, zoology, medicine, pharmacology, classical philology, theology, linguistics, and bibliography—and he wrote classic texts in each. For example, Gessner's "Bibliotheca Universalis" (1545), the first bibliography of scholarly works in the three classical languages, was the starting-point for bibliography; likewise, his "Mithridates" (1555), in 21 books, founded comparative linguistics and modern philology. Burdened by persistent financial problems, this constant work by midnight oil led to chronic illnesses and, in a weakened condition, he succumbed to an influenza epidemic, in Zürich, on 13 December 1565. Thus, this industrious man, who personified the Renaissance ideal of the humanist as universal scholar, was dead at the age of 49.

Gessner's famed work on zoology, "Historia Animalium" (in five volumes, 1551-1587; second edition 1617-1621, with numerous later reprintings and translations, most recently in 1967 and 1980), was begun in 1544. It laid the foundation for standardization of scientific terminology by listing the equivalent names of animals in a dozen languages. He combed the classical and medieval literature for information, added his own observations and those of correspondents, and then organized the whole in a very precise manner: synonymy, distribution, physical characteristics, habits, use as food and medicine, etc. By adding numerous woodcuts he produced the first illustrated work covering the entire animal kingdom, the influence of which was to continue for two centuries through the numerous reprintings and translations.

Other scholars were inspired to emulate him. Ulisse Aldrovandi of Bologna, a young correspondent of Gessner's, produced an even more extensive encyclopedia of animals (1599-1664) in 11 volumes and, somewhat later, John Jonston produced another recapitulation. These were mostly copies of Gessner, Rondelet, and other earlier authors and, except perhaps in illustration, were not an advancement in knowledge over them. The 1608 translation (reprinted 1973; second edition in 1658) of Gessner's herpetological books by Edward Topsell, an English vicar, was not an improvement either, although the latter's book was the first on reptiles in the English language, which is its main claim to fame.

Two of the volumes in Gessner's series were herpetological. Volume two ("Quadrupedibus Oviparis," 1554) covered the Amphibia and four-footed reptiles, and volume five ("Serpentium Natura," 1587), published posthumously from Gessner's notes, comprised snakes and scorpions. These works provided the most important link between the Aristotelian period, in which reptiles were first separated as a distinct kind of animal, and the Linnaean, when their systematic position was formalized in the recognition of

class Amphibia (1758), as foreshadowed in the pioneering classification of the English parson, John Ray (1693, reprinted 1978).
• *References*: L. C. Miall, 1912 (pp. 28-32, 48-49); "Conrad Gessner . . . Leben und Werk," by H. Fischer, Komm. Leemann, Zürich, 152 pages, 1966; "Conrad Gessner 1516-1565," Orell Füssli Verlag, Zürich, 240 pages, 1967; "Konrad Gesner," by P. E. Pilet, Dict. Sci. Biogr., 5: 378-379, 1972; "Conrad Gessner: a Bio-bibliography," by H. Wellisch, Jour. Soc. Bibliogr. Nat. Hist., 7: 151-247, 1975. • *Portrait*: Bibliothèque Nationale, Paris. • *Signatures* (left, Latin form): From "Conrad Gessner 1516-1565," 1967; (right, German form, 1558): Zentralbibliothek Zürich, courtesy J. P. Bodmer. • *Note*: Gessner always spelled his name with *ss* when writing in German, but with a single *s* in Latin since it usually was considered improper to write a double consonant before another consonant in Latin (see H. Schinz and K. Ulrich, 1936, Vierteljahrsschr. Naturf. Gesellsch. Zürich, 81: 285-291, and Wellisch, 1975, p. 152). Unfortunately, many later authors have translated his name from the Latin incorrectly as "Gesner." The German form of his name reproduced above, although imperfect due to loss of edge of original document, clearly shows the double *ss*.

REDI, Francesco (1626-1697).

Although best known for his experiments on spontaneous generation, Redi also performed the first scientific study on viper bites. He was born in Arezzo, Tuscany, on 18 February 1626, son of the physician to the Grand Duke, Ferdinand II. Under the Medici family, Florence, whose citizens included Dante, Galileo, Leonardo da Vinci, and Michelangelo, was intellectually at its zenith. Redi attended Jesuit schools and, later, the University of Pisa where he first studied biology (doctor of medicine and philosophy, 1647); shortly thereafter he returned to Florence to practice medicine and, in 1666, succeeded to his father's royal position. He had begun his scientific work somewhat earlier—that on vipers in 1660, following a suggestion made by the Grand Duke—and later on he also studied insects and parasites. His master work, "Esperienze Intorno alla Generazione degl'Insetti," in which he showed experimentally that maggots do not arise from corpses by spontaneous generation but from fly eggs deposited there, was published in 1668. He died during the night of 28 February-1 March 1697 at Pisa.

The means by which the bite of a viper caused death had been of great interest since antiquity and the search for an antidote had involved much credulity and even knavery. It was Redi, using controlled experiments traceable to the methodology of Galileo, who first succeeded in separating fact from fiction. He showed that viper venom was the yellowish fluid that flowed from the teeth, that the venom had to be injected beneath the skin to cause death (it had no effect when introduced into the stomach with food), and that the venom was produced in sacs in the viper's head. Through similar experiments and observations he also dispelled many myths, although his experiments on animals were for a long time regarded as cruel.

Redi's experiments were often demonstrated before the Grand Duke himself and, in 1664, were published in book form as "Osservazioni Intorno alle Vipere" (English translation in 1988). His views were severely challenged by Moyse Charas, a Parisian apothecary, in a book published in 1669, and also by others who claimed that the viper's venom was harmless and that the true poison were the spirits driven into the victim by the mind of an enraged viper! As a result, Redi was led to conduct additional research which, together with logical refutations of criticisms of his work, he published as responses in 1670, 1685, and 1686, but he was unable to fully satisfy his critics. It remained for Redi's compatriot and successor, Felice Fontana, to demonstrate the mode of action of injected venom (1767 and 1781).
• *References*: L. C. Miall, 1912 (pp. 225-228); "Vita ed Opere Inedite di Francesco Redi [part 1]," by U. Vivani, Coll. Pubbl. Stor. Lett. Aretine, 9: (xii), 1-117, 1924; "Francesco Redi (1626-1697). Physician, Naturalist, Poet," by R. Cole, Ann. Med. Hist., 8: 347-329, 1926; "Francesco Redi," by L. Belloni, Dict. Sci. Biogr., 11: 341-343, 1975; "Francesco Redi," p. 555-708. *In* M. L. Altieri Biagi and D. B. Basile (eds.), Scienziati del Seicento, R. Ricciardi Ed., Milan, 1980; "Francesco Redi on Vipers," by P. K. Knoefel, E. J. Brill, Leiden, xviii, 86 pages, 1988. • *Portrait*: From Ann. Med. Hist., 8, opposite p. 331, 1926. • *Signature*: From Cole, 1926.

SEBA, Albertus (1665-1736).

A merchant by profession, Seba possessed the largest collection of natural history objects of his era. Born on 2 May 1665 in Etzel, East Friesland, in what is now northwestern Germany, Seba was apprenticed to learn the apothecary trade and by 1697 had become a certified pharmacist in Amsterdam. He assiduously built up collections by purchasing curiosities from sailors returning from Africa, America, and the Indies, as well as from an extensive network of correspondents worldwide. His collection was so widely acclaimed that during his second visit to Holland, in 1717, Peter the Great purchased it *en toto* in order to form the nucleus of a Russian national collection in St. Petersburg (now Leningrad). Seba promptly started a second collection which, long after his death (on his birthday, 2 May 1736, in Amsterdam), was sold at auction in 1752 and dispersed widely throughout European collections; today the reptiles are in Berlin, Leiden, Leningrad, London, and Paris among other locations.

Seba published a colossal four-volume work, in folio, to describe his two collections, the famous "Locupletissimi Rerum Naturalium Thesauri," usually referred to as Seba's "Thesaurus." One can scarcely imagine its physical size, measuring 51 cm high and weighing 9 kg—each volume! The work was written by Seba in Dutch, from which Latin and French translations were made to produce two editions, issued simultaneously, in Latin-Dutch and Latin-French, both in handcolored and uncolored versions. The text is mediocre, but the 449 outstanding plates figure literally thousands of specimens in life size, including an especially large number of reptiles and amphibians, which form the largest part of volumes 1 and 2, issued in 1734 and 1735. The other volumes were published posthumously, in 1759 and 1765, edited by Arnout Vosmaer. Seba's text is not binomial and, thus, has no standing in taxonomic nomenclature, but later writers based many new species on Seba's figures, which therefore have the status of types. For example, we have Seba to thank for the misnomer *Bufo marinus*, since Carl Linnaeus based his name on Seba's figure labelled "Rana, Marina, Americana." Seba's figures provided the basis for no less than 28 species of amphibians and reptiles described by Linnaeus, and another named in his honor by J. F. Gmelin, *Python sebae* of Africa.

• *References*: "The Life of Albert Seba," by H. Engel, Svensk. Linné-Sällskap. Årsskrift., 20: 75-100, 1937; ["Albert Seba and His Contribution to the Development of Herpetology"], by K. P. Juriev, p. 109-120. *In* N. B. Ananjeva and L. J. Borkin (eds.), Proc. Zool. Inst. Acad. Sci. USSR, vol. 101, 1981. • *Portrait* (1731): From Seba, "Thesaurus", vol. 1, 1734. • *Signature* (1731): From Engel, 1937.

KLEIN, Jakob Theodor (1685-1759).

Klein, who coined the word herpetology, was an amateur naturalist and contemporary of Carl Linnaeus. He was born on 15 August 1685 in Königsberg, East Prussia (now Kaliningrad, USSR), and eventually became State Secretary to the Senate of Danzig (now Gdańsk, Poland). He had a large private collection and a broad interest in natural history, and he developed an artificial system of classification intended to rival that of Linnaeus. His many books were largely compilations, drawing heavily from Albertus Seba and early editions of Linnaeus. His nomenclature was not consistently binomial, and all but two of his books were pre-Linnaean in date of publication and thus can have no official standing in taxonomy. Klein died at Danzig on 27 February 1759.

Reptiles and amphibians naturally formed an integral part of all of Klein's classifications, beginning with his "Summa Dubiorum Circa Classes Quadrupedum et Amphibiorum" of 1743, but the book of main interest here is his 1755 "Tentamen Herpetologiae" in which the name of the discipline of herpetology was first used. Klein's definition, however, included all limbless animals—thus worms also—but excluded all quadrupeds such as frogs, turtles, and lizards.

• *References*: "Jacob Theodor Klein," by W. Hess, Allg. Deut. Biogr., 16: 92-94, 1882; "The Herpetological Works of Jacob Klein," by J. Johnson, M. L. Skaroff, H. M. Smith, and R. B. Smith, Bull. Maryland Herpetol. Soc., 20: 174-189, 1984. • *Portrait* (1743): From Klein, Système Naturel, Paris, 1754, courtesy Hobart M. Smith. • *Signature* (1724): Royal Society, London, courtesy K. Moore.

ROESEL VON ROSENHOF, A. J. (1705-1759).

Roesel (sometimes Rösel), the outstanding German artist and amateur naturalist, was born in Augustenburg, near Arnstadt, on 30 March 1705, of noble Austrian ancestry. His original name, August Johann Roesel, was emended "von Rosenhof" in 1753, perhaps to honor his uncle Wilhelm Roesel von Rosenhof, a well-known painter of animals to whom, in 1720, young Roesel was apprenticed. His father had died early, and his godmother, Auguste Dorothea, the reigning Princess of Arnstadt-Schwarzburg, encouraged his artistic career. Roesel received advanced training at the art academy in Nuremberg (1724-1726) and then became painter of portraits and miniatures to the Danish court in Copenhagen for two years. During an illness, a friend showed him a copy of the famous illustrated atlas by M. S. Merian depicting the insects, reptiles, and other natural objects of Surinam (1705), which inspired Roesel to do a similar study of the German species.

On returning to Nuremberg in 1728, Roesel spent his free time in the field, observing insects, amphibians, and reptiles, and bringing their pupae and eggs into his home to study their development and metamorphosis. His careful observations, as described in his texts and especially his detailed yet artistic drawings, were published in two great works. The first, on insects and other organisms such as crayfish and hydras, was issued in installments beginning in 1740. In it, Roesel defined insects systematically and then divided them into natural groupings, for which contribution, together with the outstanding plates, he is today regarded as the Father of German Entomology. The fourth and last volume was published posthumously, in 1761.

Simultaneously, Roesel was collecting information and drawings for his other books on amphibians and reptiles. His "Historia Naturalis Ranarum Nostratium" (alternate title, "Die natürliche Historie der Frösche hiesigen Landes"; text in parallel columns of Latin and German) was issued in several parts, the first in 1753, the last in 1758. This book includes 24 exquisitely handcolored plates, in folio, drawn and engraved by Roesel, each with a labelled (or key) plate in black and white. The value of the plates resulted in the book being reissued several times over the succeeding years, including a German-only edition ("Naturgeschichte der Froesche des Mittlern Teutschlandes," edited by J. C. D. Schreber). These plates must be among the most beautiful illustrations in all of herpetology. The complete life cycle of all species of German frogs and toads is presented in great detail—including amplexing adults and developmental stages of tadpoles—together with their anatomy and osteology.

This wealth of natural history data was incomparable for its time and stands in stark contrast to the terse one- or two-line descriptions of species in Carl Linnaeus's contemporaneous editions of "Systema Naturæ." In fact, Linnaeus based several of his binomial names on Roesel's descriptions, and the plates were copied and recopied, often poorly and without credit, in many picture books of animals well into the 20th Century. Roesel had begun a companion book on lizards and salamanders (some of his watercolors still exist, as described by Franz Leydig), but a stroke paralyzed his left side and he died in Nuremberg, on 27 March 1759.

• *References*: "Herpetologische Zeichnungen aus dem Nachlass Rösels von Rosenhof," by F. Leydig, Verhandl. Naturhist. Ver. Preuss. Rheinl. Westfal. Regierungsbez. Osnabrück, 35: 1-41, 1878; "August Johann Rösel von Rosenhof," by W. Hess, Allg. Deut. Biogr., 29: 188-189, 1889; L. C. Miall, 1912 (pp. 293-303); "August Johann Roesel von Rosenhof," by A. Geus, Dict. Sci. Biogr., 11: 502-503, 1975; "Roesel von Rosenhof—Insecten-Belustigung," ed. by E. Bauer, Müller & Schindler, Stuttgart, 80 pages, 1985; Adler, 1986 (pp. 324, 328). • *Portrait*: From Roesel von Rosenhof, "Insecten-Belustigung," vol. 4 (ed. by C. F. C. Kleemann), 1761. • *Signature* (printed form): From Roesel von Rosenhof, "Ranarum," 1758.

LINNAEUS, Carl (1707-1778).

Carl Linnaeus (sometimes Carolus Linnaeus; from 1762, Carl von Linné), the great Swedish botanist who classified the world's plants and animals and introduced binomial nomenclature, was born 13 May 1707 at Södra, in Rashult province. As a boy he was tutored in botany by his father, a country parson, who had taken his surname from a famous Swedish lime-tree—Linnæus—that grew near his home. In 1727, Carl Linnaeus entered the University of Lund to study medicine and in the following year he transferred to the University of Uppsala where he first developed his new system of plant classification based on type of sexual organs. After several field trips in Sweden, he travelled abroad (1735-1738) to England, France, and Holland, published his first important works, and met many influential persons including Albertus Seba. During this time he also finished his medical degree at the University of Harderwijk in Holland (1735), and published the first edition of "Systema Naturæ" (1735) including a description of his procedures, based partly on those of Conrad Gessner.

Despite his developing fame, no academic position awaited him in Sweden, so he practiced medicine in Stockholm. Finally, in 1741, he was offered a professorship at the University of Uppsala, which he held until his death, in Uppsala, on 10 January 1778. In 1747 he was appointed physician to the Swedish Royal family and was elevated to the nobility, in 1762. Besides his major books, Linnaeus was an incomparable teacher and supervisor of doctoral students. He extended his influence through them and their collections in distant lands: Pehr Kalm in North America, Fredrik Hasselquist in Egypt and Palestine, Pehr Forskål in Arabia, Petrus Löfling in Venezuela, Pehr Osbeck in China and the East Indies, and C. P. Thunberg in Japan, to name the principal ones.

Linnaeus was a driven man who was enormously ambitious and had a great capacity for work. He was a brilliant observer, but he was so dogmatic and philosophical in his effort to classify Nature in a simple scheme—derived, ultimately, from Aristotelian logic—that he saw himself as a prophet summoned by God to proclaim the true dogma. Egocentric, Linnaeus persecuted those, like Georges Buffon, who despised the artificiality of his system so that they never fully appreciated its practical benefits. Nevertheless, Linnaeus, inspired by John Ray, the English clergyman and naturalist, was the first to use species as a clearly-defined concept, but he saw them as fixed and unchangeable entities created by God, although his views on immutability changed over time.

Linnaeus's abhorrence of amphibians and reptiles is well known—he once wrote of them "*Terribilia sunt opera Tua, o Domine!*" or, in translation, "Terrible are Thy works, O Lord!"—yet of necessity they figured in the various editions of "Systema Naturæ" (first edition through 12th, the last prepared by him, 1735-1766). Besides naming numerous genera and species from throughout the world, Linnaeus defined the class Amphibia; although he had already done so in previous editions, officially the name dates from the tenth edition of the "Systema" (1758; reprinted 1956) since that edition was later selected by taxonomists as the starting point for zoological nomenclature.

Linnaeus's definition of "Amphibia" included not only amphibians and reptiles but cartilaginous fishes, which in later editions of his "Systema" he transferred to the Amphibia based on misconceptions about their gill structure. Numerous translations and emendations of the "Systema" appeared during and after Linnaeus's life, the most important of which, herpetologically, were those by Martinus Houttuyn (1764), P. L. Statius Müller (1774), and J. F. Gmelin (1789, the so-called 13th edition of Linnaeus, to which Gmelin added descriptions of new species and to whom authorship is credited), and John Walcott's atlas of 1788[-1789].

In addition, Linnaeus authored several works specifically on amphibians and reptiles. In 1745, he published "Amphibia Gyllenborgiana," a review of 24 species of South American, African, and European turtles, frogs, lizards, and snakes in Count Gyllenborg's collection, in which scutellation was used for the first time to define species of snakes. This work was, in fact, the doctoral dissertation of one of his students, but it was the tradition of the day for the professor to write their theses, although the student, or "respondent," had to defend it and pay for its publication! Another such thesis was on the peculiar American salamander *Siren* (1766). Linnaeus's major herpetological work was entitled "Museum S:æ R:æ M:tis Adolphi Friderici Regis Suecorum" (1754), a large folio with 33 plates, 23 of them on snakes and other "Amphibia," based on his king's private cabinet of specimens.

Some of Linnaeus's descriptions of new species were based on descriptions and illustrations in books by Seba, Gessner, Aldrovandi, Catesby, Jonston, and other authors, but most were taken from specimens examined by him. Fortunately, most of these specimens are still intact and have been reviewed in detail in three papers by the Swedish herpetologists Einar Lönnberg and Lars Gabriel Andersson (1896-1900; indexed by P. E. Vanzolini, 1969) and also by Åke Holm (1957).

• *References*: "Linnaeus," by B. D. Jackson, H. F. & G. Witherby, London, xv, 416 pages, 1923; "The Amphibia and Pisces in the First Edition of *Systema Naturae*," by K. P. Schmidt, Copeia, 1951: 2-7, 1951; Papavero, 1971 (pp. 1-3); "Carl Linnaeus," by S. Lindroth, Dict. Sci. Biogr., 8: 374-381, 1973. • *Portrait* (1774) *and signatures* (two versions, dated 1728 on left, 1765 on right): From Jackson, 1923.

LAURENTI, Josephus Nicolaus (1735-1805).

Josephus Nicolaus Laurenti (see *Note 1*) wrote the first major review of amphibians and reptiles after that of Carl Linnaeus. He was born in Vienna on 4 December 1735 and earned his medical degree from the Medical Faculty of the University of Vienna (M.D. 1768), with a thesis entitled "Specimen Medicum, Exhibens Synopsin [sic] Reptilium Emendatam cum Experimentis circa Venena." This unimposing little book of 217 pages and five plates, published in Vienna in 1768 (reprinted 1966), is of the greatest importance to herpetology. The first half is a review of the reptiles and amphibians of the world (except turtles), and the rest describes the result of careful experiments to distinguish the venomous Austrian species.

Only ten genera of amphibians and reptiles were named in the "Systema Naturæ" (1758), but Laurenti defined 30, a distinct advance over the classification of Linnaeus. Laurenti proposed some of the most familiar generic names: *Bufo*, *Hyla*, and *Salamandra* among amphibians; the lizards *Gekko*, *Chamaeleo*, *Iguana*, and *Cordylus*; *Crocodylus*; and the snakes *Natrix*, *Vipera*, *Naja*, and *Constrictor*. In addition, he first named and characterized the class Reptilia.

Historically, a few authors have attributed this book to the Hungarian scientist J. J. Winterl who, according to tradition, was a destitute student who sold the manuscript to Laurenti (see *Note 2*). It seems more likely that Winterl, being a chemist, merely collaborated in Laurenti's experiments on venoms and antidotes; Laurenti acknowledged his help on page 213 of his book. These experiments did succeed, incidentally, in deciding which Austrian species were poisonous, but made no advance in understanding the mode of action of venom, as Laurenti accepted the divergent conclusions of both Francesco Redi and Moyse Charas.

Little more is known about Laurenti's life. He attended a secondary school affiliated with the University of Vienna, graduating in 1754. There was then a long gap before his formal medical studies began, but it was normal practice at that time in Vienna for a person to perform medical duties, as a so-called field surgeon, without any higher education, and then to take a medical degree later at a somewhat advanced age (in Laurenti's case, at the age of 32). From 1769 to 1805 Laurenti was a member of the Medical Faculty of the University of Vienna, and in 1772 he passed an examination of that faculty to permit him to perform obstetrical services, implying that he also had a medical practice. Other than his thesis, he apparently published nothing else. He died in Vienna on 17 February 1805.

• *References*: "Joseph Nikolaus Laurent," p. 88. *In* J. G. Meusel, Lexikon der . . . Teutschen Schriftsteller, G. Fleischer, Jr., Leipzig, 1808; "Laurenti, medicinae Doctor," col. 1393. *In* C. G. Jöcher and J. C. Adelung, Allgemeines Gelehrten-Lexikon, vol. 3, G. Jöntzen, Delmenhorst, 1810; "Ueber die im Erzherzogthume Oesterreich verkommenden Reptilien," by L. J. Fitzinger, Arch. Gesch., Statist., Lit., Kunst, 1823: 631-634, 1823; "Joseph Nicolas Laurenti," by J.-E. Dezeimeris, Dict. Hist. Médec. Anc. Mod., 3: 409, 1836; "Joseph-Nicolas Laurenti," by A. J. L. Jourdan, Biogr. Univ. Michaud, 23: 570, 1858; "Types of the Amphibian and Reptilian Genera Proposed by Laurenti in 1768," by L. Stejneger, Copeia, 1936: 133-141, 1936. • *Signature* (1769): Archives, University of Vienna, courtesy Kurt Mühlberger. • *Note 1*: The correct spelling of Laurenti's surname has been in some dispute. The International Commission on Zoological Nomenclature ruled for "Laurenti" (direction 65, 1957), but occasional references have used "Laurent" (e.g., Meusel, 1808). The matter is further confused by the existence of two versions of Laurenti's book—and a reprint of one of them in which Laurenti's name has been critically altered—which differ only in the preliminary pages: a thesis version, issued in March 1768 on the occasion of Laurenti's doctoral defense ("Viennæ, Typis Joan. Thomæ nob. de Trattnern . . . Bibliop.," the version reprinted in 1966) and a public version, presumably issued later in 1768 ("Viennæ, Typ. Joan. Thom. Nob. de Trattnern . . . Bibliop.—1768"). The name "Laurenti" appears on the title pages of both, but it has been disputed whether this form is both genitive and nominative. At the end of the preface in the thesis version

(only), the name is given in what is believed to be the Latin nominative: Josephus Nicolaus Laurenti. Unfortunately, in the 1966 reprint of the thesis version of Laurenti's book, the final *i* in his name has been omitted. On inquiry to the Archives of the University of Vienna, Kurt Mühlberger, the University Archivist, has determined that Laurenti himself consistently used the spelling "Laurenti" in all of his university records and that it is, in fact, the nominative form.
• *Note 2*: Cloquet (1819, p. 28) and Fitzinger (1823, p. 631) seem to be the first persons to state in print that Winterl might have been the real author of the "Specimen Medicum," and numerous later authors have simply repeated the statement. Neither author provided evidence for their remark. However, the earliest published statement about Winterl's involvement presents a very different picture. Jöcher and Adelung (1810, col. 1393) stated that Professor "Winterle" [*sic*] published his [=Laurenti's] "Synopsin Reptilium." Winterl was Laurenti's fellow student and received his own M.D. degree from Vienna the previous year (1767); possibly he supervised Laurenti's book through the press or paid for its printing.

SCHNEIDER, J. G. (1750-1822).

The German philologist and naturalist Schneider was born on 18 January 1750 in Kollmen (now Collm), near Oschatz, just east of Leipzig. To his name, Johann Gottlob Theaenus Schneider, he sometimes added "Saxo" in honor of his native land, Saxony. He studied philology, especially the Greek classics, and natural history at universities in Leipzig (beginning 1769), Göttingen (1772), and finally at Strassburg, where he received his Dr.Phil. degree (1774). In 1776, he became Professor of Philology at the University of Frankfurt (Oder), where he produced a large number of translations and commentaries on classical works, including Aristotle's "Historia Animalium" (1811, in four volumes) and Nicander's "Theriaca" (1816), and became one of the foremost philologists of his day. When his university was transferred to Breslau (today Wrocław, Poland) in 1811, he moved with it and, not liking to teach, in 1814 he accepted the post of university librarian. He died in Breslau on 12 January 1822.

Even though Schneider's zoological projects were strictly secondary to his literary work, he made several important contributions. Partly this is due to the early date at which he worked, but also to the fact that he used proper Latin binomial nomenclature at a time when many others, especially the French, who were still laboring under the shadow of Georges Buffon, did not or did so only in a desultory manner. Moreover, because of his classical training, Schneider's works were well researched and referenced to the literature, with learned digressions into the classical authors. His first major herpetological title, "Allgemeine Naturgeschichte der Schildkröten" (1783; two small supplements in 1787 and 1789) was largely an historical review of cheloniology, to which he added a short systematic section including first descriptions of several well-known turtles: *Chrysemys picta*, *Chelus fimbriatus*, and *Trionyx ferox*. In 1790-1792 he published "Amphibiorum Physiologiae" which, despite its title, was devoted largely to lizard systematics.

Schneider's major herpetological title, "Historiae Amphibiorum" (in two volumes, 1799 and 1801; reprinted 1968), was a world review of amphibians and reptiles, and in it he named numerous new species and genera; to mention a few of the best known new taxa, *Crocodylus porosus*, *C. siamensis*, and *Paleosuchus trigonatus* among crocodilians and the boids *Candoia carinata*, *Eryx conicus*, *Morelia amethistina*, and *Python reticulatus*. Most of the specimens he used for this book were in the Museum Lampianum in Hannover, purchased long before by his university colleague, J. L. C. Gravenhorst, and moved to Breslau in 1811, coincidentally with Schneider himself. Later, he published several monographs on major groups, including geckos (1812) and boas (1821).

• *References*: "Memoria Joh. Theaeni Schneideri," p. 337-350. *In* F. L. C. F. Passow, Opuscula Academica, F. C. G. Vogel, Leipzig, 1835; "Johann Gottlob Schneider," by H. A. Pierer (ed.), Universal-Lexikon, Vollständ. Encyclop. Wörterbuch, 20: 60-61, 1835; "Johann Gottlob (Theaenus) Schneider," by R. Hoche, Allg. Deut. Biogr., 32: 125-127, 1891; "The Precursors in Cetacea," by G. Pilleri and L. Arvy, Investig. Cetology, vol. 12, Brain Anatomy Institute, Bern, 1981 (pp. 88-92). • *Portrait*: From Pilleri and Arvy, 1981, courtesy G. Pilleri. • *Signature*: Museum für Naturkunde der Humboldt-Universität zu Berlin, courtesy Rainer Günther.

LACEPÈDE, B.-G.-É. (1756-1825).

Bernard-Germain-Étienne de la Ville-sur-Illon, Comte (or count) de Lacepède (sometimes spelled Lacépède or de la Cépède; see *Note*), was a leading politician of Revolutionary France, a naturalist and successor to Buffon, as well as an accomplished musician. Lacepède was born to a noble family, on 26 December 1756, at Agen in southwest France. He was a recluse by instinct and his early life was devoted to philosophy and music, but at the age of 21 he went to Paris and soon formed a friendship with Georges Buffon, who sparked his interest in natural history. Buffon enlisted Lacepède's aid in his grand project to summarize all of natural history and, when he died in 1788, Lacepède, together with L.-J.-M. Daubenton, had been Buffon's choice to succeed him.

Among other duties, Lacepède became keeper of the natural history collections in the king's garden, the Jardin du Roy, but after the revolution, the first Constituent Assembly, in June 1793, transformed these collections into the Muséum National d'Histoire Naturelle and, at the same time, created 12 chairs, or professorships with tenure. Lacepède was absent from Paris during this period; because of his nobility and previous political activity, during the year-long Reign of Terror he briefly retired to the countryside near Leuville.

In the following year, however, the chair of vertebrates, then held by Étienne Geoffroy St. Hilaire, was divided, and that covering fishes and reptiles was offered to the count. Lacepède conducted some research during this period, but he became increasingly active politically, first as a Senator (1799), then as President of the Senate (1801), and later when appointed by Napoleon to be Grand Chancellor of the Legion of Honor (1803). He felt obliged to resign his professorship in 1803, and although Constant Duméril assumed the duties of that chair in 1803, he did not formally take the full title until Lacepède's death, which occurred during the night of 5-6 October 1825, at Épinay-sur-Seine, now a Paris suburb.

Lacepède published many important works on vertebrates, especially on fishes and cetaceans. Besides some shorter papers, his major herpetological *opus* was the "Histoire Naturelle des Quadrupèdes Ovipares et des Serpens," issued in both quarto (2 volumes, 1788-1789) and octavo (4 volumes, 1788-1790) editions as part of the Buffon series. This was the first comprehensive world summary of amphibians and reptiles and while it contains still-useful historical information, the illustrations are poor. And, while it far exceeded the earlier attempt of J. N. Laurenti (1768), it showed no advance in the natural arrangement of genera. The nomenclature is inconsistently binomial, and in 1987 the International Commission on Zoological Nomenclature declared that Lacepède's section on snakes (only!) is unavailable for nomenclatural purposes, except for a few conserved names.

Nevertheless, Lacepède's book was very influential, being superseded in importance only by the "Erpétologie Générale" of Constant Duméril and Gabriel Bibron half a century later. Consequently, Lacepède's work went through many printings and resettings—even as late as 1878—and two translations: a German edition by J. M. Bechstein (5 volumes, 1800-1802) and one in English by Robert Kerr (4 volumes, 1802; reprinted 1978).

• *References*: "Sketch of the Life and Character of the late Count de Lacépède," by W. Swainson, Zool. Jour., 3: 73-76, 1827; "La Vie et l'Oeuvre de Lacépède," by L. Roule, Mém. Soc. Zool. France, 27: 1-99, 1918; "Cent-cinquantenaire de la Chaire d'Ichthyologie et d'Herpétologie," by L. Bertin, Bull. Mus. Hist. Nat., 17: 353-372, 1945; "Bernard-Germain-Étienne de la Ville-sur-Illon, Comte de Lacépède," by T. A. Appel, Dict. Sci. Biogr., 7: 546-548, 1973; "Lacepède (1756-1825), l'Homme et Son Oeuvre Herpétologique," by J. L. Fischer, Bull. Soc. Zool. France, 27: 479-485 (=Bull. Soc. Herpétol. France, 7: 479-485), 1977. • *Portrait*: From Roule, 1918. • *Signature* (1825): Adler collection. • *Note*: Fischer (1977, p. 484) argued that the correct spelling is "Lacepède," without the acute accent on the first *e*. Both of the forms "Lacépède" and "Lacepède" are found in the count's own publications. However, I have seen several specimens of his signature from the period 1808-1825, and all have the form "Lacepède," which for that reason has been preferred in this biography.

SCHOEPFF, Johann David (1752-1800).

Schoepff (see *Note*), known to herpetologists for his elaborate atlas of turtles, was a German botanist, zoologist, and physician famous for his early exploration of North America. He was born on 8 March 1752 in Wunsiedel and attended the University of Erlangen (1770-1776) where his professors included the naturalists P. L. Statius Müller and the last of Carl Linnaeus's pupils, J. C. D. Schreber. He finished his medical degree in 1776 and soon departed with mercenary troops from Brandenburg, hired by the British to assist in their war with the American colonies. Schoepff was a field doctor and spent the period 1777-1784 in North America; after the war he travelled for eight months as far west as Ohio and south to Florida and made extensive natural history collections. His observations on American culture were later published in a two-volume monograph ("Reise," 1788) that became one of the most widely-read books on the new American nation.

Turtles were a special interest and Schoepff returned large numbers of them to Europe, including living specimens. Only in 1792 could he begin serious work on his turtle book, by visiting numerous public and private museums in Holland and Prussia, but the book was never finished since he died suddenly in Ansbach on 10 September 1800.

Schoepff's herpetological reputation rests solely on a single book, issued in both Latin ("Historia Testudinum") and German ("Naturgeschichte der Schildkröten") editions, each with the same plates (usually 34 of them, but occasionally 35), which were published in both uncolored and beautifully handcolored versions. Each edition was issued in four parts, apparently simultaneously, in 1792, 1793, 1795, and 1801. His former teacher, J. C. D. Schreber, who had earlier described *Rana pipiens* from specimens sent to him from America by Schoepff, issued the final part posthumously in 1801. Forty species are reviewed, each with a detailed description and a review of the literature. Seven of the new taxa described by Schoepff are still recognized and include several well-known species: the American *Pseudemys scripta*, *Malaclemys terrapin*, and *Clemmys muhlenbergii*; *Geochelone elegans* of southern Asia; and *Testudo marginata* of southeastern Europe.

• *References*: "Johann David Schöpf," by F. Ratzel, Allg. Deut. Biogr., 32: 350-352, 1891; "The Status of *Testudo terrapin* Schoepf," by M. B. Mittleman, Copeia, 1944: 245-250, 1944; "Dr. Johann David Schoepf," by H. Lang, Hofer Anz., 11 January 1958, pp. 3-4, 1958; "Johann David Schoepf—Leben und Werk," by A. Geus, Jahrb. Hist. Ver. Mittelfränken, 84: 83-161, 1968. • *Portrait (1793) and signature*: From Geus, 1968. • *Note*: During the 18th Century, spelling of the name Schoepff was not standardized and the free variations "Schoepf" and "Schoepff" existed as well as the form "Schöpf" in German Fraktur which causes the umlaut. Thus, all three variations may be found in Schoepff's own publications, although "Schoepf" has become the standard modern spelling. Based upon his signature, Schoepff preferred that form, which is therefore the one used in this biography.

MERREM, Blasius (1761-1824).

Merrem, a gifted German naturalist of unfulfilled promise, was born in Bremen on 4 February 1761. After early attempts at careers in theology and commerce, he entered the University of Göttingen in 1778 where he was converted to zoology by his teacher, J. F. Blumenbach. Merrem received his Dr.Phil. degree from Göttingen in 1781. He stayed there as a lecturer on zoology, agriculture, and mathematics while publishing his first books (on birds).

At that time, naturalists were divided into two schools: the nomenclaturists of Carl Linnaeus, concerned with the naming of immutable species created by God's hand, and the French school of Georges Buffon, involved with relationships and the modification of species, which even refused to use binomial nomenclature. Following the lead of his teacher and of P. S. Pallas, Merrem foresaw the union of these divergent schools of thought into a more correct form of natural history, based on consideration of all external characteristics of animals, not just single, isolated ones, as used by the Linnaeans. Merrem's views, however, were not appreciated by most of his contemporaries or by the book-buying public which only wanted great picture books of animals. These problems, coupled with the political disruptions flowing from the French Revolution, meant that Merrem's several attempts to produce grand monographs—on birds and reptiles—were failures and ended after only a few installments.

Because he was in dire financial straits, Merrem began to study law (1783) to increase his chance of permanent employment. Finally he got a position in 1785, at the University of Duisburg, but as a professor of mathematics, physics, and economics, and also as Rector, or president, of the university (1792 and 1800), which left him little time for zoology. In 1804 he moved to the University of Marburg where he lectured on economics and founded a zoological institute, but he was exhausted from poor health and dogged by his money troubles. His time was dissipated in areas of no real interest to him, so that the great promise of his early work was never realized. He died at Marburg on 23 February 1824, a broken man.

Merrem is better known as an ornithologist, not only for his several important books but for his discovery of the air-sac respiratory system of birds. His most elaborate herpetological book is little known today because it was published mostly at Merrem's own expense and thus in such a small edition that it is now very rare. His "Beitraege zur Geschichte der Amphibien" (only three parts appeared, in 1790, 1790 [second edition 1829], and 1821) included detailed descriptions of reptiles, with complete synonymies and numerous good, handcolored illustrations, many drawn by Merrem himself; the geographic coverage is worldwide, although in these three parts mostly snakes are described, but it is a more competent work than that of Count de Lacepède or Merrem's other contemporaries.

What a pity it was for the development of herpetology that Merrem's magnificent series had to be terminated. He also published some shorter papers, but it was his "Versuch eines Systems der Amphibien" (1820, alternate title "Tentamen Systematis Amphibiorum"), written in parallel German and Latin texts, that is best known today. In it he covered all known species, but in far more abbreviated form than in his "Beitraege"; he also separated reptiles from amphibians, although the credit for recognizing amphibians as a class apart from reptiles is generally accorded to P.-A. Latreille (1825).

• *References*: "Memoriam . . . Blasii Merremii," by C. F. C. Wagner, Acad. Bayrhoffer, Marburg, 36 pages, 1824; "Blasius Merrem," p. 438-439. *In* F. Gundlach, Catalogus Professorum Academiae Marburgensis. G. Braun, Marburg, 1927; "Blasius Merrem," p. 278-279. *In* G. von Roden, Die Universität Duisburg, W. Braun, Duisburg, 1968; Stresemann, 1975 (pp. 60-63). • *Portrait*: Courtesy Stadtarchiv Duisburg. • *Signature* (two forms): Universitätsbibliothek Göttingen, courtesy Mr. Grobe, and Museum für Naturkunde der Humboldt-Universität zu Berlin, courtesy Rainer Günther.

RUSSELL, Patrick (1726-1805).

A physician by profession, Russell was the first to study Indian snakes systematically. He was born in Edinburgh, Scotland, 6 February 1726, where he attended the university. He joined his older half-brother, Alexander Russell, in Aleppo, Syria, in 1750; he eventually succeeded him as physician to an English factory there and returned to England in 1772. In 1782 he accompanied his younger brother, Claud Russell, an officer in the East India Company, to Vizagapatam (now Visakhapatnam), on the Coromandel coast of India north of Madras. Three years later he was appointed naturalist to the company, and he largely gave up his medical practice to study the local botany, fishes, and snakes. In 1787, he issued a practical guide for distinguishing the poisonous snakes of India, which was published by the Madras Government and widely distributed throughout India. He returned to England in January 1789 and spent the remainder of his life writing books. He died in London on 2 July 1805 with many of the snakes undescribed and the book unfinished.

Russell's book "An Account of Indian Serpents" (1796) and its sequel "A Continuation of an Account of Indian Serpents" (1801-1802, but finished 1807-[1810] from his notes) are one of the great classics of herpetology and may well be the most physically imposing books ever written exclusively on a herpetological topic (they are 55 cm high and weigh 8 kg). The magnificent handcolored plates are equally remarkable and depict the snakes lifesize if rather flat. There is a total of 92 plates displayed on 87 leaves, which accounts for the widespread misinterpretation of the exact number. Russell provided native names for most species but Latin ones for only a few. Thus many of the scientific names of species he first recognized are attributed to later authors who simply based their descriptions on Russell's accounts. For example, *Vipera russellii* was well described and figured by Russell in 1796, but was not named until 1797, by George Shaw and Frederick P. Nodder, who justly immortalized Russell in their choice of a Latin name. Besides his desire to distinguish the truly poisonous species which was the practical task given to him by the East India Company, Russell had a special interest in snakebite. He was also interested in the venom delivery apparatus and hood expansion in cobras, subjects on which he contributed a few papers to the Royal Society in London, of which he was elected a Fellow in 1777.

• *References*: "Memoir of the Life and Writings of Patrick Russell," anonymous, p. ix-xv. *In* P. Russell, Continuation of an Account of Indian Serpents, G. and W. Nicol, London, 1801[-1810]; "Patrick Russell," by G. S. Boulger, Dict. Natl. Biogr., 17: 469-470, 1921. • *Portrait* (1794): Bibliothèque Nationale, Paris. • *Signature*: British Museum (Nat. Hist.), courtesy Ann Datta.

SHAW, George (1751-1813).

George Shaw, early English naturalist and prolific writer, was born on 10 December 1751 at Bierton, Buckinghamshire, where his father was a church vicar. He graduated from the University of Oxford (B.A. Magdalen College 1761; M.A. 1772), then was ordained a deacon and performed duty for several years. Later, however, he gave up church work to attend medical school at the University of Edinburgh for three years, before returning to Oxford where he was awarded a medical degree in 1787. He set up practice in London and promptly became a prominent member of the scientific establishment there, taking part in founding the Linnean Society (1788) and being elected Fellow of the Royal Society (1789). In 1791, he became Assistant Keeper of the natural history section of the old British Museum and succeeded E. W. Gray as Keeper in 1807, a position he held until his death, in London, on 22 July 1813.

Shaw was not a researcher or a museum man. His real interest lay in educating the public in science through lectures and encyclopedias, and he sought to use the British Museum for this purpose. But the museum's trustees refused his request to begin public lectures at the museum and rebuked him for dissecting and drawing museum specimens for use in his encyclopedias, yet Shaw's vision of a museum as an institution for public education was far ahead of its time. It is also true, however, that under Shaw's keepership the zoological collections reached their nadir.

Shaw never wrote anything on an exclusively herpetological topic, but "Amphibia"—in the Linnaean sense then including reptiles—were a natural part of the many comprehensive works that he produced. This, coupled with the early date at which he wrote and the fact that he used binomial nomenclature, resulted in his being the first to apply proper Latin names to many reptiles and amphibians from throughout the world. Some of the commonest and best known species such as *Ambystoma mexicanum*, *Rana catesbeiana*, and *Vipera russellii*, which had been described previously by other authors, were named by him.

Shaw's work best known to herpetologists is the Amphibia section, in two parts (1802 [1801?]) with 140 plates, from his extensive series "General Zoology or Systematic Natural History." This was the first world review of amphibians and reptiles in English. Many other amphibians and reptiles were first described by Shaw in "Naturalist's Miscellany" (1789-1813), in collaboration with Frederick Nodder, in John White's "Journal of a Voyage to New South Wales" (1790), and in several additional works.

• *References*: "George Shaw (1751-1813)," by B. B. Woodward, Dict. Natl. Biogr., 17: 1374, 1921; Gunther, 1980 (pp. 33-37). • *Portrait*: Courtesy British Museum (Bloomsbury). • *Signature* (1799): British Museum (Nat. Hist.), courtesy Ann Datta.

CUVIER, Georges (1769-1832).

The great French comparative anatomist and paleontologist Cuvier was born on 23 August 1769 in Montbéliard, then part of Württemberg but today in eastern France (he did not become a French citizen until 1793). His original given names were Jean-Léopold-Nicolas-Frédéric to which the name Dagobert was added later by his godfather, but following the early death of his oldest brother, Georges, he adopted that name, and has become generally known as

Georges Cuvier. Cuvier was precocious and displayed his enormous memory even as a boy. His father, a former French army officer, intended him to be a Lutheran minister, but fortunately for science the duke of Württemberg was seeking gifted students for his new Caroline University in Stuttgart, where Cuvier began in 1784 at the age of 15. He specialized in administrative law and economics but much preferred K. F. Kielmeyer's lectures on anatomy. At graduation (1788) he could find no suitable employment so he accepted a position in Normandy as a private tutor. Being near the sea and with much free time, his interest in natural history blossomed, especially in fishes and anatomy.

In 1795, the main events of the French Revolution over, Cuvier wrote to B.-G.-É. Lacepède and Étienne Geoffroy Saint-Hilaire, professors at the Muséum National d'Histoire Naturelle in Paris, and sent selections of his unpublished works, whereupon the latter invited him to Paris as his collaborator. Cuvier's brilliance as a researcher and teacher was immediately recognized. He shortly became a professor at the museum, at the age of 25, a member of the Institut de France in 1796, and in 1800 he succeeded L.-J.-M. Daubenton as a professor at the Collège de France. Thereafter he held numerous important government posts in addition to his academic ones, becoming in 1811 a chevalier, in 1819 a baron, and finally, in 1831, a peer of France, titles which were in keeping with his well-known vanity. He was said to have been pliant to superiors while authoritarian to others, and to have showed favoritism to those who flattered him and to friends and relatives, including his younger brother, Frédéric Cuvier, for whom he arranged a position as mammalogist at the museum. He died in Paris on 13 May 1832.

Cuvier was responsible for building the great collections of the museum, then the largest research organization in the world. He published three major works on general zoology: "Tableau Élémentaire" (1797), "Leçons d'Anatomie Comparée" (1800, 1805), in collaboration with A.-M.-C. Duméril and G.-L. Duvernoy, and his classic "Le Règne Animal," a summary of all animals (first edition in four volumes, 1817 [1816], reprinted 1969; second edition, in five volumes, 1829-1830, in collaboration with P.-A. Latreille). In the "Règne" he described many new species of amphibians and reptiles, but gave most of them vernacular, rather than Latin names, and thus authorship was claimed by later writers. Together with Achille Valenciennes, he wrote the 22-volume "Histoire Naturelle des Poissons" (1828-1849, reprinted 1969), the basis for modern ichthyology.

In addition to these works, Cuvier collaborated with Alexandre Brongniart in studies of the fossiliferous beds of the Paris environs, thus becoming leader of the field of paleontology. He also found himself in increasing disagreement with museum colleagues on the question of immutability of species, which Cuvier, perhaps partly driven by his religious beliefs, considered unchanging. He thus attributed the different forms in geological strata to destruction by catastrophic events. Geoffroy Saint-Hilaire and J.-B. Lamarck, on the other hand, believed in gradual transformation, based in part on studies of crocodilians and turtles. Geoffroy had been part of Napoléon's ill-fated Egyptian expedition (1798-1799) which, among other spoils, brought back the mummified remains of 3000-year-old crocodiles, turtles, and other organisms, and beginning in 1802 these were examined for evidence of rate of transformation. Thus began a long series of papers by both Geoffroy and Cuvier on crocodilians and turtles—fossils, mummies, and living forms—to provide evidence for their conflicting views. Coincidentally, the new taxa *Crocodylus acutus*, *C. cataphractus*, *C. rhombifer*, *Paleosuchus palpebrosus*, *Trionyx gangeticus*, and *T. subplanus* and the genera *Alligator* and *Trionyx* were named by them. After Charles Darwin, Cuvier's glory inevitably diminished.

Cuvier's herpetological work was only incidental to broader projects, yet he made important contributions. One of the earliest arose from his idea of the correspondence of body parts and his belief that none could change without altering the others. To him this meant that each part alone showed the nature of the others and led to his famous reconstructions of fossils. In 1726 J. J. Scheuchzer, who like many at that time believed in the literality of the Bible, published the description of a meter-long fossil salamander, but not knowing of any giant salamanders (*Andrias japonicus* was not described until 1837), he believed it to be the skeleton of a man who drowned in the biblical Flood, and named it "Homo diluvii testis." Scheuchzer's religious zeal was fully expressed in the date on the title page of his book, not 1726 but "MMMM XXXII"—4032 years after the Great Flood! In 1811, Cuvier examined Scheuchzer's illustration and predicted that it was a salamander based on the exposed bones, and later proved it by uncovering the rest of the skeleton. In 1837, J. J. von Tschudi finally named it *Andrias Scheuchzeri*.

Amphibians and reptiles formed a natural part of Cuvier's "Règne," although the sections treating them were only superficial reviews of the world's herpetofauna; the details were left to be filled in later by his protégé A.-M.-C. Duméril. Nevertheless, the world fame of Cuvier led to many later editions and translations of the "Règne," which therefore had enormous influence for decades. Herpetologically, the most important of these were the "Iconographie" of F.-E. Guérin-Méneville (atlas in 1829 [-1844], text 1844) and the so-called "Disciples edition" of Cuvier (reptiles by G.-L. Duvernoy, 1836-1849 [1837-1842]) in France, the German editions of H. R. Schinz (1822) and F. S. Voigt (1832), and the English edition edited by E. Griffith and E. Pidgeon (1831 [1830-1831]), with the important synopsis added by John E. Gray.

• *References*: "Le Baron George [*sic*] Léopold-Chrétien-Frédéric Cuvier," by C. L. Laurillard, Biogr. Univ. (Michaud) Ancien. Mod., new ed., 9: 590-600, 1855; "Iconographie de Georges Cuvier," by L. Bultingaire, Arch. Mus. Natl. Hist. Nat., ser. 6, 9: 1-11, 1932; "Georges Cuvier," by Y. Chatelain, Dict. Biogr. Franç., 9: cols. 1438-1442, 1961; "Georges Cuvier Zoologist," by W. Coleman, Harvard Univ. Press, Cambridge (Massachusetts), (xii), 212 pages, 1964; "Georges Cuvier," by F. Bourdier, Dict. Sci. Biogr., 3: 521-528, 1971. • *Portrait*: Adler collection. • *Signature* (1823): Royal Society, London, courtesy K. Moore.

LATREILLE, P.-A. (1762-1833).

Pierre-André Latreille, a French priest who turned naturalist, was an eminent entomologist who, among other interests, also studied reptiles. He was born on 29 November 1762 in Brives (now Brive), Corrèze, where his interest in natural history began at an early age. He was educated at the Collège du Cardinal-Lemoine in Paris, then graduated from the University of Paris in 1780 and was ordained a priest in 1786. But his ecclesiastical career was cut short by the French Revolution, during which so many Catholic clergy perished. Latreille himself was briefly imprisoned but won his freedom when he found a new species of beetle in his cell and smuggled it out to some naturalists who used their influence to have him released. Later, he abandoned the priesthood and joined the staff at the Muséum National d'Histoire Naturelle in Paris, eventually succeeding his colleague, J.-B. Lamarck.

Besides entomology, Latreille also published on animal classification and he authored one of the classics on the subject, "Familles Naturelles du Règne Animal" (1825). In this book he was the first to separate amphibians as a class from reptiles, following his colleague Alexandre Brongniart (1799) who first recognized the distinctive characteristics by which amphibians can be naturally separated from reptiles. Latreille was recognized as one of the leading zoologists of his day, like Georges Cuvier, and he received many honors, among them election to the Institute de France and the Legion d'Honneur. He died in Paris on 6 February 1833.

Latreille's herpetological work began with his little book on French salamanders (1800). In 1801 he co-authored the four-volume "Histoire Naturelle des Reptiles," with C. S. Sonnini, part of the so-called Déterville edition of Georges Buffon's great encyclopedia. This work, which covers the herpetofauna worldwide, was almost entirely written by Latreille, as Sonnini wrote only the section on salamanders and the preliminary "Discours." The importance of this book was sufficient to justify a new edition in 1826 and a reissue four years later; all three versions have 54 plates, and come in either handcolored or plain editions. By way of comparison, Sonnini and Latreille's work was far less comprehensive, both in text and illustration, than the "Histoire Naturelle . . . des Reptiles" (1801-1803) by Latreille's sometimes friend F.-M. Daudin, which was written for a rival publishing firm.

• *References*: "Pierre-André Latreille," by A. J. L. Jourdan, Biogr. Univ. (Michaud) Ancien. Mod., 23: 329-331, n.d.; "Pierre-André Latreille," by J. V., Nouv. Biogr. Gén., 29: cols. 850-854, 1862; "Pierre-André Latreille a Brive," by L. de Nussac, G. Steinheil, Paris, vii, 264 pages, 1907; Harper, 1940; "Pierre-André Latreille," by R. W. Burkhardt, Jr., Dict. Sci. Biogr., 8: 48-49, 1973. • *Portrait* (1832): Courtesy Bibliothèque Nationale. • *Signature* (1791): From de Nussac, 1907.

DAUDIN, F.-M. (1774-1804).

François-Marie Daudin, born in Paris on 24 or 25 March 1774, was a French amateur naturalist with great promise who, nevertheless, had a tragic career. A childhood illness crippled his legs and, thus, at an early age, he devoted himself to physics and natural history but the latter became predominant. This was a time of great ferment in France, both politically and scientifically. Georges Buffon had died in 1788 and now the Linnaean opposition, which wished to classify all organisms according to binomial principles, dared to emerge. During and immediately following the French Revolution, important herpetological treatises were published in Paris by L.-J.-M. Daubenton (1784), B.-G.-É. de Lacepède (1788-1789), P.-J. Bonnaterre (1789-1790), and C.-N.-S. Sonnini and P.-A. Latreille (1801), but the most comprehensive of them was that by Daudin.

Daudin's first major work, "Traite Élementaire et Complet d'Ornithologie" (1800), was the first true handbook of birds, combining Linnaean systematics with Buffonian natural history, including both anatomy and physiology. However, Daudin's greater interest was herpetology, and in his few remaining years he published three different titles on the subject. He was ably assisted in this work by his wife, Adele Daudin, who also did much of the artwork. But his books were not commercial successes and this, coupled with personal financial problems, led to despair. The Daudins eventually died of tuberculosis, he only a few weeks after his wife, both early in 1804, in Paris. He was not yet 30 years old.

Daudin's first herpetological book, written according to the same plan as his "Ornithologie," is little known and was never completed. "Histoire Naturelle des Quadrupèdes Ovipares" (1800) was to have extended to 30 parts, each with six plates and accompanying text, but only the first two parts, on frogs, appeared. Commercial circumstances interrupted publication, but the text and plates were later re-used in his less elaborate but still respectable review of the frogs of the world, "Histoire Naturelle des Rainettes, des Grenouilles et des Crapauds" (1802), issued in both quarto and folio editions and with 38 plates (either handcolored or uncolored).

Daudin's most extensive herpetological work, and the one on which his reputation rests, was the "Histoire Naturelle, Générale et Particulière des Reptiles" (1801-1803), in eight volumes with 100 plates, issued in editions with handcolored or uncolored plates or both; a reprint, in which the type was reset in most volumes, was published later. This book, part of the so-called Sonnini edition of Buffon's encyclopedia, was the standard herpetological reference of its day, acknowledged even by Daudin's grudging critics like Georges Cuvier, and was not superseded until the "Erpétologie Générale" of A.-M.-C. Duméril and Gabriel Bibron (1834-1854).

For this book, Daudin stated that he had examined over 1100 specimens belonging to 517 species, triple the number catalogued by Count de Lacepède only 25 years before; his genera were well defined, better than Lacepède's or those in Sonnini and Latreille's "Histoire Naturelle des Reptiles," but they were not yet in natural order, something that did not occur for the genera of reptiles until Duméril and Bibron. Linnaean nomenclature was used throughout. New and familiar genera—for example, *Eryx*, *Python*, *Bungarus*, *Lachesis*, *Erpeton*, and *Ophisaurus*—were named by Daudin and are often cited as from his "Reptiles," but several had been described earlier by him in Parisian journals. Thus, despite his personal handicaps and the fact that he was an amateur and not a formal member of the academic community, Daudin, while still in his twenties, managed to produce one of the great classics of herpetology.

• *References*: "Daudin (François-Marie)," by G. Cuvier, Biogr. Univers. (Michaud) Ancien. Mod., new ed., 10: 161-162, 1855; "Daudin (François-Marie)," by St. Le Tourneur, Dict. Biogr. Franç., 10: col. 266-267, 1965; Harper, 1940. • *Signature* (1803): Library, Muséum National d'Histoire Naturelle, Paris, courtesy V. Van de Ponseele.

BOJANUS, Ludwig Heinrich (1776-1827).

Bojanus, a veterinary anatomist and author of the finest iconography of the turtle, was born in Buchsweiler (now Bousville), France, in what was then the province of Alsace, on 16 July 1776, but he soon moved to Darmstadt. He attended the University of Jena (M.D. 1797) and then practiced medicine for a short time in Darmstadt. Shortly afterwards he was invited by the government of Hesse-Darmstadt to establish a school of veterinary medicine there but after many frustrating delays he accepted an offer from Russian officials to become Professor of Veterinary Medicine at the Imperial University of Vilna. He spent 18 productive years there before returning to his boyhood home in Darmstadt, in 1824, on medical advice to seek a more moderate climate, but in the meantime he had become a respected member of the Russian scientific community and was appointed Rector (or president) of the University of Vilna in 1822. He died in Darmstadt on 2 April 1827.

Bojanus published some 70 titles on anatomy and veterinary medicine, of which seven concerned the anatomy of turtles and snakes. His most famous work, and the book on which his reputation rests, was his "Anatome Testudinis Europaeae," published in two parts (1819, 1821; reprinted 1902, again in 1970), which is arguably the best atlas of any submammalian vertebrate ever published. The book includes 40 folio plates which depict the detailed anatomy of the European pond turtle, *Emys orbicularis*; the "text" is, in fact, a series of captions to these plates. Bojanus began working on this atlas shortly after arriving in Vilna and devoted a decade of work to the project before the first part was issued. He made most of the original drawings, paid for the printing himself (the original edition was limited to only 80 copies), purchased paper of the highest quality to hold the fine lines of the etched copper plates, and even imported a German artist to do the final drawings. The result was a magnificent atlas whose value continues to endure after more than a century and a half.

• *References*: "Ludwig Heinrich von Bojanus," by A. W. Otto, Nova Acta Akad. Naturforsch. Halle, 15(1): xxxix-xlv, 1831; "Memoria . . . Professoris Ludovici Henrici Bojani," by E. Eichwald, Acad. Med.-Chirurg. Typogr., Vilna, ii, 57 pages, 1835; "Ludwik Henryk Bojanus," by Z. Fedorowicz, Polska Akad. Nauk, Inst. Zool., Memorabilia Zool., 1: 1-47, 1958; "Bojanus and the Anatomy of the Turtle," by A. S. Romer, p. iii-v. *In* Bojanus's Anatome Testudinis Europaeae, repr. ed., Soc. Study Amphib. Rept., Athens (Ohio), 1970. • *Portrait and signature*: From Eichwald, 1835.

BOIE, Heinrich (1794-1827).

Boie, a German naturalist of great promise who died very young, was born on 4 May 1794 at Meldorf in Holstein. Inspired first by J. F. Blumenbach's lectures at Göttingen and later by those of Friedrich Tiedemann at Heidelberg, he was hired in 1821 as curator of the Rijksmuseum van Natuurlijke Historie in Leiden, by C. J. Temminck, the museum director. Shortly afterward he began writing a major work, "Erpétologie de Java," based upon the collections made by Carl Reinwardt, Heinrich Kuhl, and J. C. van Hasselt, members of the Natuurkundige Commissie van Nederlandsch Indië (Natural History Commission of the Dutch Indies). After Kuhl died in 1821, Boie was appointed to replace him in Java but before he could depart (1825), van Hasselt had also succumbed (1823). The rigors of the Dutch Indies eventually claimed Boie himself (who died in Java on 4 September 1827 at Buitenzorg, now called Bogor) as well as H. C. Macklot (1832) and E. A. Forsten (1843); only Reinwardt and Salomon Müller, among those commission members of herpetological interest, survived to return to Holland.

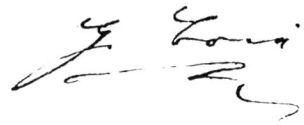

Boie's "Erpétologie," though virtually complete at the time of his departure for Java, was never published *en toto* but is frequently cited in the literature since Hermann Schlegel as well as A.-M.-C. Duméril and Gabriel Bibron made frequent references to it in their books. An extract was published by Schlegel, Boie's successor at the Rijksmuseum, in Férussac's *Bulletin* in Paris (1826), and Temminck had a few plates printed in Brussels in 1829-1830, but these were never published due to the Belgian rebellion against the Netherlands in 1830-1831. Most of Boie's few publications were in Oken's *Isis*, including a review (1826) of the collections made by J. Cock Blomhoff when he was director of the Dutch trading station on Decima Island, located in Nagasaki harbor, a paper in which many of the commonest Japanese reptiles and amphibians were first described.

Schlegel published a series of Boie's letters (in Oken's *Isis*, 1828), written just before Boie's death, in which a number of new taxa were described. Later, Boie's older brother, Friedrich Boie, a lawyer by profession, took over his dead brother's manuscripts and, as his literary executor, had some of them published. Several new genera and species were named in these posthumous papers, but Friedrich Boie is usually credited with the new names under the prevalent interpretation of the "International Code of Zoological Nomenclature" (1985). Alas, Heinrich Boie's "Erpétologie de Java" was not among those manuscripts that were published, a great loss to our early knowledge of the rich herpetofauna of that region.

• *References*: "Levensschets van Hendrik Boie," by J. A. Susanna, P. M. Warnars, Amsterdam, 288 pages, 1834; "Die Familie Boie," by K. Boie and R. Boie, Zeitschr. Gesellsch. Schlesw.-Holst. Gesch., 39: 102-106, 1909; E. Stresemann, 1975 (pp. 133-137). • *Portrait*: From Susanna, 1834, courtesy Marinus S. Hoogmoed. • *Signature*: Museum für Naturkunde, Humboldt-Universität zu Berlin, courtesy Rainer Günther.

WIED-NEUWIED, Maximilian zu (1782-1867).

Alexander Philipp Maximilian, prince of Wied-Neuwied, one of the great explorer-naturalists and ethnologists of the 19th Century, was also a specialist on reptiles. Born in Neuwied, along the Rhine River near Koblenz, Prussia, on 23 September 1782, his early interest in natural history was encouraged by his mother. He became a student of J. F. Blumenbach at the University of Göttingen, with an intended career in science, but the Napoleonic era intervened. Wied served in the Prussian army, was captured by the French at the Battle of Jena (October 1806), and later repatriated. Resuming military service, he rose to the rank of major-general and entered Paris with the Allied armies (1814), but soon resigned to resume his original interests.

Wied's first exploration was in Brazil, where he spent the period 1815-1817 studying the natural history and primitive Indian tribes of the forested areas in the coastal states north of Rio de Janeiro. His collections there were rich in reptiles, especially snakes. On returning to Neuwied he published his observations in three large works: a semi-popular account of the journey ("Reise nach Brasilien," in two volumes, 1820-1821, with appendix in 1850) which was translated into several languages, a scientific treatise on the vertebrates in a magnificent atlas of handcolored plates in folio ("Abbildungen," in 15 parts, [1822-]1823-1831), and a separate scientific text ("Beiträge," in four volumes, 1825-1833, with herpetology mainly in volume one).

Prince Max, as he was known to his contemporaries, was the first well-trained naturalist to explore Brazil, and these books established his reputation as a biologist and ethnologist of the first rank. The 90 plates in the "Abbildungen," 56 of them on reptiles, were taken from originals drawn and colored by the prince himself, and these formed the basis for his descriptions of many new species. Most of his specimens are now in New York City, at the American Museum of Natural History, which purchased his zoological collections.

In 1832, his Brazilian adventure completed, Wied embarked on his second exploration, a journey to the United States to compare the Indians and natural history of North America to those of South America. With his artist, Carl Bodmer, he spent time with the naturalists Thomas Say and Charles A. LeSueur at New Harmony, Indiana, collecting local reptiles and amphibians. Then, in the spring of 1833, Wied and Bodmer departed St. Louis for a 4400-km journey up the Missouri River, then the most treacherous of American rivers, to Fort McKenzie, near the present-day city of Great Falls, Montana. On returning, they spent the winter of 1833-1834 at a fort in central North Dakota. This was a particularly brutal winter, with temperatures sometimes below -40°C, that very nearly cost Wied his life.

Eventually, however, Wied returned to Europe (1834) and spent the rest of his life studying his North American collections. His accounts of the trip are second in importance only to those of the American explorers Meriwether Lewis and William Clark as a record of the trans-Mississippian region. Wied's report ("Reise in das innere Nord-America," published in 20 parts, 1838-1841, and translated into several languages) was justly renowned and Bodmer's watercolors are only today being recognized for their great historical and artistic value.

Wied described a number of America's commonest amphibians and reptiles—*Hyla crucifer* and *Pseudemys scripta elegans* among them—from specimens obtained on this trip. His major report on the reptiles ("Verzeichniss der Reptilien, welche auf einer Reise im nördlichen America beobachtet wurden," 1865, reprinted 1978), illustrated with seven beautifully handcolored plates of turtles and salamanders by Bodmer, was published shortly before the prince's death, in Neuwied, on 3 February 1867.

• *References*: "Max Prinz von Wied-Neuwied," by F. Ratzel, Allg. Deut. Biogr., 23: 559-564, 1886; "Maximiliano, Principe de Wied. Ensaio Bio-bibliografico," by A. do Amaral, Bol. Mus. Nac. Rio de Janeiro, 7: 187-210, 1931; "Maximilian zu Wied," by P. D. Thomas, Dict. Sci. Biogr., 14: 328-329, 1976; Adler, 1979 (pp. 22-25); "Maximilian's America," vol. 1, ed. by J. C. Porter and P. Schach, Univ. Nebraska Press, Lincoln, in press. • *Portrait*: Hunt Institute, Pittsburgh. • *Signature* (1832): Joslyn Art Museum, Omaha (Nebraska), courtesy Joseph C. Porter.

SPIX, J. B. von (1781-1826).

Johann Baptist Spix (from 1820, von Spix), German zoologist and first explorer of Amazonian Brazil, was born in Höchstadt-an-der-Aisch, Bavaria, on 9 February 1781. He had planned a theological career and spent several years at seminaries in Bamberg and Würzburg, but in 1804 he decided to pursue medicine at the University of Würzburg, where he also studied natural history, and graduated with an M.D. degree in 1806. He practiced medicine in Würzburg, then Bamberg, but in 1811 became the first Curator of Zoology at the Bavarian Academy of Science in Munich. Correspondence with Alexander von Humboldt inspired him to plan an expedition to South America, and the opportunity came as the result of an unusual circumstance. The Napoleonic invasion of Portugal (1807) caused the Portuguese royal family to seek refuge in Brazil, and in 1817 it was arranged for the crown prince to marry the daughter of the Austrian emperor. In the wedding party of this future empress of Brazil was a large group of naturalists from throughout the Austrian Empire and Bavaria: Johann Natterer, J. C. Mikan, Guiseppe Raddi, and Spix, to mention those of herpetological interest.

Spix, with his compatriot C. F. P. von Martius, a botanist, thus became the first biologists to explore the interior of Brazil, since Prince Maximilian of Wied-Neuwied had confined his earlier exploration to the coastal forests, and Humboldt had been barred from Brazil altogether during his trip to South America. Spix and Martius started at Rio de Janeiro, moved inland to the coastal mountains, which they crossed, then went north through the dry caatingas and finally up the Amazon, where they witnessed the communal egg laying of the turtle *Podocnemis expansa* and watched a family of caimans play with the head of an Indian they had killed. The dangers and hardships were great but although Spix was quite ill, he continued to collect and make observations. After eight months in Amazonia, Spix and Martius departed from Belém for Europe (June 1820) with enormous collections from vast areas never visited before by scientists. Most of the species were new to science.

Before he died on 15 May 1826, in Munich, from a disease contracted in Brazil, Spix produced eight books either alone or in collaboration with others, covering each vertebrate class plus molluscs and arthropods. Three herpetological volumes were published: "Serpentum," including caecilians and amphisbaenians (1824, written by Spix's assistant in Munich, J. G. Wagler, from Spix's notes), "Testudinum et Ranarum" (1824, by Spix), and "Lacertarum," including the crocodiles (1825, by Spix). These elaborate works, with numerous handcolored plates, were reprinted twice, once by Spix's colleague and literary executor, C. F. P. von Martius (1838-1840), and again in 1981. Together with Wied-Neuwied's works, they provide the foundation for the study of Brazil's rich herpetofauna.

• *References*: "Johannes von Spix," by J. Gistl, Gallerie Denkwürd. Naturforsch. (suppl. to Faunus, 2) 2: 7-8, 1835; "Johann Baptist von Spix," by F. Ratzel, Allg. Deut. Biogr., 35: 231-232, 1893; Papavero, 1973 (pp. 65-69, Brazilian itinerary); "Johann Baptist von Spix," by A. P. M. Sanders, Dict. Sci. Biogr., 12: 578-579, 1975; "The Scientific and Political Contexts of the Bavarian Expedition to Brasil," by P. E. Vanzolini, p. ix-xxix. *In* J. B. von Spix and J. G. Wagler's Herpetology of Brazil, repr. ed., Soc. Study Amphib. Rept., Athens (Ohio), 1981; "Johann Baptist Ritter von Spix, sein Leben und sein wissenschaftliches Werk," by E. J. Fittkau, Spixiana, suppl. 9: 11-18, 1983; "Spix and Wagler Type Specimens," by M. S. Hoogmoed and U. Gruber, Spixiana, suppl. 9: 319-415, 1983. • *Portrait*: From Gistl, 1835. • *Signature* (1810): Courtesy Bayerische Akademie der Wissenschaften, Munich.

WAGLER, J. G. (1800-1832).

A promising German naturalist with a special interest in herpetology, Johann Georg Wagler died very young from a gunshot wound. Born in Nuremburg on 28 March 1800, Wagler attended Erlangen University and received a doctor's degree in 1820. He was associated with the Bavarian Academy of Science in Munich (beginning as an assistant in

FITZINGER, Leopold (1802-1884).

Leopold Joseph Franz Johann Fitzinger, one of Austria's most prolific zoologists of the 19th Century, was born on 13 April 1802 in Vienna. He developed an early interest in nature, especially plants, and at the age of 14 became a pharmacist's apprentice. The next year he attended the University of Vienna where he studied botany and zoology under J. F. Jacquin. However, in 1817 he quit his apprenticeship to become assistant at the Naturalienkabinett (forerunner of the Naturhistorisches Museum), directed by Jacquin's son-in-law, Carl von Schreibers. For 27 years, until 1844 when he became a paid staff member, Fitzinger's work at the museum was strictly volunteer, and he supported himself by working for an Austrian political party. At first he was responsible for the collections of all lower vertebrates, but he gave up fish in 1835 and added mammals to his duties in 1844. In recognition of his important early work, he was awarded honorary degrees by the universities of Königsberg (M.D. 1833) and Halle (Ph.D. 1834). Fitzinger retired from the museum in 1861, later directed zoos in Munich and Budapest, but returned to Vienna in 1873. He died on 20 September 1884 in Hietzing, near Vienna.

Fitzinger's publications cover a broad scope, including all vertebrate classes, some invertebrate groups, and fossils. His first important herpetological book, "Neue Classification der Reptilien" (1826), was an attempt at a natural

1819) and in 1820 began to assist J. B. von Spix and C. F. P. von Martius to work up the extensive collections from their Brazilian expedition of 1817-1820. The snakes were entrusted to Wagler and resulted in his first book, "Serpentum Brasiliensium" (1824, reprinted 1981), which was written in part from Spix's notes. In 1827, he was appointed a professor at the University of Munich, where he had many students, including Louis Agassiz, who had come to Munich primarily to work with Wagler's colleagues Lorenz Oken and Martius. Wagler died in Moosach, a small village near Munich, on 23 August 1832, nine days after accidentally shooting himself in the arm.

Wagler's master work, "Natürliches System der Amphibien" (1830), with a rare folio atlas of nine plates, was the result of ten years of labor. In it he classified all amphibians and reptiles (as well as birds and mammals), but the work was uneven. He showed great insight with the lizards, for example, using tongue structure and the mode of tooth insertion, but he hopelessly mixed the venomous and non-venomous snakes. He also issued a beautiful series of illustrations of amphibians and reptiles—36 handcolored plates with accompanying letterpress—in three parts under the title "Descriptiones et Icones Amphibiorum" (1828, 1830, 1833), the last part posthumously. These major treatises, together with others on birds, represent a prodigious output for someone who lived only to the age of 32.

• *References*: "Wagler," by J. Gistl, Gallerie Denkwürd. Naturforsch. (suppl. to Faunus, 2) 1-2: 1-6, 1835; "Johann Georg Wagler," by W. Hess, Allg. Deut. Biogr., 41: 776, 1896. • *Portrait*: From Gistl, Faunus, 1, 1832. • *Signature* (1820): Archives, Bayerische Akademie der Wissenschaften, Munich.

arrangement that also included descriptions of many new genera; it was partly based on the ideas and notes of his late friends Friedrich Hemprich and Heinrich Boie, both of whom died while quite young. Among other work, Fitzinger discovered the gill openings and male copulatory organs in caecilians (1833-1834), published a classification of turtles (1836 [1835]), and made many other important contributions to herpetology, some of them in the publications of his contemporaries such as J. G. Wagler, Charles L. Bonaparte, Johann Natterer, and A.-M.-C. Duméril and Gabriel Bibron.

In 1843, Fitzinger published his most important work, entitled "Systema Reptilium" (reprinted 1973). In fact, this was only part one—covering his taxon "Amblyglossae" (the geckos, chameleons, iguanids, and agamids) with a 21-page overview of all reptiles—but the series was never continued, perhaps fortunately so, as some have written. In this book Fitzinger incorporated many ideas of the natural philosophers (including, of herpetological interest, Lorenz Oken, J. B. von Spix, and J. J. Kaup) who believed that organisms should be divided and subdivided into some predetermined and repeating number of categories. For Fitzinger, the magic numbers were five and three, and categories thus were constructed and names given simply to complete the desired arithmetical pattern. Yet, in so doing, Fitzinger named many genera, some 70 of which are in use today for amphibians and reptiles, including such well-known genera as *Pseudacris, Eretmochelys, Gonatodes, Lampropeltis,* and *Thamnophis*. In 1864, he published a popular atlas of the world's amphibians and reptiles, illustrated with 108 beautiful plates, but he never abandoned the absurd ideas of "Naturphilosophie" and became an opponent of Darwinism.

• *References*: "Autobiographische Skizze," by L. Fitzinger, Almanach Akad. Wissensch. Wien, 35: 182-190, 1885; Siebenrock, 1901 (p. 445); "Leopold Fitzinger: His Life and Herpetological Work," by R. Mertens, p. iii-vi. *In* Fitzinger's Systema Reptilium, repr. ed., Soc. Study Amphib. Rept., Athens (Ohio), 1973. • *Portrait and signature*: From Mertens, 1973.

RAFINESQUE, Constantine S. (1783-1840).

Constantine Samuel Rafinesque (see *Note*), eccentric American naturalist and archeologist, was born in Galata, a suburb of Constantinople (now Istanbul), Turkey, on 22 October 1783, of European parentage. Rafinesque lived his early life in Marseilles, where he developed an interest in botany, but during the French Revolution his family was forced to take refuge in Italy. He was taught by private tutors there and also after returning to Marseilles in 1797, but it was the Parisian naturalist F.-M. Daudin who, through correspondence, most encouraged him. His father dead and the family's fortune lost in the Revolution, Rafinesque visited briefly in America (1802-1805) where, among other activities, he collected reptiles for Daudin. He then settled in Sicily before returning to America permanently in 1815.

He had been unsuccessful in securing professorships in Palermo and Philadelphia, but in 1819 he accepted a post as Professor of Botany and Natural History at Transylvania College, founded in 1780 in Lexington, Kentucky, as the oldest college west of the Allegheny Mountains frontier. Here, Rafinesque enjoyed his only formal academic success, but it was short lived. On returning from a trip in 1825, he found that the college president, who was opposed to things scientific, had deprived him of his rooms and board. Rafinesque stayed the full term at the college, where he had been a popular teacher, but resigned in 1826. He considered joining the community of scholars at New Harmony, Indiana, but decided to return instead to Philadelphia where he spent the rest of his life. He held some teaching positions, initiated various publishing ventures, and in 1835 co-founded a savings bank. He published numerous books on natural history, especially botany, and on medical flora, American Indians, banking, and numerous other topics. He died on 18 September 1840, overworked, disease-ridden, and unappreciated in his adopted country.

Rafinesque was perhaps the most gifted American naturalist of his era, yet he missed greatness by a combination of factors, including an idiosyncratic temperament. It is said that he lived a century too soon, for he had advanced views about evolution, even acknowledged by Charles Darwin himself, and Rafinesque advocated natural classifications of organisms rather than the prevailing artificial systems of the Linnaeans. Neither endeared him to his erstwhile colleagues. But it was his mania for describing new species (he named more than 6700 organisms, mostly plants), often based on the flimsiest research and sometimes even on imaginary organisms—John James Audubon once wrote him about mythical fish and turtles which he promptly named as new species—that brought him censure and estrangement from the American scientific community. Among his harshest critics were the herpetologists Thomas Say and Richard Harlan.

Some of Rafinesque's articles were mixtures of science and fiction, such as that on water snakes and sea serpents (1819), yet others were good natural history. He published about 15 articles of a herpetological nature (1814-1832), and these included descriptions of several well-known genera (*Necturus*, *Eurycea*, and *Gopherus*) and species (*Notophthalmus viridescens*, *Desmognathus fuscus*, *Sistrurus catenatus*, and *Crotalus viridis*). Since the standard American journals were closed to him, most of his herpetological titles were contained in the *Kentucky Gazette* (1822), a local newspaper, and in two journals published by Rafinesque himself: *Annals of Nature* (1820) and *Atlantic Journal and Friend of Knowledge* (1832). Rafinesque's herpetological titles from these three serials were reprinted in 1962-1963.

• *References*: "The Life and Work of Rafinesque," by F. W. Pennell, p. 10-70, and "Herpetology and Rafinesque," by W. M. Clay, p. 84-91. *In* L. A. Brown (ed.), Rafinesque Memorial Papers. Transylvania Coll. Bull. 15, 1942; "Constantine Samuel Rafinesque," by J. Ewan, Dict. Sci. Biogr., 11: 262-264, 1975; Adler, 1979 (pp. 7, 9-12); "Fitzpatrick's Rafinesque," rev. ed. by C. Boewe, M & S Press, Weston (Massachusetts), xiii, 327 pages, 1982. • *Portrait* (1810): From Goode, 1901. • *Signature* (1819): Courtesy Library, Academy of Natural Sciences of Philadelphia, courtesy Marsha Gross. • *Note*: Rafinesque sometimes appended to his last name that of his mother's family, Schmaltz.

SAY, Thomas (1787-1834).

Thomas Say, born in Philadelphia on 27 June 1787, was one of America's leading naturalists of his day. His great uncle, the naturalist William Bartram, interested him in insects when he was a boy, but Say's father was opposed and arranged a commercial partnership with a pharmacist, John Speakman, and thus, unwittingly secured his son's future in natural history. The business shortly went bankrupt, but not before Speakman, Say, and others, meeting in a rear room of their shop in 1812, founded the Academy of Natural Sciences. The academy's tiny museum was contained in a rented room in which, after the bankruptcy, Say lived and he was totally devoted to the museum's development. It is said that he slept beneath the skeleton of a horse and fed himself on bread and milk, subsisting on 75 cents a week! Until his permanent departure from Philadelphia in 1825, Say was the mainstay of the academy: collector, curator of collections, and producer of the academy's *Journal* which he typeset and printed himself.

There were some interruptions, however. Say saw brief service during the War of 1812-1815, after which he travelled to the Sea Islands of Georgia and Spanish Florida (1817-1818). He joined the Yellowstone Expedition in 1819 as zoologist under Major Stephen H. Long, which established military posts along the Missouri River to protect the growing fur trade. The party travelled upriver to Council Bluffs in what is now Iowa, then across the Great Plains to the Rocky Mountains of present-day Colorado. Say's contributions to the early zoology of this area often are not fully recognized, since his discoveries of new species—including the well-known taxa *Bufo cognatus*, *Crotaphytus collaris*, *Scincella lateralis*, and *Elaphe obsoleta*—were published as scattered footnotes in Edwin James's account of Long's expedition (1823).

On returning to Philadelphia in 1820, Say took on additional duties as Curator of the American Philosophical Society (1821-1827) and Professor of Natural History at the University of Pennsylvania (1822-1828). In 1823, he accompanied Long on a second western trip, via Fort Dearborn (now Chicago), to explore the headwaters of the Mississippi River.

During the 1820s a communistic settlement began at New Harmony, along the Wabash River in southwestern Indiana. William Maclure, wealthy patron of the Academy of Natural Sciences, who saw the opportunity for New Harmony to become the center of education in America, persuaded Say and some other naturalists then living in Philadelphia, including Charles A. LeSueur and Gerard Troost, to join the experiment. Say became Superintendent of Literature, Science, and Education for the community and, after its failure in 1827, remained at New Harmony, except for a brief trip to Mexico (1827-1828), until his death there on 10 October 1834.

Say's seminal contributions were in entomology and conchology, in which areas he published major works; indeed, among other titles he is regarded as Father of American Entomology. Herpetology was only one of many areas of general interest, yet he did publish a catalogue of American turtles (1824 [1825]) and numerous shorter papers describing new species from the East (e.g., *Carphophis amoenus* and *Regina septemvittata*), as well as critiques of papers on American forms by other, mostly European, authors. Say was the first zoologist to visit the Great Plains and as such became the describer of many of the commonest and most widespread amphibians and reptiles of that vast region.

• *References*: Goode, 1901 (pp. 447-449); "Thomas Say. Early American Naturalist," by H. B. Weiss and G. M. Ziegler, C. C. Thomas, Springfield (Illinois), xv, 260 pages, 1931; "Thomas Say," by E. N. Shor, Dict. Sci. Biogr., 12: 132-134, 1975; Adler, 1979 (pp. 12-14, 21, 23). • *Portrait*: From Goode, 1901. • *Signature* (1819): From Weiss and Ziegler, 1931.

HARLAN, Richard (1796-1843).

Richard Harlan, American vertebrate paleontologist and comparative anatomist, was born in Philadelphia on 19 September 1796. While still a medical student, he joined the Academy of Natural Sciences, America's first natural history society, and spent a year at sea as ship's surgeon on a voyage to India, before graduating from the University of Pennsylvania (M.D. 1818). He then became a demonstrator at Parrish's private anatomical school in Philadelphia, before accepting the post of physician to the city's Dispensary (1820-1822) and Alms House Hospital (1822-1838). As a city employee, Harlan was intimately involved in emergency measures taken for the cholera epidemic in Philadelphia in 1832.

Concurrently, Harlan was a lecturer on comparative anatomy at Peale's Museum, where other lecturers of herpetological interest included Thomas Say and Gerard Troost. Throughout his time in Philadelphia, Harlan made numerous field trips, mainly in search of fossils, to the New Jersey marl pits, to the mountains and caverns of Virginia, and along the Ohio River as far west as Cincinnati. Many of America's earliest explorers and naturalists were among his friends and sent him specimens from their travels, including Stephen H. Long, Constantine S. Rafinesque, and especially John James Audubon whose work was strongly supported by Harlan.

Harlan's research in paleontology and comparative anatomy was firmly Cuvierian in tradition and, like Cuvier, Harlan believed in the immutability of species. Harlan produced substantial works on the American fauna. His first, "Fauna Americana" (1825), dealt with mammals and received critical reviews. Harlan's "Genera of North American Reptilia, and a Synopsis of the Species" (in two parts, 1826-1827, in the Academy's *Journal* [reprinted 1978]) was the first comprehensive catalogue of his country's herpetofauna; it was re-issued in 1827 as a separate, privately-printed book entitled "American Herpetology." This was later revised in his book of collected papers, "Medical and Physical Researches" (1835, herpetological section reprinted in 1978). These catalogues were an essential prelude to John E. Holbrook's great synthesis of North American herpetology a decade later.

During this period, Harlan also published many shorter journal articles in which he described numerous new species from America including the familiar *Phrynosoma cornutum* and *Opheodrys vernalis*, as well as several exotics from the West Indies and the giant *Geochelone elephantopus* of the Galapagos Islands. Other papers by him illustrated the anatomy of some of America's more peculiar species, such as the large salamanders (*Amphiuma*, *Cryptobranchus*, and *Necturus*) whose combination of larval and adult characteristics so confused European classifiers. This often brought Harlan into conflict with Continental authorities, and Harlan thus became one of the persons who helped to develop an American zoology independent from the European orbit.

Harlan visited Europe twice (1833 and 1838-1840), working with the English anatomist Richard Owen during the second trip. He moved his medical practice to New Orleans early in 1843, and died there suddenly on 30 September 1843.

• *References*: Goode, 1901 (pp. 450-451); "Richard Harlan," by W. J. Bell, Jr., Dict. Sci. Biogr., 6: 119-121, 1972; Adler, 1979 (pp. 11-13). • *Portrait*: Courtesy Historical Society of Philadelphia. • *Signature* (1823): Library, Academy of Natural Sciences of Philadelphia, courtesy Marsha Gross.

GRAVENHORST, J. L. C. (1777-1857).

Johann Ludwig Christian Carl Gravenhorst, German entomologist and zoologist, was born in Braunschweig on 14 November 1777. His interest in insects began very early, since two of his school teachers were entomologists. Thus, when he entered the University of Helmstädt to study law (1797), his mind was already on zoology and, due to his late father's fortune, he was able to indulge his passion. He then transferred to the University of Göttingen (1799) where he was trained by J. F. Blumenbach. On returning to Helmstädt in 1801 he received a Ph.D. for an entomological thesis. He then returned to Braunschweig for some years of private study, interrupted by an important few months in Paris (1802) working with Georges Cuvier, P.-A. Latreille, and Alexandre Brongniart. Later he used his financial resources to obtain a large private natural history collection, that of Lampe in Hannover, which contained many reptiles and amphibians.

Gravenhorst became a professor at Göttingen in 1805 and the next year published his famous "Monographia Coleopterorum," which went through many editions until 1840 and established his entomological reputation. Thereafter, he published numerous books and papers, the most famous of which were on parasitic wasps but also several on herpetology. He went to the University of Frankfurt (Oder) as Professor of Natural History in 1810 and in the following year moved with the university, as did his colleague, J. G. Schneider, when it became the University of Breslau. Gravenhorst became director of the museum there and donated to it his extensive personal collections, about which he wrote in many of his herpetological publications. Due to serious physical and mental illnesses, beginning in 1825 and worsening in the 1840s, he gradually reduced his scientific work; indeed, he wrote nothing new after 1842 and published only existing manuscripts. He was a comparative anatomist by training who highly valued the importance of morphology in classification. He retired completely in 1856, and died in Breslau (now Wrocław, Poland) on 14 January 1857.

Besides some short papers (beginning 1825), Gravenhorst published four major herpetological titles, the earliest of which was "Deliciae Musei Zoologici Vratislaviensis" (1829; alternate title "Reptilia Musei Zoologica Vratislaviensis"), covering the turtles and amphibians in the Breslau collection. This is an elaborate work, worldwide in coverage, and with 17 large, highly detailed plates, most of them handcolored; unfortunately the promised part on lizards and snakes was never issued. The other three titles are extensive monographs, all published in the *Nova Acta* (1833, 1837 [1838], 1847 [1851]) and concerning lizards from all parts of the world.

• *References*: "Dr. Joh. Ludw. Christian Gravenhorst," by K. Letzner, Jahr.-Ber. Schlesischen Gesellsch. Vaterländ. Kult., 35: 111-115, 1857 (1858); "Joh. Ludwig Christian Gravenhorst," by J. Carus, Allg. Deut. Biogr., 9: 616, 1879; "Johann Ludwig Christian Carl Gravenhorst," by G. Uschmann, Neue Deutsch. Biogr., 7: 12-13, 1965. • *Portrait*: Courtesy Library, Academy of Natural Sciences, Philadelphia. • *Signature*: Courtesy Library, Museum of Comparative Zoology, Harvard University.

WIEGMANN, A. F. A. (1802-1841).

Arend Friedrich August Wiegmann, German zoologist and member of a family of biologists, was born on 2 June 1802 in Brunswick, Braunschweig, where his father, the botanist A. F. Wiegmann, was Professor of Natural History. Young Wiegmann trained in medicine and philology at the University of Leipzig and then went to study with Hinrich Lichtenstein at the University of Berlin, where he was put in charge of the herpetological collection of the zoological museum directed by Lichtenstein. Wiegmann became a professor at Cologne in 1828 and at the University of Berlin two years later. Besides his major interest in herpetology, he also published on mammals and crustaceans. His son, C. A. F. Wiegmann, also followed a zoological career. A. F. A. Wiegmann published a general zoology text with J. F. Ruthe (1832, second edition 1843) and in 1835 co-founded a major German natural history journal, *Archiv für Naturgeschichte*. He died of tuberculosis at the young age of 38, on 15 January 1841.

In addition to several shorter papers in journals, Wiegmann published two major herpetological works, both in 1834. "Herpetologia Mexicana," a large folio of 60 pages and 10 handcolored plates (reprinted 1969), was the first major post-Linnaean monograph on the herpetofauna of Mexico. The book consists of two rather unrelated parts, the first being a review of the "Sauri"—in Wiegmann's classification, the lizards and crocodiles—in which he named many new genera from throughout the world. The second is a review of the specimens taken in Mexico between 1824 and 1829 by German collectors. Some 32 new species and seven new genera of lizards were described, including the first known venomous lizard, *Heloderma horridum*. A few of these taxa had been named by him earlier, in Oken's *Isis*.

Wiegmann's other major work was the herpetological section ("Amphibien," 1834 [1835]) of the extensive report on the botany and zoology of F. J. F. Meyen's voyage around the world on the Prussian cargo ship *Princess Louise* (1830-1832). This section contains ten excellent plates, half of them beautifully handcolored. Meyen visited the west coast of South and Central America, the "Sandwich Islands" or Hawaii, the Philippines, and China, returning home via the Cape of Good Hope, and thus Wiegmann described a variety of novelties from many parts of the world.

• *References*: "Arend-Frédéric-Auguste Wiegmann," anonymous, Biogr. Univ. (Michaud) Ancien. Mod., new ed., 44: 574-575, 1854-[1865]; "Wiegmann and the Herpetology of México," by E. H. Taylor, p. iii-vi. *In* Wiegmann's Herpetologia Mexicana, repr. ed., Soc. Study Amphib. Rept., Athens (Ohio), 1969; Papavero, 1971 (pp. 103-107, "Mexicana" itinerary). • *Portrait*: Instituts für Pflanzenschutzforschung, Akademie der Wissenschaften, Eberswalde, courtesy Rainer Günther. • *Signature*: Museum für Naturkunde der Humboldt-Universität zu Berlin, courtesy Rainer Günther.

BONAPARTE, Charles L. (1803-1857).

Charles-Lucien-Jules-Laurent Bonaparte, born in Paris on 24 May 1803, to a brother of Napoleon, became one of the best vertebrate zoologists of his day. His scientific career, however, was inevitably disrupted by political events surrounding the Bonaparte family. His immediate family was forced to move to Rome under papal protection, in 1804, but after the pope's deportation they had to spend 1810-1814 in England. On returning to the Papal Estates in 1814, the pope rewarded Charles's father for loyalty with the title Prince of Canino, and Charles himself was granted the papal title Prince of Musignano. About this time, Charles Bonaparte developed an interest in birds and later, after marrying his cousin (1822), visited her father, who had moved to America after Napoleon's downfall. Thus, from 1823 to 1826, he was associated with the active scientific community in Philadelphia, where he published his first paper (1824) and began his four-volume series entitled "American Ornithology" (1825-1833). After a short visit to Europe to study specimens, he spent 1827-1828 in Philadelphia, but then returned to Europe to stay.

With his extensive private collection and library housed at his estate in Rome, he began work on his vertebrate zoology of Italy, the "Iconografia della Fauna Italica," which was published in parts from 1832 to 1841. Volume two, the "Amfibi," covers herpetology and contains 54 beautiful handcolored plates, in folio size. It was the standard on the Italian herpetofauna until superseded by Lorenzo Camerano's monographs (1883-1891). Among his herpetological titles, Bonaparte published a series of about ten papers in which he proposed new classifications, especially of turtles. Most of his work, however, was in ornithology, in which he was regarded to have the most comprehensive knowledge of species of anyone in his era. Most of his books and herpetological papers are in Latin, to emphasize their international significance and his world citizenship.

Bonaparte had a modern concept of species and of their evolution, but despite his advanced scientific views, he could not escape his heritage particularly because of his striking resemblance to his famous uncle. On the death of his father (1840), he inherited the additional title of Prince of Canino, yet his liberal views, partly aroused by his experiences in democratic Philadelphia, moved him to espouse Italian republican causes (1847-1849). He became Vice-President of the Legislative Council, but when the republican cause was defeated by a French expeditionary force in 1849, he eventually lived the life of an exile, first in Leiden and finally in Paris, where he died on 29 April 1857, leaving his final project, the massive "Conspectus Generum Avium," begun in 1851, woefully incomplete.

• *References*: "Charles-Lucien-Jules-Laurent Bonaparte," by É. Franceschini, Dict. Biogr. Franç., 6: cols. 912-913, 1954; "Lucien Jules Laurent Bonaparte," by G. Petit, Dict. Sci. Biogr., 2: 281-283, 1970; E. Stresemann, 1975 (pp. 153-169). • *Portrait* (1849) *and signature*: From E. Stresemann, 1951.

SCHLEGEL, Hermann (1804-1884).

Schlegel was the most prominent of that group of German naturalists—which also included Heinrich Boie, J. J. Kaup, and Heinrich Kuhl—brought to the Rijksmuseum van Natuurlijke Historie in Leiden by its director, C. J. Temminck. Born on 10 June 1804 in Altenburg, Thuringia, young Schlegel was tutored by the pastor and ornithologist C. L. Brehm, father of Alfred Brehm who authored the "Thierleben" series. In 1824-1825 he studied in Vienna with L. J. Fitzinger, J. J. Heckel, and Carl von Schreibers before joining Temminck's staff in May 1825 as a preparator. In December 1825 he took Boie's post when the latter departed for Java; he ultimately succeeded Temminck as director in 1858. Schlegel's formal training was at the University of Leiden in 1831-1832, under Carl Reinwardt, and he received his doctorate from the University of Jena in March 1832.

Although by inclination primarily an ornithologist, most of Schlegel's early work was on reptiles. His first major work was part of the "Fauna Japonica" series edited by his friend P. F. von Siebold, who had explored Japan in 1823-1829. The herpetological volume, published in 1834 and 1838 (reprinted 1934 and again in 1972), was ostensibly co-authored with C. J. Temminck but was actually written by Schlegel. Beginning in 1837 and concluding in 1844, Schlegel issued a series of 50 outstanding folio plates of amphibians and reptiles in his "Abbildungen," together with a short text, illustrating many new species, particularly from the rich fauna of the Dutch Indies.

Schlegel's most famous herpetological work was the "Essai sur la Physionomie des Serpens," issued in 1837. This was the first truly scientific treatise on snakes. Although it contained valuable descriptions of species, based on the rich Leiden collection and Schlegel's personal examination of the major European collections, his classification scheme never gained wide acceptance and was soon overshadowed by the snake volumes published by A.-M.-C. Duméril, Gabriel Bibron, and Auguste Duméril (1844, 1853, 1854). Nevertheless, Schlegel's book was important enough to be translated into English by T. S. Traill, in 1843.

Increasingly, Schlegel's interests returned to birds, but he did co-author with Salomon Müller the zoology volumes of the "Verhandelingen" (issued over the period 1839-1844) covering the surveys of the Natural History Commission of the Dutch Indies in New Guinea, Timor, Sumatra, and Borneo. This included several folio plates of reptiles, two of them drawn by Schlegel himself, which give only a hint of his fine artistic talent better displayed in his many bird books. Also generally unrecognized was Schlegel's discovery and nurturing of three of the century's greatest natural history artists—J. G. Keulemans, Joseph Smit, and Joseph Wolf.

Schlegel was one of the most prominent naturalists of his era, concerned with fundamental questions of the definition of species, the relationship of systematics to physical geography, and the biological meaning of variation. It was he who first used trinomial nomenclature (beginning in 1844). However, like most of the older generation of German ornithologists, he was a lifelong opponent of Darwinism, and he even eschewed the use of microscopes. He died in Leiden on 17 January 1884.

• *References*: "Levensschets van Hermann Schlegel," by G. Schlegel, Jaarboek Koninkl. Akad. Wetensch., Amsterdam, 97 pages, 1884; E. Stresemann, 1975 (pp. 192-219). • *Portrait and signature* (1842): Rijksmuseum van Natuurlijke Historie, courtesy Marinus S. Hoogmoed.

DUMÉRIL, Constant (1774-1860).

André-Marie-Constant Duméril, physician and anatomist, was the greatest taxonomic herpetologist of his era. He was born in Amiens (Somme), France, on 1 January 1774, and showed an early interest in nature, especially insects and salamanders. He began his medical training in 1793, at the École Secondaire de Médecine in Rouen, specializing in anatomy, but in 1795 transferred to the École de Santé in Paris. By 1799, he had been promoted to chief of anatomical work there, and in 1801 he became Professor of Anatomy and Physiology. During this same period he continued his zoological work and became associated with the Muséum National d'Histoire Naturelle, in particular with the comparative anatomist Georges Cuvier, only five years Duméril's senior.

In 1803, Count de Lacepède, who held the professorship (or head) of ichthyology and herpetology at the museum, resigned due to his preoccupation with political work. Cuvier nominated Duméril for the post, and he reluctantly accepted but only on condition that he be regarded as temporary; eventually, on Lacepède's death in 1825, Duméril became official head, a position he relinquished, in turn, to his son, Auguste Duméril, in 1857. Thus, Constant Duméril's service to the museum spanned 54 years, during which time French domination of world herpetology reached its zenith, due largely to him.

In view of Duméril's enormous productivity at the museum it is easy to forget that he continued his medical duties throughout, only coming to the museum to teach. In fact, he was a very popular lecturer and once remarked that this was the most personally satisfying aspect of his work. Cuvier and Achille Valenciennes were responsible for the rapidly growing collections, but when Cuvier died in 1832, Valenciennes assumed the professorship of malacology, leaving Duméril without assistance. Fortunately for Duméril and for herpetology, there was already someone on the museum's staff, Gabriel Bibron, who perfectly complemented Duméril's talents. Duméril's interest was clearly at the level of higher taxa and classification, as demonstrated in his first major work, "Zoologie Analytique" (1806), which covers all animals and shows the relationships of the genera; indeed, species are not mentioned at all.

France at this time, under Napoleon, was enormously powerful, both militarily and economically, and the museum's collections were to grow as a direct result. France's victorious generals of the empire brought back large collections, even to the extent of emptying the shelves of museums in the conquered countries under the guise of "loans" or even as "gifts" of the grateful people to the French. This was also a time of great voyages and expeditions, for commercial as well as military purposes, when the European powers began to develop overseas possessions, and France was preeminent. All of the booty was returned to Paris and there came under the control of one man: Constant Duméril. Together with his primary assistant, Gabriel Bibron, Duméril proceeded to produce a project of enormous scope, a detailed review of the world's herpetofauna, based on the then-largest herpetological collection.

Duméril himself was responsible for the grand design and arrangement of genera—the first natural arrangement of genera ever undertaken for amphibians and reptiles—and thus one of the classical monuments of descriptive zoology. He was assisted in the generic arrangement by a young German, Michael Oppel, who was his student in 1807-1808. Bibron's responsibility was the description of species. The result was the "Erpétologie Générale ou Histoire Naturelle Complète des Reptiles" (1834-1854), in nine volumes (bound in ten, since volume seven is in two parts) plus an atlas of 120 plates that was issued in plain and handcolored versions. After Bibron's premature death in 1848, Auguste Duméril joined his father in writing the two remaining volumes (seven and nine).

The "Erpétologie Générale" was the first work that gave a comprehensive scientific account of all amphibians and reptiles, including their anatomy, physiology, systematics, and associated literature. In this enormous work some 1393 species are covered. Each is provided with a full synonymy and a detailed description; many are illustrated. By comparison, George A. Boulenger's later (1882-1896) review of the world's herpetofauna, also in nine volumes, covered 8469 species, but his descriptions are far less detailed.

For all the advances inherent in the "Erpétologie Générale," a major defect was the recognition of amphibians as an order of reptiles, a reversion to the arrangement of Alexandre Brongniart (1799). This was adopted in spite of P.-A. Latreille's (1825) separation of amphibians as a class and the accumulating embryological evidence of Karl E. von Baer (1828) and Johannes Müller (1831), a particularly surprising arrangement in view of the breadth of Duméril's knowledge. Besides his interests in higher taxa and anatomy, Duméril was from his boyhood an observer of living animals and, accordingly, was responsible for the creation on the museum grounds of the menagerie of live specimens (in 1838), which permitted many important observations to be made of relevance to systematics.

The death of Gabriel Bibron created a ten-year publication gap during which two separate works were issued that must be regarded as extensions of the "Erpétologie Générale": the "Catalogue Méthodique de la Collection des Reptiles" (1851), a supplement to the volumes already published (one through six; eight), which, although formally authored by both Dumérils, is generally attributed to Auguste Duméril, and the "Prodrome de la Classification des Reptiles Ophidiens" (1853), by Constant Duméril alone, giving a preview of his classification of the snakes in volume seven (published in two parts, 1854). Despite their preoccupation with this grand project, the three authors continued throughout the same period to produce other works on amphibians and reptiles, often incidental to the larger project but sometimes even quite separate from it and of a sizeable and valuable nature.

By the time the "Erpétologie Générale" was finished, Constant Duméril was 80. He had relinquished day-to-day supervision of the laboratory to his son the previous year (1853) and finally retired in 1857. During these last years, still healthy, his interests turned to his first passion, insects. He received many honors, including promotion to Commander of the Legion of Honor just two months before his death, in Paris, on 14 August 1860.

• *References*: "Éloge Historique d'André-Marie-Constant Duméril," by P. Flourens, Mém. Acad. Sci. Inst. Impér. France, 35: i-xxii, 1866; "André-Marie-Constant Duméril," anonymous, Grand Larousse, 6: 1379, 1870; "André Marie-Constant Duméril, le Père de l'Erpétologie," by J. Guibé, Bull. Mus. Natl. Hist. Nat., ser. 3, 30: 329-341, 1958. • *Portrait and signature*: Bibliothèque Nationale, Paris.

BIBRON, Gabriel (1806-1848).

Gabriel Bibron, zoologist and collaborator of A.-M.-C. Duméril, was born in Paris in 1806. As the son of a longtime employee of the Muséum National d'Histoire Naturelle, he received his education from the professors at the museum who recognized his propensity for natural history. Consequently, he was appointed to make collections of vertebrates in Italy and Sicily, but the turning point in his life came in 1832 when A.-M.-C. Duméril invited him to become his chief assistant and, later, to collaborate on the monumental treatise "Erpétologie Générale" (1834-1854). He also assisted Duméril in his teaching at the museum and Bibron himself taught natural history at a primary school in Paris, the Collége Municipal Turgot. Unfortunately, Bibron died prematurely, of tuberculosis at the age of 42, in Saint-Alban (Loire), on 27 March 1848.

While A.-M.-C. Duméril's contributions to the "Erpétologie"—overall design and arrangement of the genera—were crucial, Bibron's were no less important. He was responsible for determination of specimens, synonymy, and description of species, including many new ones. Bibron was the first to provide detailed descriptions of reptilian species, sometimes including minutiae of little value; nevertheless, the bulk of the "Erpétologie" which still has value today is its descriptive content. Besides this collaboration with Duméril, Bibron was also well positioned to determine and publish accounts of other collections of reptiles and amphibians sent to the museum by voyagers from throughout the world. The most notable of these was the lower vertebrate section of Ramón de la Sagra's "Histoire . . . de l'Ile de Cuba" (French edition, 1839-1843; Spanish 1843-1853), co-authored with his late colleague, J.-T. Cocteau; this book contains some of the most exquisite handcolored plates of reptiles ever published.

• *References*: "Allocution sur la Tome de Gabriel Bibron," by A.-M.-C. Dumeril, Rev. Mag. Zool., ser. 2, 1: 589-592, 1849; "Notice sur G. Bibron," by A.-M.-C. Duméril, 4 page notice inserted after p. vii. *In* A.-M.-C. Duméril, G. Bibron, and A.-H.-A. Duméril, Erpétologie Générale, Paris, vol. 7, 1854; "Gabriel Bibron," by Lemercier, Nouv. Biogr. Gén., 5: cols. 939-940, 1866. • *Portrait*: From Duméril, 1854 (atlas). • *Signature* (1845): Rijksmuseum van Natuurlijke Historie, courtesy Marinus S. Hoogmoed.

HOLBROOK, John Edwards (1794-1871).

Justly regarded as the Father of American Herpetology, Holbrook was born on 30 December 1794 (see *Note*) in Beaufort, South Carolina. He attended Brown University (A.B. 1815) and the University of Pennsylvania (M.D. 1818). After a short period of private practice in Boston, he received additional training at the University of Edinburgh (1819-1820) and then spent time at the Muséum National d'Histoire Naturelle in Paris where he befriended the leading French naturalists of the day—Georges Cuvier, A.-M.-C. Duméril, Gabriel Bibron, and Achille Valenciennes—associations that were to have a lasting influence on Holbrook and his work.

Returning to the United States in 1822, Holbrook established a medical practice in Charleston, South Carolina, one of the most active cities, both scientifically and culturally, in antebellum America. Later, in 1824, he became Adjunct Professor of Anatomy at the Medical College of South Carolina, which he helped to found. During the Civil War he was Chairman of the Examining Board of Surgeons for South Carolina and, despite his age (then 70), he served as a medical officer for Confederate troops in the field. He never resumed his scientific work after the war, partly because his personal papers and collections were largely destroyed when Charleston was ransacked by Federal troops, but in 1868 he was elected to the U.S. National Academy of Sciences in recognition of his contributions to herpetology and ichthyology. He died on 8 September 1871, in his boyhood town of North Wrentham (now Norfolk), Massachusetts.

Holbrook's scientific reputation rests on his books "North American Herpetology" (first edition in four volumes, 1836-1840; second, five volumes, 1842, the latter edition reprinted 1976), "Southern Ichthyology" (1847), and "Ichthyology of South Carolina" (first edition, 1855[-1857]; second, 1860). In herpetology, he published only one other title, a three-page list of the amphibians and reptiles of Georgia (1849; reprinted 1978). Despite this small number of titles, it was the massive scope of his books that secured his reputation. In his "Herpetology," which he began in the mid-1820s, he described and illustrated every American species which, at that time, meant primarily east of the Mississippi River. These books provided the first great synthesis of information on the topic, and Holbrook carefully researched the literature to determine the correct scientific names to be used, thus making a major contribution to stabilizing the nomenclature. In addition, he described many new taxa, 24 of which are currently recognized and include some of the commonest American species, among them *Emydoidea blandingii*, *Farancia abacura*, *Storeria dekayi*, *Bufo americanus*, *Gastrophryne carolinensis*, and *Ambystoma talpoideum*, as well as the genera *Crotaphytus*, *Pituophis*, and *Scaphiopus*.

The outstanding feature of Holbrook's books, however, are the illustrations. Indeed, one could say that he was obsessed with providing the most scientifically-accurate and life-like drawings possible. For his "Herpetology" he employed at least 17 artists, among them J. Sera and J. H. Richard who were particularly outstanding. Holbrook also went to great effort to obtain living examples of each species—no easy task at the time—collecting them himself from as far away as Maine and Georgia or obtaining them from his closest colleagues (James E. DeKay, Jacob Green, Edward Hallowell, Richard Harlan, J. P. Kirtland, and D. Humphreys Storer) and other correspondents.

It is largely because of Holbrook's high standards for the drawings, however, that the bibliographic history of his books is so complicated. In the first edition, species were arranged haphazardly, essentially in the order that specimens were obtained and illustrated, but he was dissatisfied with many published drawings and sometimes provided substitute plates in later volumes. In 1839, he published expanded versions of volumes one and two which had revised plates and new accounts, surreptitiously using the original title-pages, a fact that went undiscovered until 1976 when the reprint of his second edition was issued. By the time volume four was published (1840), Holbrook had become so thoroughly dissatisfied with the whole project that he abandoned it in favor of a new edition (1842), with 147 handcolored plates, in which the whole was properly organized taxonomically. Even then, however, he continued to authorize new and better plates, which were freely substituted over the years that this edition was in print. Thus, rarely do two copies agree in all particulars.

• *References*: "A Memoir of Dr. John Edwards Holbrook," by T. L. Ogier, Med. Soc. South Carolina, Charleston, 13 pages, 1871; ["Eulogy on Dr. J. E. Holbrook"], by L. Agassiz, Proc. Boston Soc. Nat. Hist., 1870-1871: 347-351, 1872; "Biographical Memoir of John Edwards Holbrook. 1794-1871," by T. Gill, Biogr. Mem. U.S. Natl. Acad. Sci., 5: 47-77, 1905; "John Edwards Holbrook," by H. L. Clark, Dict. Amer. Biogr., 5: 129-130, 1932; "John Edwards Holbrook, Father of American Herpetology," by R. D. Worthington and P. H. Worthington, p. xiii-xxvii. *In* Holbrook's North American Herpetology, repr. ed., Soc. Study Amphib. Rept., Athens (Ohio), 1976. • *Portrait* (after 1865): Courtesy Museum of Comparative Zoology, Harvard University (see also frontispiece). • *Signature* (1837): Adler collection (see also page 4). • *Note*: The year of Holbrook's birth has been in dispute. Some authorities give 1796 or even 1795. The Worthingtons, Holbrook's most recent biographers, followed Ogier in accepting 1796 and cited as supporting evidence the date given on Holbrook's tombstone. The date on the tombstone probably originated with Ogier, Holbrook's closest friend, and thus it may not be independent confirmation. Holbrook had no children and his wife had died in 1863; thus, there were no immediate family in Charleston at the time of his interment there to confirm his date of birth. Ogier also gave the day of birth incorrectly as 31 December. As Gill has pointed out, Ogier seemed to confuse Holbrook's date of birth with that of his younger brother, Silas Holbrook (born 1796). All but one of the biographies of Holbrook published during his lifetime gave his birthdate as 1794. Moreover, it seems more likely that he would have graduated from undergraduate college at the age of 20 and from medical school at age 23 (if born in 1794), rather than two years younger in age. Independent evidence is provided by dates in the Holbrook family bible, which gives Holbrook's birth date as "Born December 30, 1794" (recently confirmed for me by Claire Holbrook Cowell and Bertram Holbrook Holland, who are Holbrook's great, great niece and nephew).

GRAY, John Edward (1800-1875).

Gray is justly regarded as founder of the zoological collections at the British Museum. Although having broad interests in natural history, he gave particular attention to reptiles. Gray was born at Walsall, Staffordshire, on 12 February 1800, and began medical training at the age of 16 at St. Bartholomew's and Middlesex hospitals in London.

His early interest was botany, but piqued by his rejection for membership in the Linnean Society he turned to zoology. He gave up his medical practice in 1823 and was hired in 1824 by J. G. Children, Keeper of the Zoological Collections at the British Museum, to catalogue the reptiles. Thus began Gray's lifelong association with the museum from which he retired only in the December before his death, in London, on 7 March 1875.

Gray eventually succeeded Children as keeper (1840) and during his tenure it is said that the museum added one million specimens to its collection, for which accomplishment the University of Munich awarded Gray an honorary Ph.D. in 1852. The building of this collection was the great work of his life and it was Gray also who began the important series of catalogues of the museum's specimens. He published nearly 1200 titles, mainly on zoology, but these were often uncritical and repetitious in content. He was also an inveterate coiner of scientific names, many of which have no meaning in any language. Gray was a leading member of the London scientific establishment and was a founder or officer of several of its societies; he was elected Fellow of the Royal Society in 1832. In addition, he took an active part in questions of social and educational reform, and apparently was the first to propose a uniform rate of postage to be prepaid by means of stamps.

Gray published numerous books partly or wholly on herpetological topics. He described many new species from the Hardwicke collections, as depicted in the magnificent folio atlas entitled "Illustrations of Indian Zoology" (1830-1834[-1835]), which contains 43 herpetological plates. Unfortunately, no text or further plates were published after Thomas Hardwicke's death (1835), due to Gray's dispute with the former's executors, and the manuscript remains unpublished. Gray's "Synopsis of Species of the Class Reptilia," published originally in an English edition of Georges Cuvier's "Animal Kingdom" (1831) and reprinted separately as "Synopsis Reptilium" (also 1831), was a competent world list which foreshadowed his catalogues of the British Museum collections (turtles in 1844, lizards 1845, snakes 1849, Amphibia 1850). Later, he issued a more in-depth catalogue of the "Shield Reptiles"—the turtles, crocodilians, amphisbaenians, and tuatara, according to Gray's taxonomy—in five parts (1855-1873). In 1867 he published "Lizards of Australia and New Zealand," using plates issued previously as part of the H.M.S. *Erebus* and *Terror* reports, and in 1872 supplied the text for the re-issue of Thomas Bell's magnificent illustrations of turtles.

Gray's most spectacular novelty was the tuatara (*Sphenodon punctatus*), which he described in 1842, although he wrongly put it in the family Agamidae. Perhaps Gray's single most important act, in the development of the British Museum into the world center for herpetology at the end of the 19th Century, was his hiring of Albert Günther, in 1857, as his own replacement in charge of reptiles and fishes.

• *References*: "John Edward Gray, Ph.D., F.R.S.," p. 113-118. *In* L. Reeve (ed.), Portraits of Men of Eminence, L. Reeve Co., London, 1863; "Dr. John Edward Gray," anonymous, Proc. Linnean Soc. London, 1874-1875: xliii-xlvii, 1875; "John Edward Gray," by G. S. Boulger, Dict. Natl. Biogr., London, 8: 452-453, 1921; "The Miscellaneous Autobiographical Manuscripts of John Edward Gray (1800-1875)," by A. E. Gunther, Bull. Brit. Mus. Nat. Hist., ser. Hist., 6: 199-244, 1980. • *Portrait*: British Museum (Nat. Hist.), courtesy Ann Datta. • *Signature* (1853): Smithsonian Institution Archives, courtesy William Cox and Alan E. Leviton.

BELL, Thomas (1792-1880).

Bell, a noted zoologist and the leading British dental surgeon of his day, was born on 11 October 1792 at Poole, Dorsetshire. He entered medical college at Guy's and St. Thomas's hospitals, London, in 1813, and the Royal College of Surgeons in 1815. He was dental surgeon at Guy's Hospital for most of his professional career (1817-1861), specializing in diseases of the teeth, and was concurrently Professor of Zoology at King's College, London, beginning in 1836. Bell was a leading member of the British scientific establishment and held important positions in the Zoological, Linnean, Ray, and Royal societies, being elected a Fellow of the latter in 1828.

It was Bell, as President of the Linnean Society of London, who presided at the celebrated meeting on 1 July 1858 at which Charles Darwin's and A. R. Wallace's classic papers on natural selection were read. Later, in summarizing that year's discoveries, Bell stated: "The year which has passed . . . has not indeed been marked by any of those striking discoveries which at once revolutionise . . . it is only at remote intervals that we can reasonably expect any sudden and brilliant innovation which shall . . . confer a lasting and important service on mankind." Bell remained hostile to Darwinism throughout his life. He retired to Gilbert White's estate at Selborne, Hampshire, which he had purchased, and published a classic edition of White's "Natural History" (1877). He died there on 13 March 1880.

Besides his well-known books on British mammals and crustaceans, Bell is best remembered for his herpetological work. His *magnum opus*, "Monograph of the Testudinata," issued in eight parts from 1832 to 1842, was never finished due to his publisher's bankruptcy; it contained 40 magnificent handcolored plates, in folio size. These plates were reissued in 1872, together with 20 or sometimes 21 additional previously-unpublished plates, under the authorship of James deC. Sowerby and Edward Lear, the original artist and lithographer, with a brief text by John E. Gray; this reissue was reprinted in 1970. In 1838-1839, Bell published his "History of British Reptiles," the first synthesis of this subject; a revised edition was issued in 1849. He also contributed the herpetology volume (published in 1842-1843; reprinted 1975 and again in 1980) in the series "Zoology of the Voyage of H.M.S. *Beagle*," edited by Charles Darwin. Bell's accomplishments were primarily as an excellent compiler and popularizer, rather than as an original scientist, and as one who was more at home in his study than in the field.

• *References*: "The Late Mr. Thomas Bell, F.R.S., anonymous, Nature, 21: 499-500, 1880; "Thomas Bell," by G. T. Bettany, Dict. Natl. Biogr., 2: 175, 1921; "Bibliographic Notes on Bell's *Monograph of the Testudinata*," by E. E. Williams, p. iii-vi. *In* Sowerby and Lear's Tortoises, Terrapins, and Turtles, repr. ed., Soc. Study Amphib. Rept., Athens (Ohio), 1970; Papavero, 1973 (pp. 233-246, *Beagle* itinerary). • *Portrait*: British Museum (Nat. Hist.), courtesy Ann Datta. • *Signature* (1835): Academy of Natural Sciences of Philadelphia, courtesy Marsha Gross.

SMITH, Andrew (1797-1872).

Regarded as the Father of South African Zoology, Smith was born a shepherd's son near Hawick, Scotland, on 3 December 1797. He graduated from the University of Edinburgh (M.D. 1819) but had already joined the medical department of the British Army, which he served from 1815 to 1858. After postings to Canada and Malta, he arrived at the Cape of Good Hope in 1821, where he was to remain until his regiment was recalled to England in 1837. During this period he spent much time on the frontier, to conduct scientific investigations and to confer with native tribes on behalf of the government. He founded the South African Museum and became its first Superintendent in 1825. In 1834-1836 he headed an expedition to "Central Africa" —what is now the Transvaal—and made large zoological collections. On returning to Britain, he rose rapidly through the administrative ranks and was appointed Head of the Army Medical Department in 1851 by the Duke of Wellington. He happened to head that department during the badly mismanaged Crimean War of 1853-1855 and thus became the target of Florence Nightingale's denunciations of army medical care, but he was later vindicated and, in 1858, was knighted by Queen Victoria. He died in London on 11 August 1872.

Smith's scientific reputation largely rests on his magnificent series "Illustrations of the Zoology of South Africa," published in London in 28 parts over the period 1838-1849 (reprinted 1977). He had taken his collections and all of the drawings with him to England in 1837 (his reptilian types are in London and Edinburgh), and he continued to work on his manuscripts despite his army work. It is the beautiful illustrations which represent the most lasting feature of his books, and these depict numerous new species of amphibians and reptiles. The herpetology volume contains 78 outstanding plates drawn by G. H. Ford, who had accompanied Smith on his expedition in 1834-1836; 75 of them are finely handcolored.

Smith became a member of the London scientific establishment and counted Charles Darwin, who often quoted from Smith in his works, among his many colleagues. Smith had first befriended him when the young naturalist visited Cape Town in 1836, as H.M.S. *Beagle* was returning home from its epochal voyage. Indeed, it was Darwin who sponsored Smith as Fellow of the Royal Society in 1857.

• *References*: "Sir Andrew Smith," by W. W. Webb, Dict. Natl. Biogr., 18: 423, 1921; "Sir Andrew Smith, M.D., K.C.B. His Life, Letters and Works," by P. R. Kirby, A. A. Balkema, Cape Town, 358 pages, 1965; "Sir Andrew Smith's Life and Work," by R. F. Kennedy, p. 7-30. *In* Smith's Illustrations of the Zoology of South Africa, repr. ed., Winchester Press, Craighall (Transvaal), vol. 1, 1977. • *Portrait*: From Kirby, 1965. • *Signature*: Edinburgh University Library, courtesy Jean Archibald.

TSCHUDI, J. J. von (1818-1889).

Johann Jakob von Tschudi, Swiss naturalist, explorer, and diplomat, was born on 25 July 1818 in Glarus, Switzerland. His interest in nature began as a boy when, with his younger brother, Friedrich Tschudi, he wandered the fields in search of lizards and frogs. He attended the University of Zürich (1834-1836) where Lorenz Oken and Heinrich Schinz were among his professors, but it was his time at Neuchâtel (1836-1837) with Louis Agassiz that influenced him the most. He spent the rest of 1837 in Leiden working with Hermann Schlegel and in Paris with Gabriel Bibron, and in 1838 attained a Ph.D. from Zürich. In the meantime, he had departed on his first expedition to South America, where he spent most of his time in the high Andes of Peru, returning only in 1843. He went to the universities of Berlin and Würzburg for further study, receiving an M.D. degree from the latter in 1844, then proceeded to Vienna and Munich to study materials in preparation for his most important zoological work, "Untersuchungen über die Fauna Peruana" (1845 [1846]), which covered all vertebrate groups.

of his teacher, Louis Agassiz. Many familiar amphibian genera were described in this book (e.g., *Ambystoma, Crinia, Hynobius, Microhyla,* and *Plethodon*) and its continuing importance justified a reprinting in 1967. The herpetology section of his "Fauna Peruana" *opus* was issued in 1845 (1846), and contained 12 beautiful handcolored plates in folio; however, all the new species described in this book had been briefly defined in *Archiv für Naturgeschichte* in 1845 (reprinted 1968).

• *References*: "Johann Jakob von Tschudi," by F. Ratzel, Allg. Deutsch. Biogr., 38: 749-752, 1894; "Jean-Jacques de Tschudi" by P.-É. Schazmann, Éd. Mensch und Arbeit, Zürich, 200 pages, 1956 (1955); "Johann Jakob von Tschudi: His Life and Herpetological Work," by R. Mertens, p. iii-vi. *In* Tschudi's Classification der Batrachier, repr. ed., Soc. Study Amphib. Rept., Athens (Ohio), 1967. • *Portrait*: From Schazmann, 1956 (1955). • *Signature* (1837): Naturhistorisches Museum Basel, courtesy Lothar Forcart.

PETERS, Wilhelm (1815-1883).

Wilhelm Carl Hartwig Peters, German zoologist and explorer of East Africa, was born in Coldenbüttel, near Eiderstedt, Schleswig, on 22 April 1815, the son of a pastor. When Peters was ten years old, the family moved to Flensburg, on the Danish border, where he attended secondary school. He began his university training in medicine and natural history at the University of Copenhagen in 1834, but after six months there he transferred to the University of Berlin which, as it turned out, was to become his academic home for the rest of his life. He graduated with a M.D. degree in 1838 and, after 18 months of field work with H. Milne-Edwards in the Mediterranean region, he returned to Berlin in 1840 as an assistant to his former professor, Johannes Müller, the great anatomist.

Soon Peters began to plan what was to become the major event of his life, an exploration of Mozambique, which had the enthusiastic support of Müller and of Alexander von Humboldt, then also at Berlin. Peters departed in September 1842, travelling on a Portuguese convict ship first to Angola and finally (June 1843) to Mozambique. There he managed to explore the entire coastal region and also spent nearly a year up the Zambesi River deep in the interior. In addition, he made excursions to Zanzibar, to Madagascar and the Comoro Islands, and to Cape Town, mainly to recover from recurring illness, and he finally departed for Europe in August 1847, returning via India and Egypt. The collections he made were enormous and were written up, mainly by himself, in five elaborate quarto volumes in the series "Reise nach Mossambique" (1852-1882). This was a model faunal work for its day—comprehensive, authoritative, and well illustrated.

On returning to Berlin, Peters became a prosector at the university's Anatomical Institute, an Assistant Professor in 1849, and Assistant Director of the university's Zoological Museum, under Hinrich Lichtenstein, in 1856. Following Lichtenstein's death the next year, Peters succeeded to the

From 1857 to 1859 Tschudi returned to South America, travelling through much of the southern half of the continent, and during 1860-1861 returned again on an official mission to Brazil on behalf of the Swiss government. In 1866 he was appointed Chargé d'Affaires at the Swiss Embassy in Vienna and in 1868 was promoted to the rank of Ambassador. He held the latter position until 1888 when he retired to his country home, "Jakobshof," at Edlitz, a small town near Vienna, where he died on 8 August 1889.

Tschudi's herpetological research was a small part of his life's work, which was in the tradition of Alexander von Humboldt. Throughout his life he continued to publish major works dealing with ethnography, geography, meteorology, and medicine. His herpetological output was only 12 titles (from 1836 to 1847), but included several important works. In 1837, Tschudi named the fossil salamander (*Andrias Scheuchzeri*), which for nearly a century had been called "Homo diluvii testis"—as the name suggests, it was thought to be the remains of a human witness to the Great Flood—until Georges Cuvier demonstrated, in 1811, that it was a giant salamander.

Tschudi's most famous work, "Classification der Batrachier" (1838), was an attempt to unite into one system both fossil and Recent species, thus showing the influence directorship and to the enlargement and development of the museum into one of the world's major zoological collections. Under his leadership, the herpetological collection alone tripled in size, from about 3700 to 10,500 specimens, in its day rivalled in size only by the collections in Paris and London. From 1858 Peters was also Professor of Zoology and was a competent if unenthusiastic teacher; his real interest was the museum. Darwinism, of course, was the main ferment of the day and Peters was opposed, albeit not outspokenly so. His major contribution at the university was the synthesis of anatomy and zoology, in the tradition of his teacher, Johannes Müller. He died in Berlin on 20 April 1883.

Peters published nearly 400 titles covering many groups of vertebrates and invertebrates, but especially "Amphibia" (he was one of the last to use this term for both amphibians and reptiles), mammals, and fishes, and he often wrote about all of these groups in the same paper. Peters's taxonomic work was at the species level and his nearly 150 works in herpetology (over the period 1838-1883) included descriptions of numerous new species from throughout the world.

Peters's primary monographs include one on the peculiar uropeltid snakes of India and Sri Lanka ("De Serpentum Familia Uropeltaceorum," 1861), a major review of South American microteiid lizards ("Über Cercosaura," 1862), and "Catalogo dei Rettili e dei Batraci . . . nella Sotto-Regione Austro-Malese" (1878, with Giacomo Doria), a report on extensive Italian collections made in New Guinea, the Celebes and adjacent areas, including Queensland. His crowning achievement, published only in the year before his death, was the herpetological volume in his "Mossambique" series, based on collections he had made forty years before. This book was the first major review of the herpetofauna of East Africa, an area where Imperial Germany had territorial ambitions, and conquests began in earnest in the 1880s.

• *References*: "Peters (Guillaume-Charles-Hartwig)," anonymous, Grand Larousse, 12: 703-704, 1874; ["Wilhelm Karl Hartwig Peters"], anonymous, Science, 1: 438, 1883; "Wilhelm Karl Hartwig Peters," by F. Hilgendorf, Allg. Deut. Biogr., 25: 489-493, 1887. • *Portrait*: Archives, Zoologisches Museum Berlin, courtesy Rainer Günther. • *Signature* (1861): Adler collection.

BLYTH, Edward (1810-1873).

Blyth, describer of many of the reptiles and amphibians of the Indian subcontinent, was born in London on 23 December 1810. As a young boy he developed an interest in natural history, especially birds. He received no formal schooling in the subject, yet through self-study and diligence he quickly developed considerable skill. In 1836 he edited an edition of Gilbert White's "Selborne" classic and four years later cooperated in one of the English translations of Georges Cuvier's "Animal Kingdom," editing the sections on vertebrates (except fish). He made numerous contributions to the various learned journals in London, which brought him to the notice of the trustees of the Asiatic Society of Bengal, who were looking to hire their first salaried museum curator (in effect, director). Blyth was chosen and he sailed to Calcutta in 1841, thus beginning 22 years of service. He inherited the Asiatic Society's miscellaneous collections made by Theodore Cantor, John MacClelland, and others, to which he added assiduously from donations and his own collecting in various Indian provinces and Burma. Owing to ill health, he returned to England in 1863 and died in London on 27 December 1873.

Most of Blyth's technical publications dealt with birds and mammals, and only a few were on herpetological topics. However, over the period 1841-1863 he produced a regular series of "Curator's Reports" which included extensive descriptions of numerous new species of animals including reptiles and amphibians, not only from India and Burma but also from as far away as China, the Philippines, and other places where Blyth had correspondents. Shortly after Blyth's retirement to England the Asiatic Society's collections were transferred to the newly-created Indian Museum where they were further studied by John Anderson, the museum's first director. Besides Blyth's interest in systematics, he published on biogeography, animal domestication, and the origin of species, and his observations were important enough to be regularly quoted by Charles Darwin in his "Variation of Animals and Plants Under Domestication" (1868). Blyth himself was a staunch Darwinist.

• *References*: "Edward Blyth," anonymous, Ibis, ser. 3, 4: 465-467, 1874; ["Edward Blyth"], by A. Grote, Jour. Asiatic Soc. Bengal, 44(2, extra no.): iii-xxiv, 1875. • *Portrait* (1864) *and signature*: From Grote, 1875.

AGASSIZ, Louis (1807-1873).

Jean Louis Rodolphe Agassiz, the renowned Swiss-turned-American naturalist, was born on 28 May 1807 at his father's parsonage in Motier-en-Vuly, Switzerland. Before the age of ten he was already a collector of fishes and plants, but his family planned a career for him in commerce or medicine. Fortunately, natural history was then a proper part of a medical curriculum, and Agassiz was for a long time able to deceive his family into thinking that he was studying medicine. After preparatory school in Bienne and Lausanne, he attended the University of Zürich (1824-1826) where his professor was Heinrich R. Schinz. He then went to the University of Heidelberg (1826-1827) where he was introduced to paleontology by Heinrich Bronn and to comparative anatomy by Schinz's former student, Friedrich Tiedemann. At the University of Munich (1827-1830) he was a student of the philosopher-naturalist Lorenz Oken and the herpetologist J. G. Wagler among others. Here, he was entrusted by C. F. P. von Martius with the fishes from the Spix-Martius expedition to Brazil, which Agassiz published as his first book (1829). This also served as his Ph.D. thesis (Munich, 1829, but examined at the University of Erlangen); in 1830 he also took a M.D. degree at Munich.

His book, dedicated to Georges Cuvier, served as Agassiz's entrée to the Muséum National d'Histoire Naturelle in Paris, then the world center for natural history study, and to postdoctoral study with Cuvier himself. Agassiz had begun a work on the fossil fishes of Europe and Cuvier gave him all of his own notes for a planned similar project. This was a time of great excitement in Paris, for the Cuvier-Geoffroy debates on evolution were at their zenith, which had a marked influence on Agassiz. It was then, also, that he met Alexander von Humboldt, who became his patron and protector for many years thereafter. But the work in Paris came to an abrupt end when Cuvier died in May 1832, during a cholera epidemic. Although Agassiz had studied with Cuvier for not quite six months, he became his self-appointed successor and upheld the basic tenets of Cuvier's views on nature for the rest of his life, which had unfortunate consequences for Agassiz after Charles Darwin.

Agassiz became Professor of Natural History at the new College of Neuchâtel in Switzerland in 1832, thus beginning the most scientifically productive phase of his life. His "Poissons Fossiles," which included descriptions of 1700 species of fossil fishes, was published over the period 1833-1844. In 1837 he made his announcement of the existence of a prehistoric Ice Age, which he later published in book form (1840). These accomplishments were not without controversy, however, from colleagues and even his own students who claimed credit for much of the work and ideas.

In 1842-1846, Agassiz published "Nomenclator Zoologicus," a listing of all genera and supragenera, giving original citation and etymology (the reptile section, entitled "Nomina Systematica Generum Reptilium," was published 1844). In 1846 (octavo edition, 1848) he published the companion "Nomenclatoris Zoologici Index Universalis," an index of all generic names in zoology including fossil taxa, and in 1848-1852 his four-volume "Bibliographia Zoologiæ et Geologiæ." Agassiz was justly famous, but by 1845 he was also financially bankrupted by all the publishing ventures.

Accordingly, in September 1846 he began a lecture tour in the United States, originally intended to raise money, which resulted in a professorship at Harvard College where he stayed the rest of his career. Agassiz was an outstanding orator and teacher; audiences, professional and lay alike, were spellbound, and he enjoyed all the attention. In America he saw great possibilities for the advancement of science, even at the level of secondary schools, which he championed. At Harvard he was successful in raising money, from the college, from private donors, and from the State of Massachusetts, to build a great museum, the Museum of Comparative Zoology (still called the "Agassiz Museum" by local people), which was founded in 1859. This embodied Agassiz's concept of the combination of research, field work, publication, and graduate education at an institution of higher learning which became the model for many other American universities. His own legion of students went out to leading universities of the day. On the national scene, Agassiz was similarly influential—for example, in co-founding the U.S. National Academy of Sciences in 1863—but his influence met with increasing hostility from many who opposed his efforts to dominate the academy and to control American science.

Agassiz, like Cuvier, believed in special creation and the fixity of species and, thus, inevitably became America's chief opponent of Darwin. From that time onward, Agassiz suffered a loss of respect in intellectual circles, and even at Harvard he became isolated. He sought new challenges in two expeditions to South America (1865-1866 and 1871-1872) and in assisting with the foundation of Cornell University, where he briefly served as a professor (1868-1869). He died on 14 December 1873, in Cambridge, Massachusetts, at the age of 66. He was succeeded as Director of the Museum of Comparative Zoology by his son, Alexander Agassiz, also an accomplished biologist, who had generated substantial money from investments in copper mines and spent generously of his personal fortune to support the museum's activities.

Agassiz's primary herpetological work was contained in his magnificent series, in quarto, entitled "Contributions to the Natural History of the United States of America." Ten volumes were planned, but only five were published (volumes 1-4, 1857-1862; reprinted 1978); the fifth volume was published posthumously, in 1877, as a *Memoir* of the museum at Harvard. Volume one contained Agassiz's well-known "Essay on Classification," based on Cuvier's old plan, and gave Agassiz's account of the special and independent creation of species, which was in stark contrast to Darwin's views in "Origin of Species" published only two years later.

The last half of volume one and all of volume two (1857), together comprising Agassiz's monograph of North American turtles, covered chelonian classification, anatomy, embryonic development, physiology, distribution, and habits. Volume two contains 34 exquisitely detailed plates depicting juvenile turtles, eggs, and embryological series. In these volumes, Agassiz described new genera (*Graptemys* and *Deirochelys*) and many new species (*Kinosternon flavescens*, *Terrapene ornata*, and *Gopherus berlandieri*, among many others) of American turtles.

Agassiz had amassed an enormous collection of turtles for this work, many loaned by his friend Spencer F. Baird at the new Smithsonian Institution, but others were sent by laymen throughout the country who had heard his appeals. Such was the fame of Agassiz that one Massachusetts school teacher, learning that the professor wanted embryos not more than three hours old, camped out at a local pond for three weeks until he got fresh eggs, which he then rushed to Cambridge by a combination of a horse-and-buggy, a freight train, and finally a delivery wagon, just to play his small part in Agassiz's grand scheme.

Unfortunately, this book also illustrated a side of Agassiz's personality that was displayed long before at Neuchâtel: his inability to work with others as equals and his failure to credit collaborative work. After publication of volume four, his main assistant, H. James Clark, who was acknowledged but whose name does not appear on the titlepages, claimed to have written most of volumes two, three, and four. As far as the volume on turtles is concerned, the published correspondence between Baird and Agassiz make it clear that Agassiz was deeply involved in the actual research, if not the writing.

• *References*: "Louis Agassiz," by E. C. Agassiz, Houghton, Mifflin, Boston, 2 volumes, (xiii), 1-400 pages, and (ix), 401-794 pages, 1885; "Memoir of Louis Agassiz, 1807-1873," by A. Guyot, Biogr. Mem. U.S. Natl. Acad. Sci., 2: 39-73, 1886; "Louis Agassiz as a Teacher," by L. Cooper, Comstock Publ. Co., div. Cornell Univ. Press, Ithaca (New York), (xiii), 90 pages, 1917 (rev. ed. 1945); "Agassiz, Darwin and Evolution," by E. Mayr, Harvard Univ. Bull., 13: 165-194, 1959; "Louis Agassiz: A Life in Science," by E. Lurie, Univ. Chicago Press, Chicago, xiv, 449 pages, 1960 (abridged edition, Chicago, 1966); "Jean Louis Rodolphe Agassiz," by E. Lurie, Dict. Sci. Biogr., 1: 72-74, 1970; Adler, 1979 (pp. 17-18). • *Portrait* (1859?): Museum of Comparative Zoology, Harvard University, courtesy Roxane Coombs. • *Signature* (1873): From "Correspondence Between Spencer Fullerton Baird and Louis Agassiz," edited by E. C. Herber, Publ. Smithsonian Inst., Washington, 237 pages, 1963.

BAIRD, Spencer F. (1823-1887).

Spencer Fullerton Baird, American vertebrate zoologist, administrator, and public servant, was born in Reading, Pennsylvania, on 3 February 1823. After his father's death in 1833, the family moved to Carlisle where, in 1836, Baird entered Dickinson College at the age of 13. His interest in natural history began much earlier when he started his collection of specimens, together with his older brother William Baird, with whom he published his first papers. Before graduation (B.A. 1840, M.A. 1843) he briefly studied medicine in New York City (1841) but decided instead to devote his life to zoology. At that time there were no American institutions granting doctorates in zoology, so he engaged in extensive self instruction, supplemented by correspondence and advice from leading naturalists, especially John James Audubon with whom he began a special friendship as early as 1838, until he became Professor of Natural History at Dickinson College in 1846.

Baird's first herpetological papers, including his major paper "Revision of the North American Tailed-Batrachia" (1849 [1850]), in which the genus *Desmognathus* and several new species were named, date from his days as a professor at Dickinson College. These papers appeared at a propitious moment, both for Baird and the nation. The Smithsonian Institution, which had been founded only three years earlier, was seeking an Assistant Secretary, that is, an administrator second only to the head of the entire organization and specifically to direct the development of a research center. The authorities, impressed with Baird's publications and testimonials from leading scientists, hired him in 1850. It was a lucky choice, for there was probably no one else in America with the necessary combination of skills: solid credentials as a scientist, superb organizer, unselfish, an outstanding judge of ability in others, and enormous persuasive powers which later were used to extract large sums of money from Congress to support the museum.

The nucleus of the research center, which Baird transformed into a national museum of natural history, was twofold: the collections of the Wilkes Expedition, which had explored the Pacific Basin in 1838-1842, and Baird's own private collection. The latter was said to have required two railroad freight-cars to move to Washington. Thereafter, Baird took advantage of every opportunity to enlarge the museum's holdings. He arranged to have naturalists attached to the government's surveys in the American West and to army posts. Doubtless his father-in-law, who was Inspector-General of the Army, was helpful in these plans.

The collections returned by these surveys were enormous, and from them Baird and his assistants described a large fraction of the vertebrates of the American West. In herpetology, alone or with his chief assistant, Charles F. Girard, he named many new forms, including such genera as *Uma, Uta, Xantusia, Diadophis, Masticophis, Storeria,* and *Tantilla*. They named dozens of new species, especially in the genera *Bufo, Hyla, Rana, Cnemidophorus, Sceloporus,* and *Natrix* (sensu lato), and including many well-known lizards (*Coleonyx variegatus, Dipsosaurus dorsalis, Eumeces obsoletus, Gambelia wislizenii,* and *Sauromalus obesus*) and snakes (*Crotalus atrox, Elaphe vulpina, Heterodon nasicus,* and *Thamnophis elegans*). Baird assigned certain taxa exclusively to his assistants (*Phrynosoma* to Girard and many snake genera to Robert Kennicott), and the turtles were loaned to Louis Agassiz for his monograph of American turtles (published 1857).

All told, of his nearly 1100 publications, Baird published only 43 on herpetology (1846-1880), but those in the 1850s were of greatest importance. These included reports on the amphibians and reptiles from Stansbury's expeditions to the Great Salt Lake of Utah (1852), the Wilkes Expedition (1852-1857), the Red River exploration (1853), the several Pacific Railroad Surveys (1859), and the United States-Mexican Boundary Survey (1859) in which the Gila monster (*Heloderma suspectum*) was first illustrated (although not recognized as a new species until 1869). Most of these reports were reprinted in 1978.

Baird's major *opus* in herpetology was the "Catalogue of North American Reptiles in the Museum of the Smithsonian Institution" (1853), co-authored with Girard. Only part one, on snakes, ever appeared, due to the pressure of other duties. More than 2000 specimens were assembled for this project and were first individually labelled with locality data, then put into one huge pile before they were sorted into species. Unfortunately, Baird's monographs of the amphibians and reptiles of the United States were never published, but the manuscripts were turned over to another of his protégés, Edward D. Cope, who based two of his own books on them (published 1889 and 1900).

Herpetology was only a minor part of Baird's research. His scientific reputation was as an ornithologist and mammalogist, on which subjects he published extensive books. He also took a special interest in fisheries, and when he became concerned about the depletion of the fisheries on the Atlantic coast he lobbied successfully for creation of a U.S. Fish Commission, of which he became founding director in 1871. His establishment of a fisheries research laboratory at Woods Hole, Massachusetts, led to the development of the famed oceanographic institute. In 1878, Baird became Secretary of the Smithsonian Institution, a position he held until his death on 19 August 1887, at Woods Hole.

During his lifetime Baird received many honors, in America and overseas. The Philadelphia Medical College awarded him an honorary M.D. in 1848, and there were additional honorary doctorates. He was elected to the U.S. National Academy of Sciences in 1864, over the opposition of his friend, Louis Agassiz, who felt that Baird was merely a descriptive biologist who had made no fundamental theoretical discoveries. Although Baird recognized the enormous importance of gaining an understanding of the diversity of the newly-explored regions, he also made pioneering contributions to the problems of geographic variation and the nature of species. Unlike Agassiz, Baird was an evolutionist, and his understanding of individual and geographic variation was viewed from a Darwinian point of view.

Baird's contributions to American science increasingly overshadowed those of Agassiz. For the last decade of his life, Baird was at once Secretary of the Smithsonian Institution, Director of the National Museum, and Commissioner of Fisheries. Highly regarded by Congress, he gave frequent testimony before their committees. When Congress authorized the Pacific Railroad Surveys in 1853 to determine the most practical route for a transcontinental railway, Baird used his influence to have naturalists attached to most of the surveying parties. During the Congressional debate leading up to the purchase of Alaska from Russia in 1867, it was Baird who provided the all-important documentation as to the worth of its natural resources. This was based on a five-year survey conducted in Alaska and directed by Baird, a project in which his former student, Robert Kennicott, lost his life.

In short, as a developer of scientific institutions and as patron of collectors and aspiring young naturalists, Baird had no peer. He was the most influential figure in American zoology during the closing decades of the 19th Century, yet his modesty and unselfishness have made his many contributions comparatively unrecognized by posterity.

• *References*: "Spencer Fullerton Baird A Biography," by W. H. Dall, J. B. Lippincott, Philadelphia, xvi, 462 pages, 1915; "Spencer Fullerton Baird and the Purchase of Alaska," by E. C. Herber, Proc. Amer. Philos. Soc., 98: 139-143, 1954; "Spencer Fullerton Baird," by D. C. Allard, Dict. Sci. Biogr., 1: 404-406, 1970; Adler, 1979 (pp. 26-31). • *Portrait* (1864): Courtesy American Philosophical Society, Philadelphia. • *Signature* (1852): From Dall, 1915.

GIRARD, Charles (1822-1895).

Charles Frédéric Girard, zoologist and physician, was born in Mülhausen, Upper Alsace (now Mulhouse, Haut-Rhin, France) on 9 March 1822. He was educated at the College of Neuchâtel in Switzerland, beginning as early as 1839, as the pupil and, later, assistant of Louis Agassiz, and he joined his teacher at Harvard College in the United States in 1847. In 1850, he was hired by Spencer F. Baird, who had just joined the fledgling Smithsonian Institution in Washington, as his chief assistant. This move forever earned Girard the enmity of Agassiz, but he sought a place where his individual talents could be recognized, and Baird unselfishly provided him with unparalleled opportunities.

At that time the Smithsonian was the recipient of extensive collections made by numerous government surveying parties in the American West and the Pacific Basin. Because Baird was more occupied with mammals and birds, as well as administrative duties, the lower vertebrate groups became Girard's task to describe. During the period 1851-1859, Girard published some 25 titles on herpetology (his only other one was a short note in 1891), alone or more usually in co-authorship with Baird, in which a huge number of new genera and species were named. All told, Girard published some 81 titles, about half in ichthyology, but he also authored many papers on invertebrates.

Girard published two major works in herpetology. In 1853 he co-authored with Baird the first part (on snakes) of a catalogue of reptiles in the Smithsonian museum. This contained descriptions of numerous new genera and species and was the first comprehensive review of the snakes of North America; unfortunately, no further parts were issued. Girard's other major title was the large volume on herpetology in the famous series describing the natural history collections of the United States Exploring Expedition, a circumnavigation of the globe in 1838-1842 commanded by Charles Wilkes that surveyed the entire Pacific Basin (and coincidentally discovered Antarctica), including the coasts of the Americas, Australia, New Zealand, Fiji, Singapore, and West Africa. The new taxa had been described in a series of papers (1852-1857), but Girard's book provided fuller descriptions and a comprehensive review as well as detailed illustrations.

There were, in fact, two editions of this book, both issued in 1858. The "official" issue, which contained a separate folio atlas of 23 plates, had Baird's name on the title page, since the work had been entrusted to him and the naval authorities thus would not permit Girard's name to be used. However, with Baird's characteristic willingness to give credit where it was due, he clearly stated in his introduction that all the work was Girard's. Thus, in the "unofficial" issue (reprinted 1978), which has an enlarged atlas of 32 plates, Girard was named as author. This elaborate book won him the Cuvier Prize in 1861.

Together with Baird and sometimes alone, Girard published numerous other reports on herpetological collections made by various government parties: the Wilkes Expedition (1852-1857), the Great Salt Lake of Utah (1852-1853), the Red River (1853), the Astronomical Expedition to Chile (1854-1855), and the various reports of the Pacific Railroad Surveys (1859). Most of these reports were reprinted in 1978. During this frenzied period Girard, a person of retiring habits who worked with great industry, somehow also completed his medical training at Georgetown College in Washington (M.D. 1856).

Girard visited Europe in 1860, and while he was there the American Civil War began. His sympathies were with the South, and he accepted a commission from the Confederate Government to supply its army with medical supplies. In 1863 Girard toured Virginia and the Carolinas and after the war returned to Washington briefly, but in 1865 he departed for France. He then engaged in a medical career for the next 20 years; during the German siege of Paris (1870) he was chief physician of a military hospital. He published primarily on medical topics but later had a renewed interest in zoology (1888-1891). Girard died in Neuilly-sur-Seine, a Paris suburb, on 29 January 1895.

• *References*: "Systematic Catalogue of the Scientific Labors of Dr. Charles Girard," by C. Girard, publ. by author, Poissy, 13 pages, 1888; "The Published Writings of Dr. Charles Girard," by G. B. Goode, Bull. U. S. Natl. Mus., 41: vi, 1-141, 1891; "Charles Frédéric Girard," by H. L. Clark, Dict. Amer. Biogr., 7-8: 319, 1946; Adler, 1979 (pp. 26-27).
• *Portrait*: From Goode, 1891. • *Signature* (1857): Academy of Natural Sciences of Philadelphia, courtesy Marsha Gross.

DUMÉRIL, Auguste (1812-1870).

Auguste-Henri-André Duméril was an ichthyologist and herpetologist, collaborator with and eventually successor to his father, Constant Duméril. He was born in Paris on 30 November 1812, by which time his father had already (1803) succeeded to the professorship in charge of fishes and reptiles at the Muséum National d'Histoire Naturelle. Young Duméril received his medical training at the University of Paris (M.D. 1842; D.Sc. 1843), where in 1844 he became assistant to the professor of comparative physiology. In contrast to his father who was primarily an anatomist, Auguste Duméril's special training was in physiology, which was to characterize some of his later work.

In 1851, Duméril became Assistant Naturalist at the museum, succeeding Gabriel Bibron, who had died prematurely in 1848, and thus he took his place in collaborating with Duméril senior to complete the classic "Erpétologie Générale" (1834-1854). Although credited to father and son on the title page, Auguste Duméril wrote the "Catalogue Methodique de la Collection des Reptiles" (1851) which reviewed the museum's newly accessioned materials, and in 1854 they published the two final volumes (7 and 9) of the "Erpétologie." Auguste Duméril began, but never finished, a series entitled "Description des Reptiles . . . du Muséum d'Histoire Naturelle" (part 1, 1852; part 2, 1856).

Finally, in the last year of his life, Duméril published the first part of what became, herpetologically, his *magnum opus*: the section on reptiles of "Mission Scientifique au Mexique et dans l'Amérique Centrale," 1870-1909 (reprinted 1978); the amphibian section, by Paul Brocchi, was published in 1881-1883 (reprinted 1978). This enormous book, comprising over 1000 pages and with 99 plates in large format, resulted from collections made in Middle America by M.-F. Bocourt. Bocourt had gone to Mexico, where Napoleon III had set up a puppet government under Archduke Maximilian of Austria, but with the latter's overthrow and execution Bocourt had to collect in Guatemala and adjacent countries. On returning to Paris, Bocourt assisted Duméril in writing the text and drawing the plates, but shortly afterwards Duméril died, after a long illness, in Paris, on 12 November 1870, during the German siege of the city. Consequently, Bocourt, with the later aid of François Mocquard, Léon Vaillant, and Fernand Angel, finished the book.

Unlike his father, Auguste Duméril did not neglect fishes, which were part of their professorial responsibilities. He published two volumes (1865, 1870), thus essentially completing the monumental series on fishes left unfinished by Georges Cuvier and Achille Valenciennes. Duméril was also keenly interested in physiological aspects, and published on temperature relations of reptiles and metamorphosis in the axolotl; the proper classification of the latter was particularly vexing to systematists until he successfully induced some to metamorphose at the museum's menagerie, a feat that earned him a gold medal at the international exhibition of 1867.

• *References*: "Auguste-Henri-André Duméril," anonymous, Grand Larousse, 6: 1379, 1870; "Discours . . . de M. le Professeur Auguste Duméril," by P. Gervais, Nouv. Arch. Mus. Hist. Nat., 7(Bull.): 15-24, 1871 (reprinted, Jour. Zool., 1: 85-91, 1872). • *Portrait* (1863): George S. Myers collection, California Academy of Sciences, courtesy Alan E. Leviton. • *Signature* (1854): Adler collection.

REINHARDT, J. T. (1816-1882).

Johannes Theodor Reinhardt, Danish zoologist, was born in Copenhagen on 3 December 1816. His father, J. C. H. Reinhardt, a native of Norway, which was then under Danish control, was Professor of Zoology at the University of Copenhagen. The son, destined to follow in his father's footsteps, completed his secondary schooling in 1834 and then began his medical studies at the university and also learned zoology from his father. Despite several years of study he never completed a degree, which was not uncommon at the time in Denmark.

In 1845-1847, Reinhardt was one of the zoologists on board the Danish corvette *Galathea* which circumnavigated the globe, visiting India, the Philippines, China, Japan, the Sandwich Islands (Hawaii), and South America. On returning to Copenhagen he was placed in charge of land vertebrates at the Royal Natural History Museum and later also became Lecturer in Zoology (1856-1878) at the Polytechnic School. Concurrently he was Lecturer (1861-1865) and Associate Professor of Zoology (1865-1882) at the University of Copenhagen, where he was the first teacher of vertebrate zoology and a sympathetic early proponent of Darwinism. Reinhardt died in Frederiksberg, near Copenhagen, on 23 October 1882.

During the voyage of the *Galathea* Reinhardt had been able to explore parts of southeastern Brazil and, thus inspired, he returned in 1850-1852 and again in 1854-1856 to make extensive collections of animals on behalf of the museum. These materials formed the basis for major works on insects and vertebrates, including "Bidrag til Kundskab om Brasiliens Padder og Krybdyr" (1861 [1862]), co-authored with his university colleague Christian F. Lütken. This book, which covered the amphibians and lizards (including crocodilians and amphisbaenians) of the savanna-like cerrados, is still regarded as the best review of the herpetofauna of that part of Brazil. Numerous new species were described; unfortunately no further parts were published.

In 1862, again with Lütken, Reinhardt issued a monograph on the amphibians and reptiles of the Caribbean region based on the collections of the Royal Natural History Museum and his university's Zoological Museum. This treatise, which described many new species and had an extensive zoogeographic discussion, concentrated on the Danish West Indies located just east of Puerto Rico, islands which became the U.S. Virgin Islands upon their purchase by the United States in 1917.

Reinhardt published numerous other herpetological studies, particularly on snakes. One of these (1843), based on the snake collection at the Royal Natural History Museum which he began to curate informally in 1840, described numerous new species collected by Danish traders in West Africa ("Guinea") and other places, including the Puerto Rican *Epicrates inornatus*, *Bungarus flaviceps* of Southeast Asia, and the well-known African spitting cobra (*Naja nigricollis*). In 1863 (English translation 1955), he described the peculiar egg-eating snake of India and Bangladesh (*Elachistodon westermanni*), and other papers by him concern such diverse topics as skeletal anatomy of crocodilians and apical pits of snake scales. There is some confusion in the literature, however, since his father also published on vertebrates, including reptiles (e.g., he described the distinctive Southeast Asian colubrid *Xenodermus javanicus* in 1836, which is sometimes credited to the son).

• *References*: "Oversigt over Dansk Entomologis Historie," by K. L. Henriksen, Entomol. Meddel., 15: 1-578, 1921-1937 (see pp. 212-215, 1926); "Undervisningen i Zoologi ved Københavns Universitet," by R. Spärck, Festskr. Københavns Univ., Copenhagen, 301 pages, 1962 (see pp. 72-77); Papavero, 1973 (p. 366); "Johannes Theodor Reinhardt," by R. Spärck, Dansk Biograf. Leks., 12: 128-129, 1982. • *Portrait* (about 1880): From Spärck, 1962. • *Signature* (1860): Adler collection.

BURMEISTER, Hermann (1807-1892).

Carl Hermann Conrad Burmeister, a distinguished German entomologist who became one of Argentina's leading naturalists, was born in Stralsund, near the Baltic coast, on 15 January 1807. He was educated at the universities of Greifswald and Halle, the latter granting him the M.D. and Ph.D. in 1829. He published many works during these early years, mostly entomological, with his major title "Handbuch der Entomologie" (1832-1855), in five volumes, being a classic. Burmeister became Professor of Zoology at the University of Halle but due to the political disturbances of 1848-1850, when he sided with the "Liberals," he was forced to leave and went to Brazil, where he explored Minas Gerais (1850-1852), and also to Uruguay and Argentina (1856-1860).

In 1862, Burmeister emigrated to Buenos Aires and became founding director of the Museo Público de Buenos Aires, the national museum of natural history, where he spent the remainder of his career, retiring just a month before his death, in Buenos Aires, on 2 May 1892. During his tenure as director, the national collections grew rapidly in all branches of natural history including paleontology, an area in which Argentina is particularly rich and a subject that became a special interest to him.

Burmeister's early interest in entomology continued throughout his life, but his interests broadened to include vertebrate paleontology and physical geography. His major book on living vertebrates, "Systematische Übersicht der Thiere Brasiliens" (1854-1856), in three volumes, covered only mammals and birds. Herpetologically, besides some shorter papers and more comprehensive works which include reptiles in part, Burmeister published two books: "Der fossile Gavial von Boll," (1854), co-authored with E. d'Alton, which included a major review, with excellent plates, of living crocodilians, and "Erläuterungen zur Fauna Brasiliens" (1856), a large folio atlas of 32 plates, 22 of them beautifully handcolored, of Brazilian mammals and frogs.

• *References*: "Prof. Hermann Carl Conrad Burmeister [*sic*], Hon. F.E.S.," by R. McLachlan, Entomol. Month. Mag., 28: 221-222, 1892; "Carlos Germán Conrado Burmeister. Reseña Biográfica," by C. Berg, Anal. Mus. Nac. Buenos Aires, 4: 315-357, 1895; Papavero, 1973 (pp. 292-293). • *Portrait*: From Berg, 1895. • *Signature* (early form on left, about 1850): Museum für Naturkunde der Humboldt-Universität zu Berlin, courtesy Rainer Günther; (later form on right, 1874): Smithsonian Institution Archives, courtesy William Cox and Alan E. Leviton.

GÜNTHER, Albert (1830-1914).

Albert Carl Ludwig Gotthilf Günther was an eminent ichthyologist who also specialized in herpetology. He was born in Esslingen, Württemberg, on 3 October 1830 and entered the theological school at the University of Tübingen in 1847. Although he took holy orders in the Lutheran Church four years later, his attention quickly turned to medical studies because of his longstanding interest in natural history and the teaching of Wilhelm von Rapp at Tübingen. He received a doctorate in philosophy and arts in

1853 and then attended medical school in Berlin, where one of his professors was the celebrated zoologist Johannes Müller. During 1854, he taught in Bonn, where he developed an important association with F. H. Troschel. He finally completed his M.D. degree at Tübingen in 1855-1857.

In 1857 Günther offered his services to John E. Gray, then Keeper of the Zoology Department at the British Museum, and was hired that November; eventually he succeeded Gray as keeper in 1875. Gray had hired him to catalogue the museum's snake collection, at a salary of £75. Günther finished the job four months later, having reviewed the entire colubrid snake collection (then 3100 specimens) which resulted in his first book, "Catalogue of Colubrine Snakes" (1858; reprinted 1971, again in 1976). Gray then assigned him to do the frogs, which Günther completed in another four months and later published as "Catalogue of the Batrachia Salientia" (1858 [1859]). At this rate Gray was going broke so he then requested Günther to take on the herculean task of cataloguing the museum's fishes which thereby diverted the primary focus of Günther's research for the rest of his life, although he continued to write about amphibians and reptiles.

In 1864 Günther published "Reptiles of British India," a project that Gray was unwilling to approve, so it had to be written at home. Three years later, Günther announced his single most important herpetological discovery, his recognition that *Sphenodon* of New Zealand was not a lizard as then thought but the sole living representative of the otherwise extinct order Rhynchocephalia. He was elected a Fellow of the Royal Society that same year. In addition to some 200 papers on herpetology—his geographic areas of interest included Africa, Australia, and Asia—Günther published two other books: "The Gigantic Land-Tortoises (Living and Extinct)" (1877 [1878]) and the herpetological volume in the series "Biologia Centrali-Americana" (1885-1902; reprinted 1987). Günther's work on lower vertebrates was regularly cited by Charles Darwin in his book "Descent of Man" (1871), who depended on Günther for his information about sexual characters.

Like John E. Gray before him, Günther was a great collection builder and during his tenure as keeper the zoological collections at the British Museum grew from 1.3 to 2.2 million specimens. This was a period of expansion of the British Empire overseas and the collections expanded rapidly. Günther was also responsible for moving the collections from the old museum at Bloomsbury to the new natural history museum at South Kensington, London, in 1882, and thereafter, for building the museum's great zoology library. In 1865 Günther founded *Zoological Record*, to this day the most complete index of the world's zoological literature. Doubtless his most important contribution to herpetology, however, was his hiring of a young Belgian, George A. Boulenger, in 1879, as his successor in charge of lower vertebrates, thus assuring the long tradition of herpetological research at the British Museum. Günther retired in 1895 and died at Kew, London, on 1 February 1914.

• *References*: Papavero, 1973 (pp. 409-417, "Biologia Centrali-Americana" itinerary); "A Century of Zoology at the British Museum," by A. E. Günther, Dawsons, London, 533 pages, 1975 (see pp. 213-469); "A Sketch of the Life and Work of Albert Günther, M.D., F.R.S. (1830-1914)," by A. E. Günther, p. liii-lviii. *In* A. Günther's Biologia Centrali-Americana: Reptilia and Batrachia, repr. ed. Soc. Study Amphib. Rept., Athens (Ohio), 1987. • *Portrait* (about 1900): British Museum (Nat. Hist.), courtesy A. E. Günther. • *Signature* (1881): Courtesy A. E. Günther.

COPE, Edward Drinker (1840-1897).

Edward Drinker Cope, vertebrate paleontologist, anatomist, and America's greatest herpetologist, was born in Philadelphia on 28 July 1840. In school he was drilled in the classics and later in languages by a private tutor, but it was Cope's father, a wealthy merchant, who fostered his interest in natural history, although for an intended career in agriculture. In 1846, at the age of six, he visited the museum of the Academy of Natural Sciences in Philadelphia which, judging from his extensive notes about the trip, had a profound effect on him. During the summers of 1854-1860 when young Cope worked on the farms of various relatives, he spent his spare time in the fields and meadows collecting salamanders and snakes. By 1859 he was already a volunteer worker at the academy, busily engaged in reorganizing the herpetological collection, and he published his first paper that year at the age of 18 in which he described two new salamanders and proposed modifications in the classification of the group. Although Cope's interests later broadened to include all vertebrate groups, herpetology was his first love and favorite branch of study, and it continued to occupy him for the rest of his life.

Under the ruse that it would permit him to treat farm stock, Cope pleaded with his father to let him attend the University of Pennsylvania to study comparative anatomy with the paleontologist Joseph Leidy, and he finally was allowed to do so during 1860-1861. It was the only formal course in science that Cope ever had! It soon became clear that he would never be happy as a farmer, so his father relented and permitted him to pursue his informal studies by spending the winters of 1861-1863, during the Civil War, in Washington where he became a protégé of Spencer F. Baird.

In 1863-1864 Cope travelled throughout Europe, where he met and worked with some of the leading anatomists and herpetologists of the day, including John E. Gray and Albert Günther at the British Museum, Hermann Schlegel in Leiden, the anatomist Karl von Siebold and explorer Prince Maximilian zu Wied-Neuwied in Prussia, and Giorgio Jan in Milan. There were also extended periods at the museums in Vienna, Berlin, and Paris. The year in Europe completed Cope's student life, and the exposure set a high standard for his future research and publication.

Shortly after returning to America, Cope became Professor of Natural Science at Haverford College, near Philadelphia. Cope, of course, had no formal degree, so Haverford obliged with an honorary one (A.M. 1864). Concurrently, he was Curator of Herpetology at the Academy of Natural Sciences (1865-1873). His publications on living amphibians and reptiles continued, but in 1865 he published his first paper in paleontology—on a fossil amphibian from Illinois—and thereafter his interests shifted more to that subject. Increasingly, Cope found that his teaching duties interfered with his research and, after three years at Haverford, he resigned to pursue scientific exploration and writing, operating from two adjacent houses in Philadelphia.

From 1871 to 1893, Cope participated almost annually in extensive explorations in western North America, beginning in Kansas and later extending north to Montana, west to California, and as far south as Veracruz, Mexico. During many of these years, Cope spent about eight months of each year in the field and the remainder in Philadelphia writing up the latest results. This work was largely under the auspices of official U.S. government surveys, although Cope provided most of his own support. Travel was by stagecoach west from Philadelphia, then by horse and wagon. Cope's field parties were small, usually only himself, a guide, and a few assistants, yet they often worked in disputed areas beyond the protection of the military. The primary purpose of these expeditions was paleontological, but Cope also collected and observed living vertebrates, including amphibians and reptiles. It is not generally recognized today to what extent Cope was also a field naturalist; indeed, it is fair to say that he was the first of the great herpetologists to be thoroughly familiar with his animals in nature.

Cope's most epochal contributions were to vertebrate paleontology. He recognized the importance of the enormous collections of vertebrate fossils, especially those of dinosaurs, being unearthed mainly in the American West, and through their study foresaw his dominance of the field of paleontology. In his many books and papers on the subject, Cope described and named 1282 genera and species of North American fossil vertebrates, 510 of them amphibians and reptiles.

This work inevitably brought Cope into conflict with the other leading American paleontologist of the era, O. C. Marsh of Yale College. They had previously been friends, but beginning in 1872 their famous feud began, eventually involving dirty tricks, spying, charges of plagiarism and, in 1890, sensational headlines in a series of newspaper articles in the *New York Herald*. The competition became so fierce that Cope got into the habit of hurriedly studying his newly-collected material in the field and then promptly telegraphing his descriptions to Eastern journals, but several times the telegraph operators got the names of his new species garbled—for all posterity! Sometimes Cope's and Marsh's papers were published only one day apart. The dispute ended finally with Cope's death, but by that time the rivalry had engulfed America's leading scientific institutions and academies and many biologists and paleontologists, including the anatomist Georg Baur, Marsh's chief assistant, who was secretly Cope's ally.

Cope's father died in 1875 and left him an inheritance in excess of a quarter million dollars. With such enormous resources Cope was able to privately fund his own collections and research. In 1878 he became co-owner and publisher of the journal *American Naturalist*, which thereafter served as his personal outlet for articles, especially in his continuing competition with O. C. Marsh. Within two years, however, Cope lost most of his fortune due to bad investments in Mexican silver mines, and he had to curtail his research program. Eventually, to produce needed income, he had to go on popular scientific lecture tours, sell his extensive collection of fossils, and seek a paying job.

By 1886 Cope was on the brink of poverty, which dogged him the rest of his life. Fortunately, the University of Pennsylvania in 1889 appointed him Professor of Geology and Mineralogy and in 1895 promoted him to the chair of zoology and comparative anatomy, his teacher Joseph Leidy's old post and a position that Cope held the rest of his relatively short life. Besides his fame as an anatomist and vertebrate paleontologist, Cope was the foremost theorist of the neo-Lamarckian movement in American biology, a subject on which he published extensively. During his lifetime he received numerous prizes and other honors, including election to the U.S. National Academy of Sciences (1872) and to the presidency of the country's largest scientific society, the American Association for the Advancement of Science (1896). In 1886 the University of Heidelberg conferred upon him an honorary Ph.D. The American journal of lower vertebrates, *Copeia*, founded in 1913, was named for him.

Cope's productivity was staggering—1395 titles (of which about 170 are on Recent amphibians and reptiles), including numerous books—but this was accomplished at great physical cost. He worked at tremendous speed and, consequently, his papers are often incomplete or filled with careless errors, especially in bibliographic references. The rigors of years of hard field work, the feud with O. C. Marsh, and finally his financial failures took their toll, and by middle life he was said to be visibly aged beyond his years. He died in Philadelphia on 12 April 1897, at the age of 56 from an illness which he had waited too long to correct through surgery.

Cope's contributions to herpetology are numerous. He described many new taxa, mostly from the Western Hemisphere, but it was his major classifications which represented his most important contributions to the field. These were largely based on his profound knowledge of anatomy. The conclusions that he reached were often revolutionary but also sometimes naïve. It has been said that Cope found herpetology an art and left it a science; the field was little more than a taxonomic subject when he began his career, yet largely through his efforts it became a comprehensive entity including not only taxonomy but anatomy, geographic distribution, adaptation, and phylogenetic history. Cope was also the first to synthesize information about living and extinct taxa for all amphibians and reptiles.

His achievements often met with resistance and even antagonism, but it was in Cope's personality to cherish controversy. At the time Cope began his work it was standard practice to base classifications on external characteristics. Despite the importance of early discoveries by Johannes Müller based on internal anatomy, such characters were ignored by most workers, and dissections were strictly forbidden at major museums in Europe. Following Müller's lead, Cope broke this outdated tradition by studying the osteology and soft anatomy of vertebrates, thereby making spectacular discoveries. For example, it was he who first noted differences in the connection between the lateral halves of the sternum of frogs—the so-called arciferal and firmisternal conditions, lately re-defined—which led to his reclassification of anurans (1865). Other papers describing the skulls, vertebrae, shoulder girdles, and teeth of lizards led to the first natural arrangement of that group. Cope's classifications, with minor modifications, were adopted by George A. Boulenger in his herpetological catalogues of the British Museum, and thus Boulenger must be credited with their popularization.

Cope's major works in herpetology include his "Synopsis of the Extinct Batrachia, Reptilia and Aves of North America" (1870, supplement 1875) and its companion, "Synopsis of the Extinct Batrachia" (1875). His checklists of North American (1875) and Central American and Mexican (1887) amphibians and reptiles were important catalogues for reference. Cope was the first to use differences in the respiratory system as a basis for the classification of snakes (1894, 1895), and he paid special attention to the degree of development of the left lung and to the respiratory tissue in the trachea. His "Classification of the Ophidia" (1895, preliminary paper in 1893) introduced still another important and useful anatomical characteristic, snake hemipenes; those of lizards he published in 1896.

Cope's most widely known books on herpetology, however, were published toward the end of his career, although they had been initiated many years before. "The Batrachia of North America" (1889, reprinted 1963) was based on a manuscript and drawings by Spencer F. Baird and Charles F. Girard, to which Cope added his own material, chiefly anatomical. Cope's last work, "The Crocodilians, Lizards, and Snakes of North America" (1900), was published posthumously. It, too, was partly based on a Baird manuscript (lizard section only), but the book was never fully completed due to Cope's premature death (the turtles, however, were omitted on account of Georg Baur's intended, but never published, monograph on the subject). Both of Cope's books are still standard texts on North American herpetology, containing much information on external structure and anatomy and illustrated by numerous plates.

• *References*: "Herpetological and Ichthyological Contributions," by T. Gill, p. 2-24. *In* Addresses in Memory of Edward Drinker Cope. Amer. Philos. Soc. Mem. Vol. 1, 1897 (partly preprinted in Amer. Nat., 31: 831-863, 1897); "Biographical Memoir of Edward Drinker Cope 1840-1897," by H. F. Osborn, Biogr. Mem. U.S. Natl. Acad. Sci., 13: 125-317, 1930; "Cope: Master Naturalist," by H. F. Osborn,

Princeton Univ. Press, Princeton (New Jersey), xvi, 740 pages, 1931; "Edward Drinker Cope, Herpetologist," by W. H. Davis, Bull. Antivenin Inst. Amer., 5: 71-80, 1932; "Cope versus Marsh," by A. S. Romer, Syst. Zool., 13: 201-207, 1964; "The Fossil Feud," by E. N. Shore, Exposition Press, Hicksville (New York), xi, 340 pages, 1974; "Edward Drinker Cope," by J. M. Maline, Dict. Sci. Biogr., 15(suppl. 1): 91-93, 1978; Adler, 1979 (pp. 31-33).
• *Portrait*: Courtesy Historical Society of Pennsylvania. • *Signature* (1860): Library, Academy of Natural Sciences of Philadelphia, courtesy Marsha Gross.

STRAUCH, Alexander (1832-1893).

Alexander Alexandrovich Strauch was the first eminent Russian herpetologist. Born on 1 March 1832 in St. Petersburg (now Leningrad), in 1861 he became curator at the Zoological Museum of the Imperial Academy of Sciences in that city. Among other treasures, the academy possessed the famous collection of Albertus Seba, an Amsterdam merchant, that had been purchased in 1717 by Peter the Great himself; part of a second Seba collection was added in 1752. To this Strauch added assiduously, from various Russian expeditions and through exchanges with other museums. In 1870 he was elected Academician of the Academy of Sciences, the first herpetologist to be so honored and later, in 1879, he became Director of the museum, a post which he held until 1890, as successor to J. F. Brandt who also published on reptiles. Strauch succeeded in establishing St. Petersburg as a major world center of herpetology, and after he retired, his curatorship was taken up by A. M. Nikolsky. Strauch died in Wiesbaden, Germany, on 14 August 1893.

Strauch's scope of interest and *modus operandi* was reminiscent of the Paris school: careful systematic studies on a worldwide basis, although with an aversion to the use of internal anatomical criteria in systematics. Typical of the St. Petersburg scientific establishment of the day, most of his publications were in German. His major herpetological works included a faunal monograph on Algeria (1862), several treatises on turtles (1862, 1865), and world synopses of several groups based partly on the St. Petersburg collections and partly on his examination of specimens from European museums (crocodiles, 1866; vipers, 1869; salamanders, 1870; amphisbaenians, 1882; geckos, 1887; and turtles, 1890). His most famous work, issued in 1873, a detailed review of the snakes of the Russian Empire, was followed in 1876 by a monograph on the amphibians and reptiles collected during N. M. Przewalski's first expedition to Central Asia (1871-1873), which travelled through much *terra incognita* in Mongolia and northern China as part of a larger effort by Imperial Russia to expand its boundaries.

• *References*: "Aleksandr Aleksandrovich Strauch," *in* G. A. Kujasev (ed.), Proc. Archives Acad. Sci. USSR, 1: 99-100, 1933; Scarlato, 1982 (pp. 69, 230). • *Portrait*: Academy of Sciences, Leningrad, courtesy Natalia B. Ananjeva and Nikolai L. Orlov. • *Signature* (romanised form, 1876): Adler collection.

JAN, Giorgio (1791-1866).

Although most remembered by herpetologists for his great atlas of snake illustrations, "Iconographie Générale des Ophidiens," Jan was both a botanist and zoologist as well as a poet and literary critic. He was born in Vienna, of Hungarian ancestry, on 21 December 1791, as Georg Jan. After serving as an assistant at the University of Vienna, in 1816 he moved to the ancient University of Parma in Italy, as Professor of Botany and Director of the Botanical Garden. The Duchy of Parma had recently been assigned to the jurisdiction of Austria by the Congress of Vienna, following Napoleon's final defeat at Waterloo.

Jan's primary research at Parma was botanical but he made an enormous personal natural history collection, including fossils and minerals, which was merged with that of the Milanese nobleman Giuseppe De Cristoforis in 1831. Together, they published many catalogues of their specimens—which they offered for sale and exchange—and thereby described many new species, primarily of insects and molluscs. Cristoforis died in 1837 and his will directed that his collections would be donated to the City of Milan if the city agreed to establish a natural history museum with Jan as director. Jan also agreed to donate his own collection. Thus, the Museo Civico di Storia Naturale was founded in 1838, the oldest civic natural history museum in Italy, with Jan as founding head. He finally moved to Milan in 1845.

Jan was a latecomer to herpetology. He did not begin his work on "Iconographie Générale des Ophidiens" until 1853, when he was 62. The book was published in 50 parts, each with six plates, during the period 1860-1881, with a so-called 51st part in 1881 (1882) containing the title page, table of contents, and index (all reprinted 1961). Together with his artist, Ferdinando Sordelli, a museum assistant, Jan continued to amass snakes for the museum's collection, and those species that could not be acquired permanently were borrowed from virtually all of the major European and American museums of the day, except those in London and Berlin who refused to cooperate. Eventually Jan and Sordelli produced an atlas of 300 plates, in folio, illustrating the snakes of the world. Concurrently, Jan had separately published only the first two parts of the text in smaller format when he died, in Milan, on 8 May 1866. Thus, we must credit Sordelli for seeing the remainder of the illustrations through to publication.

• *References*: "Inaugurandosi Solennemente . . . il Busto di Giorgio Jan," by E. Cornalia, L. Giacomo Pirola, Milan, 27 pages, 1867; Siebenrock, 1901 (pp. 445-446); "Il Centenario di Giorgio Jan (1791-1866)," by C. Conci, Atti Soc. Ital. Sci. Nat. / Mus. Civ. Stor. Nat. Milano, 106: 1-94, 1967. • *Portrait*: From Conci, 1967, courtesy Wolfgang Böhme. • *Signature* (1863): Adler collection.

SORDELLI, Ferdinando (1837-1916).

Ferdinando Sordelli, naturalist and artist, is best known for his collaboration with Giorgio Jan in producing the monumental treatise "Iconographie Générale des Ophidiens." Sordelli was born in Milan on 12 December 1837 and came to work with Jan at Milan's Museo Civico di Storia Naturale in 1857, originally as a temporary assistant and later, in 1865, as permanent assistant. Later, Sordelli's own research was in zoology and paleontology, particularly on fossil plants and also including herpetology, but it was his gift as a draftsman that first brought him to Jan's attention.

Jan had begun his work on the atlas of snake illustrations in 1853. He was a good artist himself, but the task was too large for one person. Sordelli had been Jan's student prior to 1857, and his artistic abilities led Jan to entrust to him the intricate and large-scale task of illustrating his book. Apparently many of the drawings were made while Sordelli was still a teenager and before he was officially hired by the museum in 1857, but the work was not completed until about 1868, during which period it required all of Sordelli's time. Jan had died in 1866, so it was Sordelli who completed publication of the book, issued in 51 parts from 1860 to 1882 (reprinted 1961). The atlas of 300 plates, in folio, comprised the staggering total of nearly 8500 individual figures which were drawn in superlative detail and arranged on each plate with artistic skill, thus providing one of the most comprehensive series of drawings of an animal group ever issued.

Sordelli retired as museum assistant in 1915 and died in Milan on 17 January 1916.

• *References*: "In Memoria del Prof. Ferdinando Sordelli," by M. De Marchi, Atti Soc. Ital. Sci. Nat. / Mus. Civ. Stor. Nat. Milano, 55: 1-4, 1916; "Il Centenario di Giorgio Jan (1791-1866)," by C. Conci, Atti Soc. Ital. Sci. Nat. / Mus. Civ. Stor. Nat. Milano, 106: 1-94 (see p. 68-69), 1967. • *Portrait and signature* (1859): Museo Civico di Storia Naturale, Milan, courtesy Mario Schiavone.

KREFFT, Gerard (1830-1881).

Born Johann Gerhard Louis Krefft (see *Note*) in Brunswick, Braunschweig, on 17 February 1830, Gerard Krefft became one of Australia's first naturalists and the Father of Australian Herpetology. Originally intending a career in business, he emigrated to the United States in 1851 and briefly supported himself as an artist in New York City before embarking for Australia in July 1852 to join the gold rush. He was a miner until 1857 when he moved to Melbourne and was hired by Frederick McCoy at the National Museum as a collector and artist. Later, after returning for a short period to Germany (1858-1859), he became Assistant Curator at the Australian Museum in Sydney, and in 1861 was promoted to Curator (or director).

Krefft had a continuing feud with the museum's trustees, in particular with their chairman, William J. Macleay, himself a sometimes herpetologist, and was fired in 1874. He refused to leave and barricaded his office; later, he was literally carried out while still sitting in his chair and unceremoniously deposited in the street, the door locked behind him. Not to be dissuaded in an act he considered illegal, Krefft set up a rival "Office of the Curator of the Australian Museum," sued the trustees, and won a large financial settlement, but was not permitted to resume his curatorship. He never resumed serious scientific work, although he wrote natural history articles regularly for the Sydney newspapers. He died in Woolloomooloo, a Sydney suburb, on 19 February 1881.

Krefft's interests were very broad, including both vertebrates and invertebrates as well as paleontology. His single most important discovery was of the Australian lungfish (*Neoceratodus forsteri*) of the Queensland swamps. Of some 100 technical papers published by Krefft, 26 are on herpetological topics (published during the period 1862-1876). He wrote two classic books, the first of their kind for Australia: "The Snakes of Australia" (1869) and "The Mammals of Australia" (1871, reprinted 1979). The government being unwilling to fund the former title, Krefft paid for the printing of 700 copies himself. It was issued in two versions, with plain or with beautifully handcolored plates, and its continuing importance to herpetological work in Australia justified its reprinting in 1984.

• *References*: "Gerhard Krefft, ein Braunschweiger Naturforscher," by F. Grabowsky, Braunsch. Mag., 1: 36-40, 1896; "Johann Gerhard Louis Krefft," by P. Zimmerman, Allg. Deut. Biogr., 51: 373-374, 1906; "Johann Ludwig Gerard Krefft," by P. Serle, Dict. Austr. Biogr., 1: 506-507, 1949; "The Life and Work of Gerard Krefft (1830-1881)," by G. P. Whitley, Proc. Royal Zool. Soc. New S. Wales, 1958-59: 21-34, 1961 (addendum, same journal, 1967-68: 38-42, 1969). • *Portrait* (about 1857): Australian Museum, courtesy Harold G. Cogger. • *Signature* (about 1860): Museum für Naturkunde der Humboldt-Universität zu Berlin, courtesy Rainer Günther. • *Note*: German and Australian biographical sources differ on the details of Krefft's given names.

BOCAGE, J. V. Barbosa du (1823-1907).

José Vicente Barbosa du Bocage, distinguished zoologist and politician, was born on 2 May 1823 in Funchal, on the Portuguese island of Madeira off the Moroccan coast. He studied medicine and mathematics at the University of Coimbra (1839-1846) and, after some political disturbances, became Professor of Zoology at the Polytechnic School (now part of the University of Lisbon) in 1849. Later, in addition to medical duties, he became Director of the Zoological Section of the National Museum of Lisbon which in 1905 was renamed "Museu Bocage" in his honor.

The museum had been founded in 1772 by Domenico Vandelli, describer of the leatherback (*Dermochelys coriacea*), but it was Bocage who truly built the collections. Most specimens were obtained by the museum's many collectors in Portuguese colonies in Africa, Asia, and South America, but Bocage, a gifted diplomat, even obtained some by convincing the French to send many specimens to compensate for the collections stolen from Lisbon in 1807 during the Napoleonic occupation. The latter were summarily requisitioned by É. Geoffroy Saint Hilaire and taken to Paris for study and description by French naturalists including Georges Cuvier, who euphemistically referred to the specimens in print as "don du Musée de Lisbonne," a gift indeed! Besides Bocage's zoological work—he is better known as an ornithologist than as a herpetologist—he founded the Geographical Society of Lisbon and became a trusted counsellor to the Crown and Minister of State. He died in Lisbon on 3 November 1907.

Bocage authored nearly 200 papers. From 1863 to 1904 he published 58 herpetological titles, in which he described 111 new species of amphibians and reptiles, mostly from Africa, but also from Portugal, New Caledonia, and Australia, and many of these taxa are still recognized. His most elaborate herpetological work was his book "Herpétologie d'Angola et du Congo" (1895), with 20 plates; copies were issued in versions with either uncolored or handcolored plates.

• *References*: "Le Professeur Barbosa du Bocage 1823-1907," by C. França, Bull. Soc. Portug. Sci. Nat., 2: 141-194, 1909; "José Vicente Barbosa du Bocage," p. 101-102. *In* E. Pereira and G. Rodrigues, Portugal—Diccion. Hist., Biogr., Bibliogr., Lisbon, vol. 2, 1906. • *Portrait* (about 1903) *and signature*: From Bol. Soc. Portug. Ciênc. Nat., ser. 2, 12: 1, 1968-1969 (1970), courtesy Carlos Almaça.

JIMÉNEZ DE LA ESPADA, M. (1831-1898).

Marcos Jiménez de la Espada (see *Note*), Spanish naturalist and Americanist, was born in Cartagena (Murcia), on 5 March 1831. He received his bachelor's degree at the Instituto de Sevilla (1849) and attended the Universidad Central in Madrid where he received his Master's degree in 1855. Previously, in 1853, he had accepted an adjunct professorship at the university and in 1857 also became an assistant in the university's Museo de Ciencias Naturales. During these years he collected specimens throughout Spain and taught zoology and geology, but in 1862 the Spanish Government established the Comisíon Científica del Pacífico, to study the fauna and flora of the former Spanish colonies in America, and Jiménez de la Espada was named one of its zoologists. During 1862-1865, he travelled some 45,000 km throughout South America—including extensive trips in Ecuador, a crossing of the Andes, and a journey down the Amazon—making one of the most important early explorations of the continent, but not without great disappointments.

In March 1864, the admiral commanding the Spanish squadron suddenly ordered the naturalists off his ships. He then joined a larger fleet that captured some Peruvian islands which initiated a war (1864-1866) in which Spain attempted to establish a naval base. Not only had Jiménez de la Espada and the other naturalists lost their logistical support, but they were now suspected of being spies, which made their strictly scientific work all the more difficult. Ultimately, they returned to Spain laden with extensive collections, including some 1500 specimens of amphibians and reptiles, but the government refused to give proper financial support after 1872. Nevertheless, Jiménez de la Espada continued their study and published a few short papers (1871-1875) describing some novelties, but only one volume of a larger work was ever published. He also authored an important paper in 1872 on reproductive habits of the peculiar Chilean frog, *Rhinoderma darwinii*, based on his own observations of living specimens, in which he corrected Claude Gay's longstanding (1836) statement that this frog was viviparous.

In 1875, Jiménez de la Espada issued "Vertebrados del Viaje al Pacífico . . . Batracios" (reprinted 1978), the most complete catalogue of South American frogs assembled up to that time. In this book and his shorter papers he described 38 new frogs and one salamander—the latter a mislabelled Californian species—most of which are recognized today. With no real support for his herpetological work, in 1875 he turned to the study of history and culture of the Americas and became Spain's leading expert; he published only a single additional herpetological paper (1898) which was the thesis for his Ph.D., necessary to become a full professor. He died on 3 October 1898, in Madrid, his hopes for a more extensive work on the herpetology of South America sadly never realized.

• *References*: "Biografía de D. Marcos Jiménez de la Espada (1831-1898)," by A. J. Barreiro, Tip. R. C. Menor, Madrid, 43 pages, 1927; "For Science and National Glory. The Spanish Expedition to America, 1862-1866," by R. R. Miller, Univ. Oklahoma Press, Norman, xiv, 194 pages, 1968 (Spanish transl., Ed. Serbal, Barcelona, 1983); "Marcos Jiménez de la Espada, Naturalist Explorer of the Andes and Upper Amazon Basin," by J. M. Savage, p. vii-xvi. *In* M. J. de la Espada, Vertebrados del Viaje al Pacífico. Batracios, repr. ed., Soc. Study Amphib. Rept., Athens (Ohio), 1978; R. M. Gomez and B. Sanchíz, 1987 (p. 151-152). • *Portrait*: From Barreiro, 1927. • *Signature* (1869): Museo Nacional de Ciencias Naturales, Madrid, courtesy Borja Sanchíz. • *Note*: The formal family name is Jiménez de la Espada, but the shortened version "Espada" traditionally has been used for him by many herpetologists.

GUNDLACH, Juan (1810-1896).

Gundlach, an eminent Cuban naturalist, was born Johannes Christopher Gundlach in Marburg, Hesse-Cassel (now Hessen, Germany), on 17 July 1810. He was interested in insects as a boy and, following the accidental blast of a shotgun that cost him his sense of smell, became a noted taxidermist, in part due to the fact that he could work with carcasses in advanced stages of decomposition. At first hired by the University of Marburg as a preparator, he later became a student there and graduated with a Ph.D. in 1837. A Cuban fellow student invited him to collect specimens in Cuba and they arrived in 1839, setting up a base of operation in Cardenas. Later, in 1852, Gundlach moved to Havana, where he met the Cuban naturalist, Felipe Poey. Together they planned explorations in eastern Cuba, where Gundlach collected in 1856-1859. During this expedition Gundlach discovered the strange xantusiid lizard, *Cricosaura typica*, which he described in 1863 with Wilhelm Peters, to whom he had sent these and other reptiles for identification.

In 1864, Gundlach established the Cuban Museum of Natural History which was later transferred to the Havana Institute. During the Cuban insurrection (1868-1878) centered in the eastern provinces, Gundlach's field work there was blocked, so he turned his attention elsewhere: to Puerto Rico (expeditions in 1873, 1875) and to laboratory work, including studies on metamorphosis in frogs of the genus *Eleutherodactylus*. By 1884 he was collecting in eastern Cuba again, with his last trip there in 1887 at the age of 76. He received many honors during his lifetime, from organizations in Cuba and abroad, for his studies of virtually all animal groups (except fish, which were the primary area of interest to Felipe Poey), yet he remained a singularly quiet and modest person. Gundlach died in Havana on 17 March 1896.

Gundlach's work on reptiles was only incidental to his more comprehensive exploratory work in both Cuba and Puerto Rico. His special fields of interest were birds and insects, but he also published extensively on mammals and molluscs. His few papers on herpetology were published between 1863 and 1878, and his major work, "Contribucion á la Erpetologia Cubana," was published as a book in Havana in 1880.

• *References*: "Sketch of John Gundlach," by J. Vilaró, Pop. Sci. Month., 50: 691-697, 1897; "Vida y Exploraciones Zoologicas del Dr. Juan Gundlach en Cuba (1839-1896)," by C. T. Ramsden, Mem. Soc. Cubana Hist. Nat. "Felipe Poey," 3: 146-168, 1918; "Johannes Gundlach, Naturforscher aus Cuba," by R. Mertens, Lebensbild. Kurhessen Waldeck (Marburg), 3: 103-133, 1942; "Juan Cristóbal Gundlach," by I. Castellano Rodiles, Publ. Soc. Cubana Hist. Nat. "Filipe Poey," 1: 1-11, 1960; Papavero, 1971 (pp. 181-183). • *Portrait and signature* (1864): Academy of Natural Sciences of Philadelphia, courtesy Marsha Gross.

STEINDACHNER, Franz (1834-1919).

Franz Steindachner, a noted Austrian ichthyologist and herpetologist, was born in Vienna on 11 November 1834. At the University of Vienna he originally studied law, although his main interest was science. One of his teachers, Eduard Seuss, a geologist, interested him in studying fossil fishes and thus his career in ichthyology was set. Steindachner began work at the Naturhistorisches Museum in Vienna in 1860, originally with fishes. When L. J. Fitzinger retired in 1861, Steindachner took over his curatorship which included herpetology. Eventually he became head of the zoology section (1874) and director of the museum (1898), a position he held until his death, in Vienna, on 10 December 1919.

Steindachner was responsible for building the Vienna collection into one of the world's largest, not only through collections by Austrian military personnel and purchase of specimens from collectors (often using his personal funds), but from his own expeditions to North and South America, Africa, and the Middle East, at a time when museum curators rarely conducted field work at all. He superintended the Red Sea Expedition in 1895 and, in 1903, a voyage to Brazil, his last major trip. Because of his increasingly more time-consuming administrative duties, the herpetological collections were eventually curated by Friedrich Siebenrock.

Steindachner is best known as an ichthyologist, but he published nearly 60 herpetological titles (1862-1917) including one major work, the amphibian and reptile sections of the circumnavigation by the Austrian frigate *Novara* (1867, republished 1869). This monograph reviewed a collection of nearly 1000 specimens from along the ship's route in South America, Africa, Ceylon, Java, China, Australia, and New Zealand, as well as other collections purchased from or donated by Austrian nationals overseas (including Johann Natterer in Brazil and Ferdinand Stoliczka in India), some of it obtained by the museum many years before.

Steindachner's other major herpetological works include a monograph on the snakes and lizards of the Galapagos Islands (1876), which he had visited personally, and another on Formosan snakes (1913). His papers are systematic and faunal in nature, including descriptions of numerous new genera and species, and are worldwide in coverage. Perhaps his most spectacular novelty was the unique lizard *Lanthanotus borneensis*, which he described (1877 [1878]) and correctly related to the American helodermatids. This paper also nicely demonstrates the work of one of the best natural history artists, Eduard Konopický, who regularly illustrated Steindachner's works. He had a special ability to catch life-like poses as well as details of scutellation, in lithographs that in many respects are far superior to modern illustrations.

• *References*: Siebenrock, 1901 (pp. 446-449); "†Franz Steindachner," by V. Pietschmann, Ann. Naturhist. Mus. Wien, 33(Jahresber.): 47-48, 1920; "Localities of the Herpetological Collections Made During the 'Nova Reise'," by C. Gans, Ann. Carnegie Mus., 33: 275-285, 1955; "Intendant Dr. Franz Steindachner, sein Leben und Werk," by P. Kähsbauer, Ann. Naturhist. Mus. Wien, 63: 1-30, 1959 (1960); Papavero, 1973 (pp. 286-288, *Novara* collections). • *Portrait* (1912): Courtesy Oliver G. Dely. • *Signature* (1912): British Museum (Nat. Hist.), courtesy A. F. Stimson.

ANDERSON, John (1833-1900).

Anderson, a physician by training and field naturalist by inclination, was born in Edinburgh, Scotland, on 4 October 1833. He graduated from the University of Edinburgh (M.D. 1861) and, after a few years on the staff of a college in Edinburgh, joined his brother in Calcutta (1864). In 1865, he became Curator, or director, of the newly-created Indian Museum and, partly from his own collections made in remote parts of India, he built up the museum's holdings which were begun by Edward Blyth at the Asiatic Society of Bengal. In 1868 Anderson joined an expedition, as medical officer and naturalist, to Burma and to Yunnan in southwest China; a second expedition, in 1875, was turned back at the Chinese border and thus collected only in Burma.

Anderson retired from the Indian Service and returned to England in 1876. Following some field work in Algeria in 1889, he shifted his attention to Egypt, where he spent his winters due to health reasons. The Egyptian fauna had been neglected since the work of the French naturalists who accompanied Napoleon's army at the beginning of the century and, with the encouragement of George A. Boulenger, Anderson amassed large collections which led to his "Zoology of Egypt" series. Anderson did not live to complete the series, however; he died at Buxton, England, on 15 August 1900.

Unlike Albert Günther, George A. Boulenger, and other contemporaries, Anderson was foremost a field naturalist. Thus, his publications are full of natural history data and information on the local geography. Besides some 20 herpetological papers, he published three major works. "Anatomical and Zoological Researches . . . of the Two Expeditions to Western Yunnan in 1868 and 1875" (1878[-1879]) was the first extensive monograph on the fauna of Burma and Yunnan, and included the first review of the rich turtle fauna of southeast Asia, illustrated with 21 beautiful handcolored plates of turtles. Anderson's modest "Contribution to the Herpetology of Arabia" (1896; reprinted 1984) was a potpourri of miscellaneous chapters, but still the pioneer and, to this day, the only complete synthesis of Arabian herpetology. His most comprehensive herpetological work, however, was the Reptilia and Batrachia volume (1898; reprinted 1965) in his "Zoology of Egypt" series, containing some of the finest color drawings of reptiles ever published, drawn by P. J. Smit and John Green. This volume remains one of the most competent faunal works in herpetology.

• *References*: "John Anderson," by W. T. Blanford, Nature, 62: 529-531, 1900; "John Anderson," by G. A. Boulenger, Proc. Royal Soc. London, 75: 113-116, 1905; "John Anderson (1833-1900): A Zoologist in the Victorian Period," by A. E. Leviton and M. L. Aldrich, p. v-xxxv. *In* Anderson's Herpetology of Arabia, repr. ed., Soc. Study Amphib. Rept., Athens (Ohio), 1984. • *Portrait and signature* (1894): From Leviton and Aldrich, 1984.

BOULENGER, George A. (1858-1937).

George Albert Boulenger was the leading taxonomic herpetologist of his era. Born in Brussels, Belgium, on 19 October 1858, he had an interest from boyhood in animals and flowers, especially roses. He attended the Université Libre de Brussels, graduating in 1880, and in the same year was appointed to the staff of the Musée Royal d'Histoire Naturelle de Belgique as Assistant Naturalist. The museum's collections and library were limited, however, and Boulenger thus had to make frequent trips to Paris and London. In Paris, Fernand Lataste, a naturalist specializing in herpetology, became Boulenger's mentor and his alter ego for life.

During Boulenger's trips to London he so impressed Albert Günther, keeper of zoology at the British Museum (Nat. Hist.), that Günther hired him in 1881 as assistant in charge of lower vertebrates, much as he himself had been hired by John E. Gray during a similar visit in 1857. Günther assigned him the task to prepare a new edition of the catalogues of amphibians and reptiles, which Boulenger produced in nine volumes over the period 1882-1896. In addition to this monumental task, which still stands as the most comprehensive review of the world's herpetofauna, covering 8469 species, Boulenger published several other books and hundreds of papers, not only on amphibians and reptiles, but also on freshwater fish. It may safely be said that no person ever was as familiar with the world's species of lower vertebrates as he was.

His work earned him numerous honors: election as Fellow of the Royal Society (1894), honorary doctorates from St. Andrews, Giessen, and Louvain, and in 1935 the American Society of Ichthyologists and Herpetologists elected him their first Honorary Foreign Member. The King of Belgium conferred upon him the Order of Leopold in 1937. Boulenger had three sons, two of them biologists; the youngest (Edward George Boulenger) became Director of the Aquarium and Curator of Reptiles at the Zoological Society of London, and his numerous publications are sometimes credited to his father—and vice versa. On retirement in 1920, Boulenger senior returned to Brussels and, taking an honorary appointment at the Jardin Botanique de l'Etat, devoted himself to the study of roses, a subject of interest in his youth and in his free time throughout his life. His major work on this subject was the two-volume book "Les Roses d'Europe" (1924, 1933), but the sequel on American species was unfinished at the time of his death, on 23 November 1937 at St. Malo, Brittany (now part of Ille-et-Vilaine) in coastal France.

Boulenger's contemporaries have noted his incredible memory for specimens, species descriptions, and for the associated literature, which was the key to the great speed at which he worked. He could read six languages well and converse in most of them, and wrote with equal facility, but his manuscripts were never typed and went straight to the typesetter in longhand. Boulenger was a tireless worker with strict self discipline, and he also had great charm: a distinguished figure in the London scientific establishment and a proper Victorian gentleman. Even on collecting trips he would dress in tweed suit, homburg, and cloth cape. Yet apropos to the age in which he began his work, Boulenger refused to use the binocular microscope, employing a small hand lens instead to count scales, and he counted teeth by inserting a needle into the mouth, drawing it along inside the jaw and counting the clicks. He rarely dissected specimens and even then he usually made just a small slit through which only gross features could be seen, quite in keeping with the longstanding British Museum policy at that time that specimens were not to be cut up. Understandably, he did not make fundamental anatomical discoveries such as those made by Edward D. Cope, his great contemporary, which led to new arrangements of families, but Boulenger was quick to capitalize on these innovations and improve on them. His arrangement of reptiles is still essentially current, although that of amphibians has been very largely revised.

Boulenger's habit of not fixing types and of failing to indicate the specific specimens from which descriptions were drawn was custom at the time in Europe. He also did not observe the rules of the "International Code of Zoological Nomenclature" which, in fairness, only came into prominence late in his lifetime. To him scientific names were meant to define and describe, regardless of the legalities involved. Despite Boulenger's mastery of the world's herpetofauna, he had little understanding of geographic distribution or of subspecific variation. Nevertheless, Boulenger's methods and conclusions dominated European herpetology long after his retirement, although his influence in North America, where a quite independent school of herpetology developed, was never as great, and coverage of its fauna was the weakest part of Boulenger's catalogues.

Boulenger published numerous books, beginning with the British Museum catalogues: amphibians (two volumes, 1882), lizards (three volumes, 1885-1887), turtles, crocodilians, and the tuatara (1889), and snakes (three volumes, 1893-1896), were all published within a 15-year span by the age of 38, an incredible feat for an individual; these volumes were reprinted 1961-1966. Boulenger also published the herpetological volume in the "Fauna of British India" series (1890), a two-volume "Tailless Batrachians of Europe" (1896-1897; reprinted 1978) with its exquisite illustrations, "Les Batraciens . . . d'Europe" (1910), the herpetological section of the series "Vertebrate Fauna of the Malay Peninsula" (1912), "Snakes of Europe" (1913), and finally his two-volume "Monograph of the Lacertidae" (1920-1921, reprinted 1966), which summarized a vast body of meristic data. Besides these books, he published nearly 900 scientific papers, most of which were on herpetology, and served as section editor of *Zoological Record* from 1880 to 1914.

• *References*: "Dr. G. A. Boulenger, F.R.S.," by J. R. Norman, Nature, 141: 16-17, 1938; "George Albert Boulenger, 1858-1937," by M. Smith, Copeia, 1938: 1-3, 1938; "George Albert Boulenger. 1858-1937," by D. M. S. Watson, Obit. Notices Fellows Royal Soc., 3: 13-17, 1940; "Notice sur George-Albert Boulenger Associé de l'Académie," by M. Poll, Annu. Acad. Roy. Belg., 133(Not. Biogr.): 191-228, 1967; "G. A. Boulenger: His Life and Herpetological Work," by J. C. Battersby, p. vii-xii. *In* Boulenger's Contributions to American Herpetology, vol. 1, repr. ed., Soc. Study Amphib. Rept., Athens (Ohio), 1971. • *Portrait* (1919): Courtesy Gaston F. de Witte. • *Signature* (1916): Adler collection.

BOETTGER, Oskar (1844-1910).

Boettger (sometimes spelled Böttger), Germany's leading herpetologist and malacologist at the turn of the century, was responsible for establishing the Senckenberg Museum as a world center for herpetology. He was born in Frankfurt on 31 March 1844 and developed an interest in paleontology as a boy. He entered the University of Frankfurt in 1863 and later attended the school of mines in Freiberg (Saxony), intending to become a mining engineer, but on graduation in 1866 he could find no job due to political disturbances. Instead, he took a Ph.D. at the University of Würzburg in 1869, with a paleontological thesis, and became a teacher, first in Offenbach and finally in Frankfurt, where he died on 25 September 1910.

Boettger's posts at the Senckenberg Museum—as paleontologist (beginning 1870) and later as herpetologist (1875)—were unsalaried. Indeed, his curating was all the more remarkable in that for the period 1876-1894 he could not even enter the museum! He remained at home due to a nervous disorder (agoraphobia) and the spell was broken only when a relative tricked him into coming out of his house to get a desirable postage stamp for his prized collection, whereupon he resumed his teaching, curating, and even took overseas trips. His most productive period was during his confinement at home, when assistants had to bring every specimen to him.

Senckenberg's herpetological collections were begun by Eduard Rüppell, famous for his expeditions in northeast Africa, and numbered 370 species when he catalogued them in 1845. Near the end of Boettger's curatorship, the number of species was 1436, as listed in his three-part "Katalog" (1892, 1893, 1898), and the collections included a large number of types. Boettger obtained most new accessions from business friends and former students, over 200 of them, who travelled the world and who sent specimens (and stamps!) to Boettger. His interests were cosmopolitan—the Mediterranean lands, Southwest Africa and the Congo, Madagascar, China, the Philippines, and Central and South America being only the main areas of emphasis—and he described many new genera and species, a research program much like that of George A. Boulenger. Boettger worked at the species level, in taxonomy, and developed faunal catalogues for remote, little-known regions of the world. In addition to these strictly scientific interests, he co-authored the herpetological volume for the third edition of Alfred Brehm's "Tierleben" (1892).

• *References*: "Oskar Boettger," by W. Kobelt, Ber. Senckenb. Naturf. Gesellsch., 42: 74-83, 1911; "Prof. Dr. Oscar Boettger," by F. Haas, Jahresber. Mitt. Oberrhein. Geol. Verein., new ser., 1: 19-20, 1911; "Zum 100. Geburtstag Oskar Boettger's," by R. Mertens, Natur und Volk, 74: 93-95, 1944; Mertens, 1967 (pp. 8-15). • *Portrait and signature:* From Kobelt, 1911.

MÜLLER, Fritz (1834-1895).

Friedrich Müller (generally known as Fritz Müller), Swiss zoologist and physican, was born in Basel on 8 May 1834. His first studies were in the humanities, but in 1852 he briefly studied theology before finally deciding on a career in medicine. He studied at the University of Basel (1852-1854), later at Würzburg and Prague, before graduating with an M.D. degree in 1857. After a year and a half of additional training in Vienna, Berlin, and Paris, he returned to Basel and set up a medical practice. He was an active member of the medical community, helping to found the local medical society in 1860 (he became its president in 1876), but in 1872 he gave up his practice to become part of the Basel city government with responsibility for sanitary control. During 1868-1869 he lectured on zoology at the University of Basel, but most of his official duties after 1872 were governmental. In 1875 he became a member of the Commission for the Natural History Collection, thus beginning his herpetological work, and later he also studied crustaceans and spiders. Müller was plagued with chronic illnesses since at least 1873, which necessitated travel to the Mediterranean during the winter months to recuperate. He died in Basel on 10 March 1895.

Müller's few herpetological titles span the period 1877-1895. His major work, a catalogue of the amphibians and reptiles in the Basel Museum (1878), was followed by seven supplements from 1882 (1880) to 1892; an eighth and final supplement was published by his colleague, Ehrenfried Schenkel, in 1901. These works were worldwide in scope, based primarily upon collections made by Swiss nationals overseas and largely in tropical countries; many new taxa were described. Prominent among these collections were those made in Ceylon and the Celebes by Paul B. Sarasin and his cousin, Fritz Sarasin, amateur naturalists who published a classic monograph on *Ichthyophis glutinosus* (in four parts, 1887-1890) that, a century later, is still the most comprehensive study of any caecilian species. Müller's last herpetological publication (1894, in two parts) was a report on the Sarasin collections from the Celebes.

• *References:* "Dr. Fritz Müller †," by T. Lotz, Correspond.-Blatt Schweiz. Aerzte, 1895(11): 1-5, 1895; "Dr. Friedr. Müller," by T. Lotz and L. Rütimeyer, Verh. Naturf. Gesellsch. Basel, 11: 259-283, 1897. • *Portrait*: Universitäts-Bibliothek Basel, courtesy Christine Stocker. • *Signature*: Adler collection.

BEDRIAGA, Jacques von (1854-1906).

Jacques Vladimir von Bedriaga (in its Russian form, Jacob Vladimirovich Bedriaga, sometimes Bedryagha), a specialist on Eurasian herpetology, was born of Russian nobility in 1854 at his mother's estate in the village of Kriniz, near Voronezh, nearly 500 km south of Moscow. He entered the University of Moscow in 1872 where he studied with A. P. Bogdanov, but had to leave Russia in 1873 because of serious health problems. In Germany he had the good fortune to work with two eminent comparative anatomists, the leaders of the Darwinian revolution in Germany. At the University of Jena, Bedriaga studied with Ernst Haeckel, a former pupil of Johannes Müller and Carl Gegenbauer. He received his Ph.D. degree in 1875, for a thesis on the urogenital organs of reptiles, and then proceeded to Heidelberg for postdoctoral work with Gegenbauer whose broad anatomical interests included the anatomy of amphibians and reptiles.

Bedriaga's earliest publications (beginning in 1874) dealt with captive animals, but soon he began his frequent trips to Italy, Greece, and other Mediterranean areas where he made extensive collections, particularly of lizards. In 1881, to improve his health, he moved to Nice, along the French Riviera, which became his home base for many years, and later he moved to Florence, Italy. He died in 1906.

Bedriaga's first major work was "Die Amphibien und Reptilien Griechenlands" (1880, published in Moscow in three parts; reprinted 1882, and corrections in 1883), the first monograph of the herpetology of Greece. His special interest for many years was lacertids and in 1886 he published, in Frankfurt (Main), his book "Beiträge zur Kenntnis der Lacertiden-Familie." Thereafter his interests turned more to amphibians, resulting in an extensive monograph, published in five parts in Moscow, entitled "Die Lurchfauna Europa's" (frogs in 1889, reprinted 1891; salamanders in 1896-1897, reprinted 1897), which contained a wealth of life history information, much of it personally collected by him.

From time to time Bedriaga visited Russia and in particular St. Petersburg (now Leningrad) where he was informally associated with A. M. Nikolsky and the Zoological Museum. There he was asked by the museum authorities to study the enormous numbers of amphibians and reptiles collected by Russian explorers in "Central Asia"—that is, Mongolia and northern China—as an adjunct to their military expeditions to expand the Russian Empire. Alexander Strauch had reported on the earliest collections (1876), but much more had accumulated, particularly from the several expeditions by N. M. Przewalski and his lieutenants in 1876-1888, which Bedriaga covered in two short papers full of new taxa, published by the Academy of Sciences in St. Petersburg (1905 [1906]), and monographed in a large book of nearly 800 pages, in quarto (published in four parts, 1898-1912). The book is in parallel German and Russian texts (Nikolsky translated it into the Russian) and the ten highly-detailed plates were drawn by Italian artists.

• *References*: "Esboço Biographico do . . . Dr. Jacques Wladimir de Bedriaga," by A. M. Seabra d'Albuquerque, Imp. Univ., Coïmbra, 8 pages, 1891; "Jacob Vladimirovich Bedryaga," *In* A. P. Bogdanov, Materials for the History of Scientific Activity in Russia on Zoology. Newsbull. Imper. Moscow Univ., 70 (Trans. Dept. Zool., 6), Moscow, 1891. • *Portrait*: From Bogdanov, 1891, courtesy Natalia B. Ananjeva. • *Signature* (1876): Carnegie Museum, courtesy C. J. McCoy.

VAILLANT, Léon (1834-1914).

Léon-Louis Vaillant, noted French ichthyologist and herpetologist, was born in Paris on 11 November 1834. He graduated from the Collège d'Arras in 1854, where he showed an aptitude for drawing and natural history. Thus when he returned to Paris and began his medical studies he naturally attended courses in zoology as well. In 1861 he finished his M.D. degree but then continued his zoological research under the guidance of Henri Milne-Edwards, himself a student of Georges Cuvier. Vaillant received his doctorate in natural sciences in 1865. Except for four short papers on the anatomy of amphibians (1862-1863), all of his publications before 1872 were on invertebrates, but this quickly changed. Auguste Duméril, who held the professorship in reptiles and fishes at the Muséum National d'Histoire Naturelle, died in 1870. Vaillant was invited to assume his duties; however, he could not do so formally because Émile Blanchard, who held the position temporarily, wished to keep it. Blanchard had authored an atlas of vertebrate anatomy (1852-1864), which included a series of 20 plates on reptiles, and in 1871 described and named the Chinese giant salamander. Eventually, however, Vaillant prevailed (1875).

Vaillant thus came to hold the chair of Count de Lacepède and the Dumérils and to inherit responsibility for the then-largest collections of reptiles and fishes in the world, but at a time of grave peril. Because of the German siege of Paris, most of the specimens had to be removed from their showcases and many of them, especially the all-important types, were dispersed to basements and cellars around Paris for safekeeping. Vaillant supervised this work and their later reinstallation in the museum, but scarcely had this been completed than it was recognized that the collections had outgrown their quarters and new galleries were needed. The new buildings were occupied in 1889; Vaillant again supervised the transfer of the collections. In addition to these administrative duties, he was in charge of the reptile menagerie and the aquarium. Despite these many responsibilities, Vaillant continued his research, now devoted to fishes and reptiles, and received many honors. He died in Paris on 24 November 1914, ironically during a new German bombardment, just as when his career at the museum had begun in 1870.

Vaillant published over 260 titles, of which nearly 90 are herpetological. His special interests were the systematics and anatomy of living and extinct turtles and crocodiles, but as a result of the living collections in the menagerie, he also published on reptilian behavior and physiology. His most magnificent work in herpetology, co-authored with Guillaume Grandidier, was the volume on turtles and crocodiles (1910) in the "Histoire Physique, Naturelle et Politique de Madagascar" series, with its exquisite colored plates. Regrettably, no other herpetological volumes were issued in this series.

• *References*: "Léon Vaillant," by E.-L. Bouvier, Bull. Soc. Philomath. Paris, ser. 10, 9: 53-56, 1917; "Le Professeur Léon-Louis Vaillant (1834-1914)," by L. Roule, Arch. Mus. Hist. Nat., 4: 1-14, 1929. • *Portrait*: Courtesy Brooks M. Burr. • *Signature* (1893): Smithsonian Institution Archives, courtesy William Cox.

DUGÈS, Alfredo (1826-1910).

Dugès, the Father of Mexican Herpetology, was born Alfred Auguste Delsescautz Dugès, in Montpellier, France, on 16 April 1826. He attended the university at Montpellier where his father, Antoine L. D. Dugès, author of several herpetological works, including a major study on amphibian osteology (1834, republished 1835), had been a professor. Later, he completed his M.D. at the University of Paris (1852). Dugès emigrated to Guanajuato, Mexico, in 1853, where he established a gynecological practice. Concurrently he was Professor of Natural History and director of the museum at the State College of Guanajuato. Despite his isolation from other research centers and some personal handicaps, including lifelong deafness, Dugès established himself as the leading vertebrate biologist in Mexico, with special emphasis on herpetology. His brother, Eugenio Dugès, also emigrated to Mexico and became a well-known entomologist. Alfredo Dugès made collections together with his brother, who also lived in Guanajuato, and often they were accompanied by Alfredo's students, especially on Sundays.

Alfredo Dugès was the first person to summarize the Mexican herpetofauna in Linnaean terms and was virtually the sole link between the early explorers and the modern era. He described 40 new reptiles and amphibians, of which nearly half are recognized as valid today, and also published on all other vertebrate groups, insects, and botany. His herpetological collections are still largely intact at the museum, later renamed "Museo Alfredo Dugès," at the University of Guanajuato. He died in Guanajuato on 7 January 1910.

Although he had published several herpetological papers while in France, beginning in 1865 he began a nearly unbroken stream of papers until his last one in 1907. All told, he published 184 titles of which 94 were herpetological; most were illustrated by him as well. The herpetofauna of Mexico is one of the world's most diverse, with over 1200 species and subspecies recognized, yet virtually alone Dugès made good progress in working out distributions and describing that diversity. Many of his papers were short and sometimes published multiple times in different journals, presumably to ensure wider distribution, but the repetition often creates confusion in the literature in terms of correct citation of original descriptions of his new taxa. Dugès's most important and comprehensive works were "Erpetología del Valle de México" (1888), whose coverage extends beyond the valley itself, and "Reptiles y Batracios de los E[stados] U[nidos] Mexicanos" (1896; reprinted 1978).

• *References*: "Alfredo Augusto Delsescautz Duges Ensayo Biografico," by R. Martín del Campo, Anal. Inst. Biol. Mexico, 8: 437-455, 1937; "The Second Century of Alfredo Dugès, Father of Mexican Herpetology," by H. M. Smith and R. W. Reese, Herpetol. Rev., 1(7): 5-7, 1969; "Early Foundations of Mexican Herpetology," by H. M. Smith and R. B. Smith, Univ. Illinois Press, Urbana and Chicago, [ix], 85 pages, 1969. • *Portrait and signature* (1888): Adler collection.

BAUR, Georg (1859-1898).

Georg Hermann Carl Ludwig Baur, vertebrate morphologist and specialist on reptiles who died tragically at a young age, was born on 4 January 1859 in Weisswasser, Bohemia (now Bílá Voda, Czechoslovakia, near the Polish border). He had the good fortune to study with some of Germany's most outstanding comparative anatomists and paleontologists, including Karl von Siebold and Karl von Zittel at the University of Munich and, at the University of Leipzig, Rudolf Leuckart and J. V. Carus. Baur received his doctorate at Munich in 1882 and stayed on for a short period as assistant in histology, but in 1884 he was hired as assistant by O. C. Marsh at Yale College in the United States. There Baur remained until 1890 when he resigned because of increasing disputes with Marsh who had failed to support his bid to become Professor of Anatomy at Yale and because Baur had sided with Edward D. Cope in Cope's feud with Marsh. Baur then joined the staff at Clark University in Massachusetts where he began his life's major projects, a monograph of North American turtles and his investigations on the faunas and floras of oceanic islands. In 1891 he spent five months in the Galapagos Islands where he made extensive collections.

In 1892, the newly-founded University of Chicago, looking to start a department of biology, raided Clark University and hired the entire biology staff, including Baur, who thus became Assistant Professor of Comparative Osteology and Paleontology at Chicago. Apparently due to incessant work on too many projects and a highly nervous disposition, Baur developed motor paralysis. In 1897, friends persuaded him to visit relatives in Munich for recuperation, but there his condition only worsened. He had to be transferred to an asylum, where he died on 25 June 1898, at the age of 39.

Baur was a brilliant zoologist in the pattern of Edward D. Cope, whose successor he had hoped to become. Like Cope, he had broad interests and many diverse projects on which he sometimes published too hastily, and he, too, was a confirmed neo-Lamarckist. Baur's special interest was the morphology of the vertebrate skeleton, about which he published some important papers on the carpus and tarsus and the temporal region. Reptiles, and especially turtles, were his favorite group, both living and extinct forms, but his opinions on their proper classification often brought him into conflict with Cope, George A. Boulenger, and others.

At the time of his death, Baur had virtually completed a monograph of North American turtles, which was to have been published as a companion to Cope's books on the other groups of amphibians and reptiles. His manuscript was turned over to Leonhard Stejneger, who had intended a similar monograph of his own, but his book never appeared. Fortunately, Baur's drawings of fossil turtles were published by Oliver P. Hay (in 1908), who had worked with him for four years. Altogether, Baur published about 150 titles in a very short period (1882-1899), most of them on reptiles. Included were descriptions of many familiar and still valid taxa of turtles (in the genera *Geochelone*, *Graptemys*, *Kinosternon*, *Pseudemys*, and *Trionyx*).

• *References*: "George Baur," by O. P. Hay, Science, new ser., 8: 68-71, 1898; "George Baur's Life and Writings," by W. M. Wheeler, Amer. Nat., 33: 15-30, 1899. • *Portrait*: From Wheeler, 1899. • *Signature* (1888): Academy of Natural Sciences of Philadelphia, courtesy Marsha Gross.

GARMAN, Samuel (1843-1927).

Samuel Walton Garman (or "Garmann" as he sometimes styled himself), noted American ichthyologist and herpetologist, was born in Indiana County, Pennsylvania, on 5 June 1843. He left home as a boy, drifted westward and, while working for a Union Pacific Railroad surveying party, fought Indians and shot game for the work crews. In 1868 he joined John W. Powell's expedition to the Colorado mountains, then followed him to Illinois State Normal University where Garman enrolled and graduated in 1870. Garman then taught school for two years, during which time he corresponded with Edward D. Cope. He accompanied Cope briefly during the summer of 1872 to collect fossils in Wyoming but was fired for making unreasonable demands for pay. This, fortuitously, led to the most important event in his life. Garman happened to be in San Francisco later that summer when the ship *Hassler* docked, carrying Louis Agassiz and his scientific party which had just completed a journey through the Straits of Magellan. Agassiz recognized his potential and invited Garman to join him as a pupil.

Garman thus arrived at Harvard College, where he worked for the remainder of his career in the Museum of Comparative Zoology. In 1874, he joined Alexander Agassiz, Louis Agassiz's son, on his expedition to Lake Titicaca in the Andes and, later, on several of his expeditions to the West Indies. Harvard College awarded him honorary degrees (B.S. 1898, A.M. 1899) for his scientific work. Garman shunned publicity and, as time went on, he became a recluse at the museum. Apparently deeply affected by the Cope-Marsh feud, he was very secretive about his research, refusing to talk about it or to show specimens even to most of his museum colleagues. He worked in dimly-lighted quarters at the museum and when someone knocked on his door, he would cover all specimens on which he happened to be working before going to the door, and even then cracked it open only enough to talk with the visitor. Garman's primary interest was fishes, but he also had responsibility for the herpetological collection until 1910 when Thomas Barbour effectively took control. Garman died at Plymouth, Massachusetts, on 30 September 1927.

Most of Garman's titles were about fishes, a subject on which he was an acknowledged authority. His special interest was sharks, and he published a major monograph on this group in 1913 entitled "The Plagiostoma." Garman authored nearly 50 works in herpetology (1874-1917), with numerous papers describing new species (including two sea turtles, *Chelonia depressa* and *Lepidochelys kempii*), as well as faunal lists from collections made throughout the Americas and Australia.

Garman also authored two book-length monographs on reptiles. In 1883, he published "The Reptiles and Batrachians of North America. Part 1. Snakes," an extensive work with detailed descriptions of genera and species from north of the Isthmus of Tehuantepec in Mexico, and also including extralimital species from as far afield as South America; no further parts appeared. "The Galapagos Tortoises," Garman's last herpetological publication (1917), described and figured in 42 quarto plates the large series of tortoises amassed by Louis Agassiz on the *Hassler* expedition and also those collected by Georg Baur, among others.

It should be noted that Garman's younger brother, Harrison Garman, also published on herpetology (1890-1897), including the first summary of the amphibians and reptiles of Illinois (1892).

• *References*: "Samuel Garman—1843-1927," by D. S. Jordan and T. Barbour, Science, 67: 232-233, 1928; "Samuel Garman," by H. L. Clark, Dict. Amer. Biogr., 7-8: 154, 1946. • *Portrait and signature*: Museum of Comparative Zoology, Harvard University, courtesy Roxane Coombs.

STEJNEGER, Leonhard (1851-1943).

Leonhard Hess Stejneger (before 1870, Steineger), the dean of American herpetologists after the death of Edward D. Cope, was born in Bergen, Norway, on 30 October 1851. He had an early interest in birds and later, in Berlin, he briefly studied medicine, but turned to law (graduating from the University of Kristiania in 1875) in order to assist in his father's mercantile business. The business went bankrupt in 1880, so in 1881 Stejneger, who by this time had published 22 papers on birds, the study of which was then merely a hobby, emigrated to the United States. He proceeded to the Smithsonian Institution where he was promptly hired by Spencer F. Baird to assist in ornithology. Baird soon arranged for him to accompany a government party to Alaska and adjacent Siberia (1882-1883) to set up weather stations, but Stejneger's main function was to collect vertebrates. His important monographs on birds (1885) and fur seals (1896, 1899) resulted from this and several later expeditions to the region, the last in 1922 at the age of 70. Through these expeditions, which included visits to the Japanese fur seal islands, Stejneger developed a major interest in Japanese birds and, later, reptiles.

In 1889, when Henry C. Yarrow resigned as Honorary Curator of Amphibians and Reptiles at the National Museum, Stejneger was asked to assume charge. Thus, at the age of 37, with an excellent reputation as an ornithologist, he switched to herpetology and began a new career. His first paper on reptiles was published later that year. The need to completely reorganize a department that had been without regular curation since that of Charles F. Girard in the 1850s led to overwork, and he was assigned to join a field party under C. Hart Merriam that was conducting a survey of the San Francisco Mountains in Arizona, where Stejneger quickly recuperated. This was also his first serious field experience with amphibians and reptiles.

Although Stejneger continued to publish on birds (146 of his 411 titles are on ornithology, 152 on herpetology), he quickly immersed himself in his new area of study and published numerous careful taxonomic studies in which he described many new genera and species. Some of the most distinctive North American genera were named by him, including the tailed frog *Ascaphus*, the blind cave salamanders *Typhlomolge* and *Typhlotriton*, the skink *Neoseps*, and the snake *Phyllorhynchus* of southwestern deserts. Most of his papers dealt with North America, but other areas of special interest included Japan, China, the Philippines, Central and South America, the West Indies, and Africa.

Stejneger's studies were characterized by a critical, scholarly approach, in which he reformed the practices of Edward D. Cope and of George A. Boulenger by introducing detailed descriptions of identifiable individuals and careful designations of type specimens and type localities. Besides his museum work, Stejneger participated in several expeditions, including Puerto Rico (1900), which had recently come under American control following the brief Spanish-American War (1898), the American West, and Japan.

In 1911 Stejneger was promoted to Head Curator of Biology at the National Museum, and thereafter administrative duties increasingly occupied his time. Previously, he had published his major works in herpetology, "Poisonous Snakes of North America" (1895; reprinted 1975), "Herpetology of Porto Rico" (1904), and the classic "Herpetology of Japan and Adjacent Territory" (1907), which is still a regularly-consulted monograph on Japan, Korea, Taiwan, and coastal mainland China. In 1917, together with Thomas Barbour, he issued the first of five editions of "A Check List of North American Amphibians and Reptiles" (later editions in 1923, 1933, 1939, and 1943). These had a profound influence on stabilizing nomenclature and, at the same time, revealed numerous problems for further study which, in turn, contributed to the rapid growth of herpetological study in North America.

Stejneger's interest in nomenclature was probably based on his early schooling in the classical languages and to his later formal legal training. In 1898 he was elected a member of the International Commission on Zoological Nomenclature and became a strong proponent of universal rules and procedures, although his critics argued that he was sometimes overly legalistic. As an example of the extreme positions that he sometimes took, he practically required the author of a 1809 paper, in which the type species of *Trionyx* was designated, to follow the exact working of the 1930 "Code," in order for that designation to be valid!

Stejneger remained a productive scientist to the end of his very long life. When in 1921 he attained the normal age (70) for retirement from government service, he was requested to continue his research and administrative duties. Frank N. Blanchard had joined him as Aide in 1918, to assist with the herpetological collections, but resigned the next year and was replaced by Doris M. Cochran, who eventually succeeded Stejneger as curator. In 1932 Stejneger was exempted from mandatory retirement by order of President Hoover, and he continued in office until his death in Washington on 28 February 1943, at the age of 92.

During his lifetime Stejneger was honored by organizations in his adopted country as well as abroad. Norway made him Knight First Class of the Order of St. Olav (1906), and the University of Oslo (formerly Kristiania) awarded him an honorary Ph.D. in 1930. In 1923 he was elected to the U.S. National Academy of Sciences, and in 1931 the American Society of Ichthyologists and Herpetologists, which he had served as President during the period 1918-1923, made him Honorary President for life. Despite his many duties, Stejneger always gave strong encouragement and advice to many aspiring young herpetologists, among them Thomas Barbour, Frank N. Blanchard, Emmett R. Dunn, and Alexander G. Ruthven, who in their turn provided the main impetus for scientific study of amphibians and reptiles in the 1920s and 1930s in the United States.

• *References*: "Leonhard Stejneger," by A. K. Fisher, Copeia, 1931: 74-83, 1931; "Dr. Leonhard Stejneger," by P. Bartsch, Science, 98: 51-54, 1943; "A Herpetological Bibliography of Leonhard Stejneger (1851-1943)," by W. L. Necker, Herpetologica, 2: 87-92, 1943; "Leonhard Stejneger," by A. K. Fisher, Copeia, 1943: 137-141, 1943; "Biographical Memoir of Leonhard Hess Stejneger," by A. Wetmore, Biogr. Mem. U.S. Natl. Acad. Sci., 24: 142-195, 1947; Schmidt, 1955 (pp. 608-610); "Leonhard Stejneger," by W. L. Schmitt, Syst. Zool., 13: 243-249, 1964; "Leonhard Hess Stejneger," by E. N. Shor, Dict. Sci. Biogr., 13: 25, 1976. • *Portrait* (1902): Courtesy Academy of Natural Sciences of Philadelphia. • *Signature* (1916): Adler collection.

PHILIPPI, Rodolfo Amando (1808-1904).

Born Rudolph Amandus Philippi on 14 September 1808, in Charlottenburg, near Berlin, where his father had a position at the Prussian court, Philippi became one of Chile's leading naturalists. He developed an interest in natural history as a boy and studied medicine at the University of Berlin, graduating in 1830, where his teachers included the naturalists Hinrich Lichtenstein, H. F. Link, A. F. A. Wiegmann, and even Alexander von Humboldt. Beginning in 1835 he was a professor at a school in Cassel but this ended due to his connection with the political disturbances of 1848-1850.

Philippi emigrated to Chile in 1851 to join his brother, becoming Professor of Natural History at the University of Chile in 1853 and, concurrently, Director of both the National Museum and the Botanic Gardens. In such positions he strongly influenced the study of natural history in Chile and arranged numerous expeditions to remote regions of Chile including the Strait of Magellan and Chile's oceanic islands. He participated in the expedition to the Atacama Desert of northern Chile in 1853, about which he later published "Viaje al Desierto de Atacama" (1860). He retired from most of his administrative positions in 1883 and, finally, at the age of 88, as Director of the National Museum (1896). He died in Santiago on 23 July 1904.

Herpetology was only one of Philippi's diverse interests which covered nearly every branch of natural history in some 350 publications. His herpetological works, which began in 1861, cover Chilean and Argentinian species, especially snakes and turtles, but his most extensive herpetological treatise, "Suplementos a los Batraquios Chilenos" (1902), intended as an update to Claude Gay's review (1848-1849) of Chile's amphibian fauna, was written at the end of Philippi's very long life and unfortunately is replete with error; the plates to this work, drawn by Philippi himself, remained unpublished until issued by José M. Cei in 1958. Somewhat earlier (1899) Philippi had published his major work on Chilean snakes, listing 45 species; once his specimens were found again by Roberto Donoso-Barros (1965), it turned out that most had been misidentified and that only three species, in fact, were Chilean.

• *References*: "El Doctor Rodolfo Amando Philippi. Su Vida i Sus Obras," by D. Barros Arana, Impr. Cervantes, Santiago, vii, 248 pages, 1909; "Dr. Rudolph Amandus Philippi," by M. E. McLellan, Auk, 44: 158-159, 1927; Papavero, 1973 (pp. 275-281). • *Portrait and signature* (1895): From Barros Arana, 1904.

BERG, Carlos (1843-1902).

Friedrich Wilhelm Carl Berg, originally a Russian entomologist, was born of German ancestry in Tuckum, Kurland (now Tukums, Latvian SSR), Russia, on 21 March 1843, and became one of Argentina's leading naturalists. After working some years in business, in 1865 he moved to Riga, as curator of entomology at the museum, and then to the university there. In 1873, he was hired by Hermann Burmeister, director of the natural history museum in Buenos Aires, and he emigrated to Argentina. Beginning with an expedition to Patagonia the next year, Berg collected natural history specimens for the museum from throughout Argentina and also from Chile and Uruguay. Except for the period 1890-1892, during which he was on staff at the Museo Nacional de Montevideo in Uruguay, Berg spent the rest of his career in Buenos Aires and eventually succeeded Burmeister as director, after the latter's death in 1892.

Berg's major research interests were in entomology but these expanded to encompass paleontology and vertebrates. His herpetological publications span the period 1884-1901 and include two major works, "Batracios Argentinos" (1896) and "Contribuciones . . . Fauna Erpetologica Argentina" (1898). Berg died at Buenos Aires on 19 January 1902.

• *References*: "Carlos Berg. Reseña Biográfica," by Á. Gallardo, Anal. Mus. Nac. Buenos Aires, 7: ix-xl, 1902; "Carlos Berg," by V. O. Cutolo, Nuevo Dicc. Biogr. Argent., 1: 422, 1968. • *Portrait*: From Gallardo, 1902. • *Signature* (1901): Academy of Natural Sciences of Philadelphia, courtesy Marsha Gross.

GOELDI, Emílio Augusto (1859-1917).

Goeldi (sometimes Gœldi or Göldi), one of Brazil's leading naturalists, was born Emil August Göldi in Ennetbühl, Canton St. Gall, Switzerland, on 28 August 1859. He studied at the university and marine station in Naples before completing his doctorate in 1883 at the University of Jena, under the celebrated evolutionary biologist Ernst Haeckel. He then assisted Haeckel for a while, but emigrated to Brazil in 1884, by the invitation of the Emperor, Dom Pedro II, to become assistant head of the Section of Zoology at the Museu Nacional in Rio de Janeiro.

Pedro II was an enlightened monarch who was interested in developing Brazil's natural resources and, because of his personal interest in the natural sciences, he initiated several scientific explorations in the country. In 1889, however, he was deposed by the army, and in the political turmoil that followed Goeldi lost his museum position (1890). Goeldi lived in Rio for some years, but in 1894 he was invited by the governor of Pará, the large state encompassing the mouth of the Amazon, to become Director of the Museu Paraense, which also included zoological and botanical gardens, and see to its reorganization and development. Goeldi remained as director until 1907 when he resigned due to health problems and returned to Switzerland. The museum was later renamed Museu Paraense Emílio Goeldi in his honor. In 1908 he became Professor of Zoology at the university in Bern, a position he held until his death, in Bern, on 5 July 1917.

Goeldi's scientific work encompassed a large scope, but it emphasized vertebrates and, due to his practical medical concerns, parasitic insects. His best known book was on birds ("As Aves do Brasil," 1894-1900, with an atlas in 1900-1906). He published about 15 titles on herpetology (1885-1907), primarily on systematics and distribution but also some papers on reproductive biology in which he had a special interest. It was Goeldi who discovered the frog, later named *Hyla* (now *Flectonotus*) *goeldii* by George A. Boulenger in 1895, in which the female carries the eggs attached to her back. Unfortunately, Goeldi's major work on herpetology, "Reptis do Brazil," which was completed in 1892-1894, was never published as a single volume but in sections, those on lizards and turtles being issued in 1902 and 1906, respectively; these are mostly compilations from the literature but contain many field notes by Goeldi.

• *References*: "Dr. Emil August Goeldi," by T. S. Palmer, Auk, 34: 510, 1917; "Professor Dr. Emil August Goeldi, 1859-1917," by T. Studer, Verhandl. Schweizer. Naturforsch. Gesellsch., 1918: 36-59, 1918; Papavero, 1973 (p. 374); "Emílio Augusto Goeldi—(1859-1917)," by O. R. da Cunha, Ciênc. e Cult., 35: 1965-1972, 1983. • *Portrait* (1895) *and signature* (1897): Museu Paraense Emílio Goeldi, courtesy Osvaldo Rodrigues da Cunha.

MÉHELŸ, Lajos (1862-1952?).

One of Hungary's eminent zoologists, Ludwig von Méhelÿ (in German form, as he sometimes used) was born in Bodrogszegi (now Kisfalud-Szögi), Hungary, on 24 August 1862, of an aristocratic family. He attended Müegyetem (or Technical) University in Budapest, graduating in 1880, and thereafter taught at schools in Budapest and Brassó until he joined the staff of the Hungarian National Museum, in charge of amphibians and reptiles, in 1896. His earliest papers were on insects, but in 1882 he published the first of about 90 titles on herpetology. Most of the latter were systematic, distributional, and faunistic in nature and dealt with Hungary, but there were also papers on distant places: Armenia, Ceylon, Paraguay, Persia, the expedition of Baron Eugen Zichy to Siberia, Mongolia, and China, and a series of papers on New Guinea. Vipers and lacertid lizards were topics of special interest, on which he published major monographs and described numerous new taxa, although his views on the proper phylogenetic position of lacertid species brought him into public conflict with George A. Boulenger and Franz Werner. Unfortunately, his most elaborate herpetological work, entitled "Herpetologia Hungarica," for which he won an important prize from the Hungarian Academy of Sciences in 1897, was never published.

In 1913, Méhelÿ was promoted to head of the museum's zoology section, and in 1915 he became Professor of Zoology and Anatomy at Pázmány Péter University of Budapest (now Eötvös Loránd University). His interests gradually turned more to mammals, paleontology and, at the end of his career, to invertebrates. His last major herpetological paper appeared in 1911, but in the meantime his compatriots S. J. Bolkay and G. J. von Fejérváry took over the study of Hungarian amphibians and reptiles. Méhelÿ was a staunch disciple of Charles Darwin and helped to disseminate his views, not only in his scientific publications but also in his semi-popular writing. He retired in 1932 and published his last paper in 1942. He is believed to have died in 1952, but the exact date is not known because he was held as a political prisoner, near Budapest.

• *References*: "Die wissenschaftliche und literarische Tätigkeit von Ludwig Méhely auf dem Gebiete der Zoologie," by O. G. Dely, Vertebr. Hungar., 9: 21-64, 1967, and supplement by I. Boros and O. G. Dely, same volume: 65-165, 1967. • *Portrait and signature*: Courtesy Oliver G. Dely.

PERACCA, Mario G. (1861-1923).

Conte (or count) Mario Giacinto Peracca, noted Italian herpetologist, was born in Turin on 21 November 1861. Interested in zoology by his father, an amateur ornithologist, Peracca enrolled as a medical student at the University of Turin but soon switched to zoology. He obtained his doctor's degree in 1886 under the guidance of Michele Lessona, author of several major monographs on the Italian herpetofauna. Earlier, as a student, Peracca had been an assistant to Lessona and, in 1887, officially became Assistant at the Zoological Institute of the university, a position he retained when the directorship of the institute passed from Lessona to Lorenzo Camerano. The latter was also an eminent Italian herpetologist, who published the first modern monographs of the amphibians and reptiles of Italy (in five volumes, 1883-1891).

Until he retired from his assistantship in 1920, Peracca curated the herpetological collections and added specimens purchased by him from collectors around the world or from exchange with other museums. Many specimens also came from his estate, where he built an enormous temperature-controlled vivarium adorned with luxuriant tropical plants and which he stocked with giant salamanders, Galapagos tortoises, and other local and exotic specimens. Unfortunately, most of his studies on living animals were never published. Despite his nobility, Peracca was said to have lived modestly, and he devoted his fortune to zoology and to helping others. He died in Turin on 23 May 1923 after a long, debilitating illness.

Peracca's research was clearly in the tradition of George A. Boulenger, who was a longtime friend. He published about 55 papers (1886-1917), virtually all of them on herpetology. Some are on Italian species, but his most extensive series of papers dealt with collections obtained in South America and Africa, including Madagascar, but there were also titles on the Middle East, China, and Australia. Many new genera and species were described, and the collections were eventually donated to the Zoological Institute at Turin.

• *References*: "Mario Giacinto Peracca," by L. C. de Martiis, Boll. Mus. Zool. Anat. Comp. Univ. Torino, new ser., 38 (12): 1-7, addenda, 1923; "La Collezione Erpetologica del Museo di Zoologia dell'Università di Torino," by O. Elter, Catal. Mus. Region. Sci. Nat. Torino, 5: 1-116, 1981 (1982). • *Portrait*: From Martiis, 1923, courtesy Olindo Bortesi and O. Elter. • *Signature* (1907): Adler collection.

ROLLINAT, Raymond (1859-1931).

Pierre André Marie Raymond Rollinat, French amateur naturalist, became his country's leading authority on the life histories of native reptiles. He was born in Saint-Gaultier, near Argenton-sur-Creuse (Indre), on 2 September 1853, and developed a love for animals as a young boy, under the influence of his great uncle who was an amateur ornithologist. Rollinat attended high school in Châteauroux but resisted scholastic discipline and much preferred his sessions, every Thursday, with a local taxidermist who taught him his technique. He declined to go to Paris for advanced study in zoology and decided instead to stay in Argenton. Except for military service, he remained there until his death, on 27 December 1931.

During most of his life, Rollinat made detailed studies of the vertebrates of central France, in particular the reptiles and amphibians. He built a laboratory in his house and in the large garden adjacent to it he erected various enclosures where he could study animals at close hand. The lizards were often so tame that they would take food from his fingers or even from his lips. The structures in his garden included terraria installed in flower beds where temperature and humidity were most favorable for hibernation, breeding, and hatching of eggs, all topics on which he published. Besides these subjects, Rollinat made a special study of embryonic development in several species. The local turtle, *Emys orbicularis*, which he maintained in two artificial ponds, was a favorite species for study. He even kept aquaria next to his bed, so that he could observe the nocturnal egg-laying activities of some species. The result was a long series of about 30 papers (over the period 1892-1932), in which he detailed the life histories of all the local species, and his book, "La Vie des Reptiles de la France Centrale," published posthumously in 1934 (reissued in 1937, 1946, and 1980).

Rollinat was the most prominent of a small group of French herpetologists who were interested in the local fauna at a time when their museum-based colleagues in Paris

LÖNNBERG, Einar (1865-1942).

Axel Johan Einar Lönnberg (sometimes spelled Loennberg or Lœnnberg), naturalist and conservationist, was born in Stockholm on 24 December 1865. His interest in nature was encouraged by his father, a wealthy landowner and member of the Swedish Parliament, who had taken a doctorate in botany at Uppsala University. Einar Lönnberg himself earned degrees at Uppsala (B.A. 1887, M.A. 1890, D.Phil. 1891). During 1891-1903 he was Inspector of Fisheries and then, following a six-month period at the museum in Göteborg, in 1904 he became professor in charge of the vertebrate department at the Royal Natural History Museum in Stockholm, until he retired in 1933. He continued some zoological research up to his death in Stockholm, on 21 November 1942.

Lönnberg's early work was on botany and parasitology, but his main interest became vertebrates. Best known for his research on mammals and fishes, he published numerous titles on amphibians and reptiles over the period 1894-1938. His first papers on reptiles resulted from his expedition to Florida in 1892-1893. On this trip he also collected plethodontid salamanders and discovered that they were lungless, although Harris H. Wilder and Lorenzo Camerano had previously and independently recognized that other species of salamanders lacked lungs; Lönnberg's papers on this subject were published in 1896 and 1899.

concentrated on the faunas of the far-flung French colonies. Among the others were Fernand Lataste, the mentor of George A. Boulenger, and L. F. Héron-Royer, an amateur naturalist who specialized on frogs and who, during the last 15 months of his life, made several trips to Argenton in 1890-1891 to advise Rollinat on setting up his outdoor laboratory. Rollinat also profited from routine correspondence with Boulenger, M. G. Peracca, and Willy Wolterstorff, among others, who advised him and who received specimens and research information in return. Despite his amateur status, Rollinat became justly famous for his work—carried out in the tradition of the great French amateur entomologist, J.-H. Fabre, with whom Rollinat is often compared—and he received several prizes from the scientific community.

• *References*: "Rollinat et le Monde des Reptiles," by A. Legros, La Nature, 40: 173-176, 1932; "Raymond Rollinat (1859-1931). Notice Nécrologique," anonymous, Bull. Soc. Natl. Acclim. France, 79: 397-403, 1932; "Biographie du Raymond Rollinat 1859-1931," by P. Rangde, Bull. Soc. Herpétol. France, 6: 5-25, 1978 (entire issue devoted to Rollinat); "Le Naturaliste Raymond Rollinat," by P. Rangde, Bull. Soc. Herpétol. France, 9: 10-25, 1979 (entire issue devoted to Rollinat). • *Portrait* (1918): From Rollinat, 1934. • *Signature* (1902): Adler collection.

In 1899, Lönnberg participated in an expedition to the Caspian basin and, later, to British East Africa (1910-1911). He published on the reptiles and fishes from the latter in 1911, and many of his herpetological papers described collections made by Swedish expeditions to New Guinea, Australia, Siam, the Congo, Brazil, and Sumatra; several of these were co-authored by his junior colleague, Lars Gabriel Andersson, or his successor at the museum, Hialmar Rendahl. In 1899, Lönnberg published an important paper re-describing Linnaean types of amphibians and reptiles and noting their present locations; Andersson published two companion papers (1899, 1900), and P. E. Vanzolini issued an index to the three in 1969.

Besides his extensive zoological research, Lönnberg was also an early conservationist. He was particularly active in establishing legislation to protect waterfowl and reindeer. In 1906 he founded the magazine *Fauna och Flora*, which served to popularize natural history in Sweden, and he edited it for the rest of his life.

• *References*: "Einar Lönnberg," anonymous, p. 569. *In* Vem är Det.—Svensk Biograf. Handb., Stockholm, 1934; "Axel Johan Einar Lönnberg," by T. S. Palmer, Auk, 61: 177-178, 1944; "Einar Lönnberg," anonymous, Ann. Mus. Civ. Stor. Nat. Giacomo Doria Genova, 62: 5-6, 1945; "Axel Johan Einar Lönnberg," by T. Flensburg, p. 172. *In* Svenska män och Kvinnor—Biograf. Upsalgsbok, Stockholm, 1942-1955; "Einar Lönnberg," by C. L. Hubbs, Copeia, 1950: 70, 1950. • *Portrait* (1936): British Museum (Nat. Hist.), courtesy A. F. Stimson. • *Signature* (1926): Natural History Museum, Stockholm, courtesy Göran Nilson.

ANDERSSON, Lars Gabriel (1868-1951).

Lars Gabriel Andersson, Swedish zoologist, was born the son of a priest in Vagnhärad, south of Stockholm, on 22 February 1868. He attended Uppsala University (B.A. 1890, M.A. 1897, D.Phil. 1909) and during most of his life was a school teacher, except for the periods 1894-1895 and 1897-1902 when he was an employee in the vertebrate zoology section at the Royal Natural History Museum in Stockholm. In 1906 he became Assistant Master at a secondary school in Stockholm, then at another, and from 1912 he was Senior Master at a third school in Stockholm; from 1920 until his retirement in 1933 he was Headmaster at a high school in Lidingö just north of Stockholm. Throughout this period he was a volunteer worker at the museum and on 13 February 1951, having put in a full day's work there, he died suddenly, in Lidingö, at the age of 83.

Besides several popular zoological works, Andersson published many papers on amphibians and reptiles (1898-1945). Among his first were two papers (1899, 1900) describing Linnaean herpetological types at the Royal Museum; his senior colleague at the museum, Einar Lönnberg, had published an earlier paper on the subject in 1899. The rest of Andersson's titles were careful and often extensive reviews of collections made by Swedish expeditions to many parts of the world and included descriptions of numerous new species. The geographical coverage was vast, including many parts of Africa, Madagascar, New Guinea, Borneo, Southeast Asia, and even eastern Siberia. His special interest was South America, which occupied nearly half of his publications and covered areas from the high Andes to Tierra del Fuego at the southern tip of the continent. His most comprehensive monograph (1945), also his last publication, reviewed the amphibians of eastern Ecuador.

• *References*: "Lars Gabriel Andersson," anonymous, p. 34. *In* Vem är Det.—Svensk Biograf. Handb., Stockholm, 1934; "Lars Gabriel Andersson," by S. Lindman, p. 98. *In* Svenska män och Kvinnor—Biograf. Uppsalgsbok, Stockholm, 1942-1955; "Dr. Lars Gabriel Andersson," anonymous, Copeia, 1951: 184, 1951. • *Portrait*: Courtesy British Museum (Nat. Hist.). • *Signature* (1914): British Museum (Nat. Hist.), courtesy A. F. Stimson.

NIKOLSKY, A. M. (1858-1942).

Alexander Mikhailovich Nikolsky was the leading Russian herpetologist at the turn of the century and successor to Alexander Strauch as custodian of his nation's largest collection. Nikolsky was born on 18 February 1858, in Astrakhan', near the Caspian Sea, as the son of a military surgeon. He studied natural sciences at St. Petersburg Imperial University (1877-1881), defended his graduate dissertation in 1887, and then became an associate professor at the university. In the meantime (about 1880), he was appointed curator of the university's zoological collection and in 1895 became zoologist and head of the herpetological

• *References*: ["From the Memoirs of Zoologist Alexander Mikhailovich Nikolsky"], by B. E. Raikov, Mem. Hist. Nat. (Inst. Hist. Nat. Sci. Tech.), 1: 79-104, 1966; "Aleksandr Mikhaylovich Nikol'skiy," anonymous, p. 415. *In* Who Was Who in the USSR, Munich, 1972; ["Aleksandr Mikhailovich Nikolskij,"] by B. N. Mazormovich, Nauka, Moscow, 77 pages, 1983. • *Portrait*: Academy of Sciences, Leningrad, courtesy Ilya S. Darevsky and Natalia B. Ananjeva. • *Signature* (romanised form, 1906): Smithsonian Institution Archives, courtesy William Cox and Alan E. Leviton.

SIEBENROCK, Friedrich (1853-1925).

Siebenrock, reptilian anatomist and turtle taxonomist, was born on 20 January 1853 at Schörfling, near Attersee (Lake Atter) in Upper Austria. He received his university education in Innsbruck and Vienna. For many years Siebenrock was Demonstrator at the Zootomical Institute of the University of Vienna, under Carl B. Brühl who had authored a well-known atlas of crocodile osteology (1862), and he started voluntary work at the Naturhistorisches Museum in Vienna in January 1886. In December of that year he became an assistant to Franz Steindachner and thus worked with fishes as well as amphibians and reptiles. As Steindachner's duties at the museum became more administrative, Siebenrock gradually assumed responsibility for the collections of lower vertebrates and he was promoted to curator in 1895. He spent the rest of his career at the museum until retiring in 1920, when he was succeeded by Otto von Wettstein. Siebenrock then returned to the place of his birth, where he died in poverty on 28 January 1925.

department at the Zoological Museum of the Academy of Sciences; he held that position until 1903 when he became a professor at Kharkov University in the Ukraine, where he spent the rest of his career. Earlier, from 1881 to 1891, he participated in numerous expeditions (Sakhalin, Lake Balkhash, Japan, Transcaspia, Iran, and the Caucasus). Among his students at Kharkov was S. A. Chernov, who later held Nikolsky's former post at the Academy of Sciences.

Nikolsky was interested in all vertebrate groups but especially lower vertebrates. His work emphasized systematics and faunal surveys and, unlike his Russian predecessors, also zoogeography and life histories. His primary works included books on the vertebrates of the Crimea (1891), "Herpetologia Turanica," which concerned the Fedchenko expedition to Turkestan (1899), "Herpetologia Rossica" (1905), which was his first major herpetological treatise, and "Herpetologia Caucasica" (1913). His most comprehensive herpetological work, the reptilian (1915, 1916) and amphibian (1918) volumes in the series "Fauna of Russia and Adjacent Countries," was of such continuing importance that it was republished in English in 1962-1964. Nikolsky also authored a text on animal geography in 1909. His career was honored by his election as Academician of the Ukrainian Academy of Sciences in 1919. He died in Kharkov on 8 December 1942.

Siebenrock's numerous herpetological publications cover the period 1892-1924. His earliest papers were on the anatomy and osteology of lizards, turtles, and the tuatara, but his interests shifted to the systematics of turtles at the turn of the century. He described and classified turtles from all parts of the world, with special emphasis on those from South America, Australia, Asia, and Africa. He named 27 taxa of turtles, including the genus *Pseudemydura*, the Madagascan *Acinixys planicauda*, the soft-shelled tortoise of East Africa (*Malacochersus tornieri*), and numerous emydids, kinosternids, and trionychids. He also published a few papers on crocodilians. His *magnum opus* was "Synopsis der rezenten Schildröten" (1909), the first major review of the world's turtles since that of Boulenger (1889).

• *References*: "Friedrich Siebenrock," p. 507-508. *In* A. Handlirsch and R. von Wettstein (ed.), Botanik und Zoologie in Österreich in den Jahren 1850 bis 1900. Festschr. Zool.-Bot. Gesellsch., Vienna, 1901; "Friedrich Siebenrock (1853-1925)," by F. Tiedemann and G. Grillitsch, Herpetozoa (Vienna), 1, in press. • *Portrait*: Courtesy Heinz Wermuth. • *Signature* (1908): Smithsonian Institution Archives, courtesy William Cox and Alan E. Leviton.

WERNER, Franz (1867-1939).

The zoologist and explorer Franz Joseph Maria Werner was born in Vienna on 15 August 1867. His father interested him in amphibians and reptiles when he was six. As a precocious high school student, he received instruction and encouragement through correspondence with George A. Boulenger and Oskar Boettger whose scope and style of research Werner came to emulate. After obtaining his Ph.D. from the University of Vienna (1890) and a year of further training in Leipzig, he returned to the university where he became an instructor at the Zoological Institute and rose through the ranks to become a full professor in 1919; he retired in 1933 and died on 28 February 1939, in Vienna.

During his tenure at the university, Werner supervised numerous students, not only in herpetology but in entomology, as he was equally interested in insects, especially Orthoptera. Among his herpetological students were Georg Haas, Walter Mosauer, and Otto von Wettstein. Werner was also an avid explorer, especially of Southeast Europe, Asia Minor, northern Africa, and the islands of the eastern Mediterranean. Despite rough seas and difficult landings, he explored 45 of the Greek islands, the last of them (Thasos) at the age of 71 and less than a year before his death.

Although situated near the Naturhistorisches Museum in Vienna, with its extensive herpetological collections, the personal animosity of then-director Franz Steindachner prevented Werner from working with those collections. In his youth, Werner's fervent wish had been to be appointed to the museum's staff, but Steindachner rejected him, probably seeing in him a serious competitor. Thus, until Steindachner's death (1919), Werner had to base his research on collections in other European museums and on his own enormous private collection, which was later donated to the Vienna museum by his sons.

Despite this handicap, Werner published over 550 titles, the vast majority on herpetology. These works emphasize taxonomy and faunistics and cover the world fauna, but there are also papers on anatomy and behavior. In his eagerness to describe new taxa (he named 24 genera of amphibians and reptiles, and over 400 species and subspecies), he occasionally erred and named as new some well-known species of snakes from specimens whose continent of origin had been mixed up or had no locality data at all, but his other publications were quite competent.

Werner's first major faunal work, in 1897, was on Austria-Hungary and the so-called occupation territories (part of the Balkans), followed by one on the Bismarck Archipelago east of New Guinea (1900) and, much later, on Greece (1938). His primary taxonomic monographs include one on chameleons (1902), a three-part series covering colubrid snakes (1923, 1924, 1929), and several issues in the *Das Tierreich* series. Werner also wrote the section on Amphibia for the *Handbuch der Zoologie* (1930) and co-authored a standard work on venoms and venomous snakes (1931), with Rodolpho Kraus, a former colleague in Vienna who later became Director of the Instituto Butantan in Brazil. Unlike most of his contemporary fellow scientists, he was not adverse to popular or semi-popular writing and contributed significantly to the popularity of terrarium-keeping in the German-speaking countries of Europe by his little volume entitled "Amphibien and Reptilien" (1910) and also by another book that was the major treatise of his career, the two herpetological volumes in the fourth edition of Alfred Brehm's "Tierleben" series (1912-1913).

• *References*: Siebenrock, 1901 (pp. 452-456); "Prof. Franz Werner," by B. C. Mahendra, Nature, 143: 711-712, 1939; "Franz Werner 1867-1939," by W. Mosauer, Herpetologica, 1: 178-183, 1940; "Franz Werner als Mensch und Forscher," by O. von Wettstein, Ann. Naturhist. Mus. Wien, 51: 8-53, 1941. • *Portrait*: Naturhistorisches Museum Wien, courtesy Josef Eiselt and Franz Tiedemann. • *Signature* (1927): Adler collection.

WAITE, Edgar R. (1866-1928).

Edgar Ravenswood Waite, eminent Australian ichthyologist and herpetologist, was born in Leeds, England, on 5 May 1866. He had a boyhood interest in natural history and after brief training at Victoria University (now University of Manchester) he became, in 1888 at the age of 22, Sub-Curator and, later, Curator (or director) of the Leeds Museum. His main interests then were ornithological, but after he emigrated to Australia in 1892 he became Zoological Assistant (or curator) in charge of mammals, reptiles, fishes, and osteology at the Australian Museum in Sydney. Thereafter, his publications covered all vertebrate groups, especially fishes, and he participated as naturalist aboard H.M.C.S. *Thetis*, which trawled off the coast of New South Wales in 1898.

From 1906 to 1914, Waite was Curator (or director) of the Canterbury Museum in Christchurch, New Zealand, and participated in two expeditions to Antarctic waters. In 1914, he became Director of the South Australian Museum in Adelaide, a position he held until his death. During those years he accompanied several expeditions, including one to Central Australia and, later, at the age of 50, to New Guinea during which he contracted malaria that debilitated him for the rest of his life. He died on 19 January 1928 in Hobart, Tasmania, where he was attending a science congress.

Waite's international reputation was primarily as an ichthyologist, but of his 144 titles, about 30 were herpetological (published 1893-1929) and these included two books. The first of his books, "Popular Account of Australian Snakes" (1898), illustrated with 16 color plates drawn by Waite himself, was probably the first pocket-sized guide on a herpetological subject—Australians have a special need to promptly recognize their numerous species of poisonous snakes. Waite's best known herpetological book was "Reptiles and Amphibians of South Australia," published posthumously in 1929. In addition to his scientific contributions, Waite founded and edited major museum journals in New Zealand and Australia, pioneered new museum exhibition techniques, and was an excellent museum administrator.

• *Reference*: "Obituary and Bibliography of Mr. Edgar R. Waite," by H. M. Hale, Rec. S. Austr. Mus., 3: 345-359, 1928. • *Portrait and signature* (1900): Australian Museum, courtesy Harold G. Cogger.

WALL, Frank (1868-1950).

Frank Wall, a British medical officer and leading student of the snakes of the Indian Empire at the turn of the century, was born in Nuwara Eliya, in the central mountains east of Colombo, Ceylon (now Sri Lanka), on 21 April 1868. His father, a prominent political figure in Ceylon, first interested him in natural history. Wall qualified in medicine (M.R.C.S., L.R.C.P., Middlesex Hospital, London, 1892), entered the Indian Medical Service in 1893, and after completing his training at Netley, departed for India in 1894. Until he retired and returned to England in 1925, he was at various times stationed throughout most of the Indian peninsula as well as Ceylon and Burma, and collected snakes wherever he went. During the First World War, he joined the Mesopotamian Expeditionary Force in Iraq and also served in France, work for which he was decorated with the C.M.G. He was promoted to Colonel, his final rank, in 1920.

On retiring to England, Wall donated his holotypes and collection of skulls to the British Museum (Nat. Hist.) and gave up further herpetological work. Wall was a field worker, by choice not a museum scientist, and he was interested in all aspects of ophiology including taxonomy, distribution, mode of life, and anatomy. Once living animals were no longer available to him, he turned to other interests. He died at his home in Bournemouth, Dorset, on 19 May 1950.

Wall published about 215 titles over the period 1898-1928. His first major work, issued in serial form, was "A Popular Treatise on the Common Indian Snakes." This was published in 29 parts, with 28 beautiful colored plates, in the *Journal of the Bombay Natural History Society*, over the period 1905-1919 and, unfortunately, was never reissued in book form. In 1907, the first edition of Wall's widely-used "The Poisonous Terrestrial Snakes of Our British Indian Dominions" was published. This manual for identification and treatment of snake bite went through three more editions (1908, 1913, 1928). Perhaps his book best known to scientists is "Ophidia Taprobanica or the Snakes of Ceylon" (1921). He also published "A Hand List of Snakes of the Indian Empire," in five parts (1923-1925), which was a systematic list with full synonymies.

• *References*: "Colonel Frank Wall," p. 412. *In* A. Lawrence (ed.), Who's Who Among Living Authors of Older Nations, vol. 1. Golden Syndic. Publ. Co., Los Angeles, 1931; "Frank Wall, 1868-1950," by M. Smith, Copeia, 1951: 113-114, 1951; "Bibliography of the Herpetological Papers of Frank Wall," by S. M. Campden-Main, Smithson. Herpetol. Informat. Serv., pp. 1-7, 1969. • *Portrait* (1935) *and signature* (1926): British Museum (Nat. Hist.), courtesy A. F. Stimson.

VAN DENBURGH, John (1872-1924).

As Holbrook's "North American Herpetology" (1842) did for eastern North America, Van Denburgh's *magnum opus*, "The Reptiles of Western North America," in two large volumes (1922), laid the foundation for present-day herpetological studies in the American West. Born of Dutch-English ancestry in San Francisco on 23 August 1872, Van Denburgh showed an early interest in birds and later attended newly-founded Stanford University (A.B. 1894, M.A. 1895, Ph.D. 1897) where his major professor, Charles H. Gilbert, interested him in reptiles. In 1894, he became associated with the California Academy of Sciences as an assistant; he was promoted to curator the next year, a position he held until his death except for a period when he attended medical school (M.D. Johns Hopkins University 1902). In addition to the curatorship, he also had a medical practice in San Francisco. His death in Honolulu, Hawaii, on 24 October 1924, was officially recorded as a suicide.

Van Denburgh's main interests were life histories, distribution, and taxonomy, as exemplified in his first major works, a three-part series on the herpetology of Baja California, Mexico (1895-1896, reprinted 1978), and "Reptiles of the Pacific Coast and Great Basin," issued in 1897 (reprinted 1971). He was an accomplished field naturalist, and also skilled in laboratory technique, as shown in his early studies on the venom of *Heloderma*. He was an avid collection builder and amassed one of the largest collections of amphibians and reptiles in North America when, in April 1906, the great earthquake and resulting fire destroyed the academy. Of the 8100 specimens, Van Denburgh, at great personal risk, saved only 13, but these included most of the holotypes. Despite this calamity, Van Denburgh rebuilt, beginning with the substantial Galapagos Islands collections (4506 specimens) made by his assistant, Joseph R. Slevin, in 1905-1906.

The museum existed in temporary quarters until the move to its present location in Golden Gate Park in 1915. By the time of Van Denburgh's death, the collections had been built to more than 50,000 specimens, largely through field work by Slevin but also by Van Denburgh himself (to Arizona, the Channel Islands and Sierra Nevada Mountains of California, and the San Pedro Martir Mountains in Mexico). Another assistant curator, Joseph C. Thompson, like Van Denburgh a medical doctor, made extensive collections in the Far East, but he was terminated by the academy (in 1912) after a series of disputes with Van Denburgh. Thus began one of the celebrated controversies in herpetology, with claims and counterclaims for authorship of new taxa, which had serious and even still-continuing consequences for the nomenclature of some East Asian amphibians and reptiles.

• *References*: "A Most Regretable [*sic*] Tangle of Names [Van Denburgh-Thompson Controversy]," by T. Barbour, Occas. Pap. Mus. Zool. Univ. Michigan, 44: 1-9, 1917; "John Van Denburgh 1872-1924," by B. W. Evermann, Science, 61: 508-510, 1925; "Holotype Specimens of Reptiles and Amphibians in the Collection of the California Academy of Sciences," by J. R. Slevin and A. E. Leviton, Proc. Calif. Acad. Sci., ser. 4, 28: 529-560, 1956. • *Portrait*: By Sarah Atsatt, from the Howard K. Gloyd collection, courtesy Kathryn J. Gloyd and Roger Conant. • *Signature* (1922): Adler collection.

DITMARS, Raymond L. (1876-1942).

Raymond Lee Ditmars, America's first great popularizer of reptiles, was born on 22 June 1876 in Newark, New Jersey, descended from Dutch immigrants named van Ditmarsen. He had an interest in animals, especially reptiles, from boyhood, but his life's chief interest was apparently set when, at the age of 12, he caught his first snakes at Gravesend Bay near Brooklyn, New York City. He attended secondary school at Barnard Military Academy in the city, but was an indifferent student, preferring instead the vacation periods when he could collect specimens or study those he kept at his family's apartment.

In 1893, Ditmars was hired by the entomology department at the American Museum of Natural History. He resigned in 1897 to take higher-paying positions, first as a stenographer at an optical instrument company and later, in July 1898, as a reporter for *The New York Times* newspaper, for which one of his beats was the newly-formed New York Zoological Society, shortly due to open its zoo in the Bronx section of the city. Ditmars was hired as Assistant Curator (in charge) of Reptiles, beginning July 1899, at the age of barely 23. He donated his private collection which formed the nucleus of the zoo's reptile displays. Thus began his lifelong association with the Bronx Zoo, to which duties he added responsibility for mammals (1926) and insects (1940).

Ditmars published eight herpetological books, plus a dozen more on other animal groups or autobiographical titles about his work or travels, but it was "The Reptile Book," published in 1907, that established his reputation and also the seeds of controversy. This book, reissued in an expanded version in 1930, together with several others ("Reptiles of the World," 1910, revised version 1933; "Snakes of the World," 1931; "Reptiles of North America," 1936; and "Field Book of North American Snakes," 1939), had an enormous public influence for fully half a century, making Ditmars responsible, more than any other single person, for kindling a widespread interest in reptiles. Indeed, several generations of professional herpetologists, including many still active today, can trace their interest to one or another of Ditmars's books.

Despite these achievements, even today Ditmars is often dismissed as simply a showman. He wrote highly popular books for the general public, beginning with "Strange Animals I Have Known" (1931), followed by two autobiographical books, "Confessions of a Scientist" (1934) and "The Making of a Scientist" (1937), but he also made substantive scientific and public contributions. He pioneered techniques to cure diseases of reptiles. He was a major organizer of the first effort to establish an antivenin institute in America; prior to that, Ditmars himself was the country's distribution center for antivenins produced in Brazil. He was a producer of influential nature films, the first in 1910 ("The Jungle Circus"), and led the development of the Bronx Zoo into the first rank of the world's zoos.

Lincoln Memorial College in Tennessee awarded Ditmars an honorary Litt.D. degree in 1930 (although he had carried the unofficial title of "doctor" for many years), but academic herpetologists still gave him grudging credit for his accomplishments. They objected to his cursory knowledge of the scientific literature and especially to his habit of exaggerating information for effect, such as maximum lengths for species, and his attribution of data on reproduction to authorities who were almost certainly fictitious. He wrote some technical papers and conducted field work in many places, including South Carolina, Florida, California, Central America, Trinidad, and Brazil, although most of his collecting was done within 100 km of New York City.

To be sure Ditmars was a showman, knowing exactly how to stage an event to maximize publicity, and he was also a dynamic speaker. To supplement his zoo income he lectured regularly to large and small audiences, resulting in chronic throat ailments which, complicated by pneumonia, eventually led to his death in New York City on 12 May 1942.

• *References*: "Raymond L. Ditmars," by Alexander G. Ruthven, Copeia, 1942: 131, 1942; "Raymond L. Ditmars, Aspects of an Extraordinary Career," by H. F. Osborn, Anim. Kingdom (Bull. New York Zool. Soc.), 45: 82-84, 1942; "Raymond L. Ditmars. His Exciting Career," by L. N. Wood, J. Messner Inc., New York, x, 272 pages, 1944; "Raymond Lee Ditmars," by C. H. Pope, Dict. Amer. Biogr., suppl. 3: 228-229, 1973; "The Herpetologists—Part I," by R. Pinney, Herp, Bull. New York Herpetol. Soc., 15(2): 27-43, 1980. • *Portrait* (1930s): Carnegie Museum, courtesy M. Graham Netting. • *Signature* (1929): Courtesy Roger Conant.

RUTHVEN, Alexander G. (1882-1971).

Alexander Grant Ruthven, one of America's most influential herpetologists who later became President of the University of Michigan, was born in Hull, Iowa, near the South Dakota border, on 1 April 1882. After graduation from Morningside College in Sioux City, Iowa (B.S. 1903), he entered the University of Michigan to study with the ecologist Charles C. Adams. His doctoral thesis, for which he earned a Ph.D. in 1906, was later published as "Variations and Genetic Relationships of the Garter-Snakes" (1908), a work that had a profound influence in herpetology. Ruthven's scope was a single, well-defined genus and he thereby set the standard for a new style of Ph.D. thesis. Moreover, since many species of *Thamnophis* are common, he had a very substantial sample size which required the development of statistical methods in order to visualize geographic changes in scutellation. This enabled Ruthven to pioneer the use of biometric methods in herpetology.

After finishing his doctorate, Ruthven joined the staff of the Museum of Zoology at the University of Michigan as Curator of Herpetology and, concurrently, Instructor in Zoology (1906). Eventually, he became Director of the Museum (1913), Professor (1915), and Chairman of the Department of Zoology (1927). The main structure of the present museum building was erected under his supervision, and under his directorship the museum became the leading center for graduate training in herpetology. The program of research, teaching, publication, and field work was clearly patterned after the plan laid down for the Museum of Comparative Zoology at Harvard College by Louis Agassiz; indeed, the principal herpetologists at the two institutions, Ruthven and Thomas Barbour, maintained the closest personal relationship and had a regular weekly correspondence for nearly two decades. As part of this development of herpetology at Michigan, Ruthven was instrumental in having editorship of the journal *Copeia* headquartered in the museum, beginning in 1930.

While he worked at the museum, Ruthven had numerous students, including Frank N. Blanchard, M. Graham Netting, Arthur I. Ortenburger, and Olive G. Stull, but he became President of the University of Michigan in 1929 and thereafter he had less time for herpetology and supervision of students. Fortunately, he had an able assistant in Helen T. Gaige, who joined Ruthven's staff in herpetology in 1910 and who became Curator of Amphibians in 1923. When Ruthven was named president, Gaige assumed responsibilities for the herpetology program and the day-to-day supervision of Ruthven's Ph.D. students after 1929: Joseph R. Bailey, Charles E. Burt, Norman E. Hartweg, James A. Oliver, Laurence C. Stuart, and Charles F. Walker. Three of these students—Hartweg, Stuart, and Walker—were to become staff members at the museum, and with an earlier Ruthven student, Frank N. Blanchard, as a professor in Michigan's zoology department, the tradition of herpetological study at the university was assured.

As University President from 1929 to 1951, Ruthven was generally acknowledged as the person most responsible for the emergence of the University of Michigan as one of the great American universities. During his 22 years as president, the university experienced an enormous growth despite the Depression and a world war, and throughout Ruthven had a record of a highly democratic administration.

Ruthven published about 130 titles of which about 75 are herpetological. Many of these report results of his expeditions to the western United States, Mexico (Veracruz), and Colombia, the last resulting in one of Ruthven's major treatises, "Amphibians and Reptiles of the Sierra Nevada de Santa Marta, Colombia" (1922). Besides a primary interest in systematics, all of these studies have an ecological outlook. His publications also dealt with collections from Venezuela, British Guiana, Panama, the West Indies, Brazil, and Bolivia.

Despite this major focus on the Neotropics, Ruthven also produced a local handbook, "Herpetology of Michigan" (1912), co-authored with Helen T. Gaige and her sister, Crystal Thompson (second edition 1928). Ruthven's last paper in herpetology was published in 1936. Indeed, it was remarkable that he continued research at all after assuming the presidency in 1929, although many of the later papers were co-authored with Mrs. Gaige. He died alone in Ann Arbor, on or about 19 January 1971, at the age of 88.

• *References*: "A Naturalist in a University Museum," by A. G. Ruthven, Alumni Press, Univ. Michigan, Ann Arbor, 143 pages, 1931; Schmidt, 1955 (p. 612); "Naturalist in Two Worlds: Random Recollections of a University President," by A. G. Ruthven, Univ. Michigan Press, Ann Arbor, 162 pages, [1963]; "Dr. Alexander G. Ruthven Dies," anonymous, New York Times, 20 January 1971, p. 38, 1971; "Alexander Grant Ruthven," anonymous, Copeia, 1971: 587, 1971; "Alexander Grant Ruthven of Michigan: Biography of a University President," by P. E. Van de Water, W. B. Eerdmans, Grand Rapids (Michigan), 251 pages, [1977]. • *Portrait* (1928): Howard K. Gloyd collection, courtesy Kathryn J. Gloyd and Roger Conant. • *Signature*: Adler collection.

WRIGHT, Albert H. (1879-1970).
WRIGHT, Anna A. (1882-1964).

Albert Hazen Wright and Anna Allen Wright, co-authors of the handbooks of American frogs and toads and of snakes, were both born in New York State—Albert in Hamlin (Monroe County) on 15 August 1879 and Anna in Buffalo, 4 March 1882. Albert graduated from Brockport State Normal School in 1899 and taught public school for a year before entering Cornell University in 1900; Anna also taught school, in Buffalo, before attending Cornell. Albert completed his A.B. degree in 1904, M.A. in 1905, and took his Ph.D. degree in 1908 under the supervision of Burt G. Wilder, one of Louis Agassiz's principal students. Anna graduated with an A.B. in 1909. They were married the next year and became companions in field and laboratory for the next 55 years, until Mrs. Wright's death, in Ithaca, New York, on 5 December 1964. Albert Wright served on the Cornell faculty until his retirement in 1947; he died at the age of 90 on 4 July 1970, in Ithaca, after 70 years of association with Cornell.

Albert H. Wright's publications on amphibians and reptiles spanned the period 1908-1962, but comprised relatively few titles. He was a general naturalist, however, and published on all vertebrate classes and botany, as well as history. His major herpetological works included "Life-Histories of the Anura of Ithaca, New York" (1914), "Synopsis and Description of North American Tadpoles" (1929), and "Life-histories of the Frogs of Okefinokee [*sic*] Swamp, Georgia" (1932) which encompassed his major research interest, the life histories of amphibians, for which he provided the first comprehensive studies of American species.

Wright also authored or co-authored a series of three long papers (1915-1926) on the herpetology of the Okefenokee Swamp, based on the Cornell University expeditions from 1909 to 1922 by Wright and his students Sherman C. Bishop, Alvin R. Cahn, W. D. Funkhouser, and Francis Harper. These studies, together with others by them on the fauna and flora of this great swamp on the Georgia-Florida border, were instrumental in its becoming a national wildlife refuge in 1937. Wright's pioneering work in ecology was later recognized in 1955 by his election as Eminent Ecologist by the Ecological Society of America.

Albert H. Wright's major books, however, were produced and co-authored with his wife and resulted from years of extensive field work throughout North America, during which they collected notes and specimens and took over 12,000 photographs of amphibians and reptiles. As a result, it has been said that the Wrights had more first-hand experience in the field with more species and subspecies of

North American amphibians and reptiles than anyone else. Their photographs were used in the "Handbooks of American Natural History" series, published by Cornell University Press and edited by Albert Wright.

This series of volumes included the well-known handbooks on salamanders (by Sherman C. Bishop, 1943), lizards (by Hobart M. Smith, 1946), and turtles (by Archie F. Carr, 1952). In this same series appeared the handbooks of frogs and toads (by Anna and Albert Wright, 1933, second edition in 1942, third edition, with authorship reversed, in 1949) and of snakes, in three volumes (by Albert and Anna Wright, text volumes in 1952, bibliography volume in 1962, the latter reprinted 1979). All of these books have gone through numerous reprintings and have become the standard natural histories for the North American herpetofauna. The Wrights's own books were packed with vast collections of information and anecdote, often presented in an unsynthesized and sometimes haphazard and repetitious manner.

Professor Wright was one of Cornell's outstanding teachers of vertebrate natural history and comparative anatomy, and he and his wife were beloved by generations of students. His most lasting legacy to herpetology resulted from the supervision of numerous students, many of them herpetologists, who went on to found important schools of research throughout the United States and also in China. Among Wright's most prominent students were Sherman C. Bishop, Raymond B. Cowles, Kenneth L. Gordon, William J. Hamilton, Jr., James Kezer, Ch'eng-chao Liu, Robert L. Livezey, Karl P. Schmidt, and Harold Trapido.

• *References*: "Mrs. Anna A. Wright," by W. J. Hamilton, Jr., Copeia, 1965: 124, 1965; "Albert Hazen Wright," by W. J. Hamilton, Jr., Copeia, 1971: 381-382, 1971; "Albert Hazen Wright," by H. E. Evans, E. C. Raney, and W. J. Hamilton, Jr., Necrol. Fac. Cornell Univ., 1969-1970: 52-54, 1971; "Amphibian and Reptile Biology at Cornell University," by K. Adler and F. H. Pough, Herpetol. Rev., 15: 40-42, 1984. • *Portrait* (about 1956): Courtesy William J. Hamilton, Jr., and Cornell University Department of Manuscripts and University Archives. • *Signatures* (1955): Adler collection.

ROUX, Jean (1876-1939).

Jean Roux, Swiss zoologist, was born in Geneva on 5 March 1876. At the University of Geneva he studied under Emile Yung and completed his doctoral thesis in 1899. His research at this time was on protozoans, and after postdoctoral work in Berlin, he became curator at the Naturhistorisches Museum in Basel (1902). Ciliates not being proper museum animals, he was asked to concentrate on herpetology and thus fell heir to the important collections assembled at the museum by his predecessor, Fritz Müller.

Roux began his new work in earnest and published numerous papers on amphibians and reptiles, beginning in 1904. His interests later broadened to include crustaceans, fishes, and mammals. He participated in two expeditions which resulted in his major herpetological works. In 1907-1908 he accompanied Hugo Merton to the Aru and Kei (or Kai) islands, situated between western New Guinea and Australia and which are now part of Indonesia; the results were published in 1910. In 1911-1912 he joined Fritz Sarasin on an expedition to New Caledonia and the Loyalty Islands in the South Pacific; the reptiles were published in a large monograph in 1913. Beyond these interests, Roux published numerous careful taxonomic studies on amphibians and reptiles collected in Africa, South America, Indonesia, and New Guinea, as well as other parts of the world. In total, he published about 35 papers on herpetological subjects, the last in 1939.

Roux turned down an offer to move to the Senckenberg Museum in Frankfurt (Main) in 1913, electing to stay in Basel. He retired from the museum in 1937 and died on 1 December 1939. He was succeeded as curator by Lothar Forcart.

• *References*: "Jean Roux 1876-1939," by F. Sarasin, Verh. Naturf. Gesellsch. Basel, 50: 1-8, 1940; "Jean Roux 1876-1939," by P. Revilliod, Verhandl. Schweizer. Naturforsch. Gesellsch., 120: 487-493, 1940. • *Portrait*: From Revilliod, 1940. • *Signature* (1927): Adler collection.

ROOIJ, Nelly de (1883-1964).

Petronella Johanna de Rooij (see *Note*) authored the definitive survey of the reptiles of the Indo-Australian region. She was born in Weesp, just southeast of Amsterdam, on 30 July 1883, the daughter of a local physician. She had attended a high school for girls and, therefore, under existing Dutch law, because she lacked a grammer school education, she could not obtain an undergraduate degree from Dutch universities. She did attend lectures at the University of Amsterdam, however, and received a teacher's certificate (1904), with which she was permitted, under different laws in Switzerland, to enroll at the University of Zürich. She completed her Ph.D. degree there in 1907, although the research was done in Amsterdam, with a thesis on the cardiovascular system of *Andrias*, the giant salamander (published 1906 [1908]-1907). From 1907 she held the position of Curator of Reptiles and Amphibians in the Zoological Museum, University of Amsterdam, but resigned in 1922, at the age of 39, because of a change in museum administration which was not to her liking, and she did no further scientific work. She married in 1929, thereafter being known as P. J. Breijer-de Rooij, and died at Arnhem, Netherlands, on 10 June 1964.

During her very short professional career, De Rooij published only a dozen papers, primarily on the herpetology of Dutch possessions in the East and West Indies, but she is long remembered for her book, "The Reptiles of the Indo-Australian Archipelago" (in two volumes, 1915 and 1917; reprinted 1970). This monograph, still the definitive and most comprehensive work on the region, covers 624 species of reptiles; the companion volume on amphibians was published by P. N. van Kampen in 1923. Both titles cover the region from Sumatra through New Guinea and include Borneo. These books were the herpetological culmination of the Dutch effort to explore their territory in the East Indies, begun early in the 19th Century, which consumed vast sums of money and even the lives of outstanding young naturalists such as Heinrich Boie, Heinrich Kuhl, and J. C. van Hasselt.

• *References*: "In Memoriam," anonymous, Vakblad voor Biologen, 44(7): 132, 1964; "Mrs. Dr. P. J. Breyer-de Rooy [*sic*]," by W. S. S. van der Feen-van Bethem Jutting, Copeia, 1964: 598, 1964; "Dr. P. J. de Rooij," by L. D. Brongersma, unpublished manuscript. • *Portrait* (about 1929): Rijksmuseum van Natuurlijke Historie, courtesy Marinus S. Hoogmoed. • *Signatures* (maiden name, 1916): Naturhistorisches Museum Basel, courtesy Christine Stocker; (married name, 1931): Courtesy Leo D. Brongersma. • *Note*: "Nelly" is the diminutive of Petronella, a call name, but De Rooij regularly used it in her publications. In some of her papers her name was spelled "Rooy"; the 25th letter of the Dutch alphabet, *ij*, is sometimes incorrectly written as *y*.

MÜLLER, Lorenz (1868-1953).

The German herpetologist Lorenz Müller was born in Mainz on 18 February 1868. Originally self trained as an artist, he attended the Academy of Art in Munich. He had been a terrarist as a boy, with a special interest in amphibians and reptiles, so after he returned to Munich from art studies in Paris, Belgium, and Holland, he developed important contacts with Oskar Boettger in Frankfurt and Willy Wolterstorff in Madgeburg, both of whom encouraged his herpetological interests. Müller worked at the Zoologische Staatssammlung in Munich as a scientific illustrator, but since the herpetological collections were without a curator, he took on those duties beginning in 1903. These historically-important but long-neglected collections, which had their beginnings in the 1820s with the work of J. B. von Spix and J. G. Wagler, began to grow rapidly. In 1909-1910 Müller was a member of the museum's expedition to the Lower Amazon of Brazil, which had been first explored by von Spix in 1817-1820, and this stimulated his lifetime interest in the herpetofauna of South America.

By 1912, the museum finally established a curatorship for Müller. During the First World War he served with the army in the Balkans, but spent much of the time collecting specimens. In 1928 he became Chief Curator of Zoology at the museum and held this post until retirement, at which time his research was recognized with an honorary doctorate from the University of Munich. He continued his research and was called back to active duty during World War II, to supervise packing of the collections and their temporary removal for safekeeping, unfortunately without complete success.

The museum's crocodilian collection, which had not yet been removed, was destroyed in a bombing raid in April 1944 and one third of the main collection, which had been stored in the cellar of a restaurant in Planegg, south of Munich, was lost in another raid on one of the last days of the war, in April 1945. Only quick action by Müller the morning after the raid saved the collection from further losses. Even his own large private collection and library were lost when his home was destroyed during the bombing of Munich, yet he resumed his work and built collections anew. He died on 1 February 1953.

Müller published more than 100 herpetological titles, beginning in 1900. He maintained a lifelong interest in living animals and often published notes about them in the terrarium journals. His scientific papers emphasized the descriptions of new taxa, particularly of South America but also of the Mediterranean Region, Africa, and even China and Australia. He was also interested in zoogeographic and evolutionary questions, thus recognizing the importance of variability and the need for large series of each species, a view that was uncommon among European herpetologists at the time.

Müller's major works include monographs on the herpetofaunas of the Cameroons (1910) and Brazil (1927), and the first two editions of the checklist of European amphibians and reptiles (1928 and 1940, with Robert Mertens). He also wrote a book on the amphibians, turtles, and crocodiles collected by the German expedition to the Gran Chaco, a vast prairie region in the interior of south-central South America (1936, with Walter Hellmich, who in 1960 published the lizard section). Müller was responsible for completing the book "Die Europaeischen Schlangen" (1913-1931) after the death of the original author, Fritz Steinheil, in 1926. He also published a long series of papers on Chilean lizards (1932-1938) with Hellmich, who had made large collections in Chile and then joined the museum staff in 1932; eventually he succeeded Müller as curator.

• *References*: "Prof. Dr. Lorenz Müllers Reisen und Veröffentlichungen," by W. Hellmich, Isis-Festschr. L. Müller und K. Lankes, Munich, pp. 1-8, 1948; "Prof. Dr. h. c. Lorenz Müller," by R. Mertens, Natur und Volk, 83: 105-107, 1953; "Prof. Dr. h. c. Lorenz Müller," by W. Hellmich, Aquar. Terrar. Zeitschr., 6: 80-81, 1953 (English translation in Copeia, 1953: 133-134, 1953); "Lorenz Müller†," by W. Hellmich, Zool. Anz., 18 (suppl.): 471-473, 1955. • *Portrait*: Courtesy Heinz Wermuth. • *Signature* (1911): British Museum (Nat. Hist.), courtesy A. F. Stimson.

ANNANDALE, Nelson (1876-1924).

Thomas Nelson Annandale, zoologist, anthropologist, and founder of the Zoological Survey of India, was born in Edinburgh on 15 June 1876, the son of a distinguished surgeon. He attended the University of Edinburgh and the University of Oxford (Balliol College) and in 1905 earned a D.Sc. from Edinburgh. However, the year before he was awarded the degree, he went to Calcutta as Deputy Superintendent of the Indian Museum, and in 1907 was promoted to Superintendent, a post once held by John Anderson. Under his active leadership the museum's program grew rapidly; collections and studies were made throughout the Indian Empire, the *Records* and *Memoirs* were initiated, and he trained a generation of Indian zoologists.

In 1916, Annandale founded the Zoological Survey of India and was its director until his sudden death, in Calcutta, on 10 April 1924, at the age of 48. He had been nominated to be a Fellow of the Royal Society just before his death. Among those students of herpetological interest trained by Annandale were C. R. Narayan Rao, Baini Prashad, and Sunder Lal Hora, the latter taking responsibility for lower vertebrates in 1921 and, in 1947, succeeding to Annandale's former post as director.

Annandale's earliest interest was anthropology, although his first paper on herpetology was in 1902. In India his work was largely zoological and covered nearly every major group including invertebrates. Thus, reptiles and amphibians were only one of many interests. He was an enthusiastic field collector, penetrating all corners of the Indian Empire and beyond (Malay Peninsula, Palestine, China, Japan, Morocco). His special interest was in Asiatic lakes, but his major comparative review of that subject was unfinished at his untimely death.

Most of Annandale's nearly 70 papers on amphibians and reptiles were taxonomic—he described numerous new species from everywhere he travelled—but he also conducted faunal surveys. Some of his most significant herpetological works concerned the desert tracts in southern India (1906), soft-shelled turtles (1912), and the Far East (1917). His interest in lake faunas naturally led him to the study of frog tadpoles, which were the subject of several of his papers. In 1912 he published an important herpetological survey of Abor (now Arunachal Pradesh), a frontier region north of Assam, territory today occupied by India but largely claimed by China. Annandale also published a report on lower vertebrates from Nepal and the western Himalayas (1907), co-authored with George A. Boulenger and Frank Wall.

• *References*: "Nelson Annandale, 1876-1924," anonymous, Jour. Bombay Nat. Hist. Soc., 30: 213-214, 1924; "Nelson Annandale 1876-1924," by S. Kemp, Rec. Indian Mus., 27: 1-28, 1925; "Thomas Nelson Annandale—1876-1924," by W. T. Clausen, Proc. Royal Soc. London, ser. B, 97: xviii-xxi, 1925; "(Thomas) Nelson Annandale," p. 25. *In* Who Was Who, 1916-1928. A. & C. Black, London, 1929. • *Portrait*: From Kemp, 1925, courtesy T. S. N. Murthy. • *Signature* (1908): British Museum (Nat. Hist.), courtesy A. F. Stimson.

ŌSHIMA, Masamitsu (1884-1965).

Masamitsu Ōshima, Japanese herpetologist and ichthyologist, was born on 21 June 1884 in Sapporo. After graduating from Tokyo Imperial University in 1908, where he worked with the early Japanese herpetologist Motokichi Namiye, whose own research Leonhard Stejneger initiated during his visits to Japan, Ōshima became a technician at the Central Research Institute of the Japanese colonial government in Formosa (now Taiwan). One of the functions of the institute was to investigate poisonous snakes, and Ōshima studied their systematics, which resulted in his first paper, a list of Formosan snakes with descriptions of several new species (1910). He published some additional papers on snakes (particularly sea snakes) and lizards, but his next major work was a monograph of the venomous snakes of Formosa and the Ryukyu Islands (1920).

In the meantime, Ōshima had gone to Stanford University in the United States and earned his M.A. degree (1919) under the renowned ichthyologist David Starr Jordan. Ōshima returned to Japan, where he completed his Ph.D. degree at Tokyo in 1920, and thereafter most of his work was on fishes. In 1924 he became a lecturer at Tokyo Women's University, affiliated with the Misaki Marine Experiment Station, and later he had positions with the Tokyo District High School (1930), Palau Tropical Research Institute (1940) and the Army-Navy Medical School.

During World War II, Ōshima was called upon to write his most extensive work on reptiles, a review of the poisonous snakes of the so-called Greater East Asia Co-Prosperity Sphere, which was intended to stretch from Manchuria to Australia. This book, which includes keys and many illustrations, with detailed island-by-island distributional charts, was produced for use by Japanese troops during the war. After the war Ōshima became a technical advisor to General Headquarters and completed a second Ph.D. (1958) in agriculture, for research on freshwater trout.

Ōshima published many titles on fishes, including a textbook entitled "The Fish" (1940). Some of his other books served to popularize zoology. His last herpetological paper (1952, co-authored with Carl Gans) was on the egg-eating habits of a Japanese rat snake. Ōshima died in Sapporo on 26 June 1965.

• *References*: ["A Remembrance of Dr. Masamitsu Ōshima,"] by Y. Okada, Acta Herpetol. Japon., 2: 37-38, 1967; Hasegawa, 1967 (pp. 19-20). • *Portrait* (1933): Courtesy Toshijiro Kawamura. • *Signature* (1915): Smithsonian Institution Archives, courtesy William Cox and Alan E. Leviton.

HEWITT, John (1880-1961).

John Hewitt, South African zoologist, archeologist, and specialist on the herpetofauna of the Eastern Cape Province, was born on 23 December 1880 in Dromfield, Sheffield, England. Interested in science as a schoolboy, he later studied natural science at the University of Cambridge (B.A. Jesus College 1903). During 1905-1908 he was Curator of the Sarawak Museum in Kuching, in the portion of Borneo that is now part of Malaysia.

In 1909, Hewitt went to South Africa where he became Assistant for Lower Vertebrates at the Transvaal Museum in Pretoria, but the next year he was named Director of the Albany Museum, a small provincial museum in Grahamstown in the eastern portion of Cape Province. Here, despite severely limited funds, he built the exhibit and study collections almost single-handedly and supervised extensions to the building in 1920 and 1938. In September 1941 a disastrous fire gutted the main building and much of his 31 years of work went up in flames, including many specimens as well as the accessions register. Hewitt received many honors for his research, including an honorary D.Sc. degree in 1935 from Rhodes University. He retired in 1958 and died on 4 August 1961, in Grahamstown.

Hewitt was a broadly-interested naturalist, as circumstances required at the two museums he directed. His first major research interest was arachnids, on which he published numerous papers. His first herpetological paper was a semi-popular one on Sarawak snakes (1905); later he authored about 45 technical papers on amphibians and reptiles (over the period 1909-1938), most of them taxonomic and distributional in nature. The greatest number concerned South Africa, and he described numerous new taxa. Hewitt had a particular interest in the biogeographic affinities of the herpetofaunas of South Africa and Madagascar, and in 1911 he and Paul A. Methuen collected in Madagascar (two papers, 1913).

Beginning in 1931 Hewitt published several papers on tortoises. The Albany Museum had acquired James E. Duerden's large collection of tortoises to which Hewitt added greatly. Based on these large samples he distinguished many local populations and recognized most with formal names. In the species *Psammobates tentorius* alone, Hewitt accepted 27 different subspecies, of which he had named 16, but Arthur Loveridge and Ernest E. Williams, in their later monograph of African turtles and tortoises (1957), recognized only three.

Hewitt published one book, an extensive illustrated guide to the amphibians and reptiles (and other vertebrates) of the Eastern Cape Province, in two parts (1918, 1937). After the 1941 fire, however, he gave up herpetological research altogether and concentrated on archeology.

• *References*: "John Hewitt," anonymous, South Afr. Jour. Sci., 57: 312, 1961; "Dr John Hewitt," by J. Greig, Cape Herpetol. News, 3: 29-31, 1980; "The Remarkable Dr Hewitt," by D. Sülter, Elephant's Child (Albany Mus. Publ.), 103: 12-14, 1987. • *Portrait*: Albany Museum, courtesy Wouter Holleman. • *Signature* (1915): Adler collection.

CAMP, Charles L. (1893-1975).

Charles Lewis Camp, American paleontologist, anatomist, and historian, was born in Jamestown, North Dakota, on 12 March 1893. His father, a pioneer lawyer, had participated in the framing of the constitution for the new state of North Dakota. The family moved to southern California in 1895 where young Camp met Joseph Grinnell, who involved him in an excavation of fossils from the Rancho La Brea tar pits and who was later to found the Museum of Vertebrate Zoology of the University of California at Berkeley. Recognizing Camp's interest in natural history and fossils, Grinnell some years afterward arranged for Camp, then aged 15, to join a museum team surveying Mt. San Jacinto in Southern California.

By the time Camp entered Berkeley as an undergraduate (1911) he had amassed a large herpetological collection which he donated to the museum. His studies of the museum's collections and further field work led to nine papers (1915-1917) with descriptions of new taxa of salamanders, frogs, lizards, and a fossil toad from the tar pits. Camp's most spectacular discovery occurred quite by accident when two plethodontid salamanders got caught in a trap he had set for mice. Thus he found the first *Hydromantes* taken in the Western Hemispere. He named the new species *Spelerpes platycephalus*, thinking it was related to the Mexican species of that genus, but in 1923 Emmett R. Dunn demonstrated its European affinities. In 1953, Camp, together with Joe Gorman, announced the discovery of a second Californian species, which they named *H. shastae*.

After Berkeley, Camp began graduate studies at Columbia University in New York as a student of William K. Gregory. However, the First World War intervened, and after two years in the army Camp returned to complete his Ph.D. thesis at Columbia and at the American Museum of Natural History, where Mary C. Dickerson arranged for him to work. Gregory's other students at that time included G. Kingsley Noble and Alfred S. Romer, who, like Camp, were working on questions concerning the myology of amphibians and reptiles.

Camp's thesis was later published as "Classification of the Lizards" (1923, reprinted 1971), a far-sighted work which provided the basis for modern lizard systematics and phylogeny. His classification was based upon all available evidence—morphology, geographical distribution, ontogeny, fossil record—and not on key characters. The various states of each character were discussed in detail and the primitive and derived conditions inferred, methods that have earned him recognition as an early cladist.

After finishing his thesis, Camp joined the zoology staff of the University of California at Berkeley in 1922 to teach comparative anatomy, but his research interests were already shifting to paleontology. In 1930 he transferred to Berkeley's Department of Paleontology and simultaneously became Director of the Museum of Paleontology (until 1949). In 1960 he retired and on 14 August 1975 he died in San Jose, California.

Camp's primary work was in paleontology, a subject on which he wrote major monographs (e.g., on the phytosaurs in 1930, mosasaurs in 1942, and dicynodonts in 1956). He conducted numerous collecting trips in the American West, South Africa (1935, 1947-1948), China (1935), and Australia (1960); on the latter trip, together with John Cosgriff, he discovered the first Triassic amphibians on the continent. His best-known work on fossils was the eight-volume "Bibliography of Fossil Vertebrates" (1940-1968), in succession to Oliver P. Hay's earlier bibliographies of the same subject. Camp was also a widely recognized authority on Western American history, about which he wrote several books.

• *References*: "Charles Lewis Camp," anonymous, News Bull. Soc. Vertebr. Paleontol., 14: 1-3, 1945; "Charles Lewis Camp, Biographer of the West," by F. P. Farquhar, Pacific Hist., November 1963 issue (6 p.); "Charles Lewis Camp 1893-1975," by J. T. Gregory and S. P. Welles, News Bull. Soc. Vertebr. Paleontol., 105: 47-49, 1975; "Charles Lewis Camp," by J. T. Gregory, G. P. Hammond, G. R. Stewart, and S. P. Welles, In Memoriam: Publ. Univ. Calif. Acad. Senate, pp. 37-38, 1977; "Tracy I. Storer and Charles L. Camp," by D. B. Wake, Copeia, 1978: 196-197, 1978. • *Portrait* (about 1926): Courtesy Charles M. Bogert. • *Signature* (1969): Adler collection.

GAIGE, Helen T. (1890-1976).

Helen T. Gaige, born Helen Beulah Thompson on 24 November 1890, in Bad Axe (Huron Co.), Michigan, was influential in the development of modern herpetology in the United States. She attended the University of Michigan (A.B. 1909, M.A. 1910), and in 1910 was appointed Scientific Assistant in the herpetology division of the Museum of Zoology at Michigan which was then under the curatorship of Alexander G. Ruthven. When he became director of the museum in 1913, the responsibilities of running the herpetology program fell more and more to Gaige, and in 1918 she was promoted to Assistant Curator of Reptiles and Amphibians and to full Curator in 1923. As Ruthven took on heavier administrative duties—eventually becoming university president in 1929—Gaige finally took charge of the herpetology program, including curation of the collections and supervision of graduate students. Unlike Ruthven, she was not concurrently a professor and thus could not officially sponsor graduate students, but she supervised a large number of them, including those who eventually succeeded her in charge of the museum's herpetology program, although Ruthven officially was their major professor.

Gaige's first publications, in 1912-1913 (under her maiden name, Helen B. Thompson), dealt mostly with herpetological collections from the Midwest, primarily Michigan, but in 1912 she went on a museum expedition to Nevada, then to the Davis Mountains of Texas (1916), the Olympic Mountains in Washington State (1919), Panama (1923), and to Florida and Colorado (1925). In 1917 she described a peculiar new salamander from Washington which she thought was related to the Siberian species of *Ranodon*, but today is known as *Rhyacotriton olympicus*, an ambystomatid. During her 1919 trip she was successful in finding in abundance the tailed frog (*Ascaphus truei*), a species that had previously been known from only five specimens since its discovery in 1897. She provided the first detailed life history information on this remarkable species and also described the tadpole. Thereafter her interests shifted to the Neotropics, especially to frogs, although the quantity of her papers was small. She collaborated with Ruthven and her sister, Crystal Thompson, who also worked for the museum, in writing the book "Herpetology of Michigan" (1912, second edition 1928).

Gaige's primary impact on American herpetology, besides her curation of a major collection and the day-to-day training of doctoral students, was as an editor. At the 1929 meeting of the American Society of Ichthyologists and Herpetologists, held in Ann Arbor and only the second (after Chicago in 1922) to be held away from the East Coast, Ruthven persuaded the officers of the society to move their main offices to the University of Michigan, including the editorship of the journal, *Copeia*. Accordingly, in 1930 Gaige became the Herpetological Editor, and Carl L. Hubbs, an ichthyologist also on the museum's staff, Ichthyological Editor. The journal was instantly transformed from a consecutively-numbered series of somewhat unprofessional looking pamphlets (although previously well edited by Emmett R. Dunn) to the larger-format quarterly journal that it is today. Gaige's role in this metamorphosis was paramount. In 1937 she became Managing Editor of *Copeia* (Karl P. Schmidt took over her duties as Herpetological Editor), then Editor-in-Chief in 1946, holding that post through 1949, several years after she retired from the university. She thus served as editor for 20 years, a record still unmatched for that journal, and a grateful society elected her Honorary President for life in 1946.

Personally shy and self-effacing and always preferring to be inconspicuous, Gaige nevertheless had a major influence in developing the herpetology program at the University of Michigan and also in the maturation of *Copeia* into the premier journal of herpetology in the world. She retired as Curator of Amphibians in 1945 and was succeeded by her former student, Charles F. Walker. She later moved to Gainesville, Florida, where she died on 24 October 1976.

• *References*: "Helen Thompson Gaige," by A. G. Ruthven, Herpetologica, 1: 1-3, 1936; "Helen Thompson Gaige . . . 1890-1976," by J. R. Bailey, Copeia, 1977: 609-611, 1977. • *Portrait* (about 1935) *and signature* (1932): Adler collection.

TAYLOR, Edward H. (1889-1978).

Edward Harrison Taylor, American herpetologist, field collector, and taxonomist, was born in Maysville (De Kalb Co.), just north of Kansas City, Missouri, on 23 April 1889. At the age of five, his mother took him to visit relatives in Colorado, an "unbelievable wonderland," he once said, compared to life in a small Missouri town. This experience and his love for travel books inspired Taylor to a life of exploration coupled with an interest in reptiles that began in early boyhood. He attended the University of Kansas (B.A. 1912), where he studied zoology and paleontology, but on graduation he obtained a Civil Service position in the Philippine Islands, which in 1898 had become an United States-administered territory.

Taylor's first adventure began on Mindanao Island, where he was a school teacher for children of the primitive Manobos tribe. Later, he was transferred to Negros Island for similar duties. In his free time he eagerly collected vertebrate specimens and, in 1915, published his first herpetological paper. He briefly returned to the University of Kansas to complete his Master's thesis (M.S. 1916), and then soon went back to the Philippines, as Chief of Fisheries at the Bureau of Science in Manila. In this new position Taylor could freely explore and collect specimens throughout the Philippines, and this resulted in numerous papers.

At the end of World War I, Taylor was sent to Siberia for secret intelligence duties on behalf of the U.S. Government, and in 1920 he returned to the United States. From April 1920 to July 1921, Taylor made an extensive collecting trip throughout the Philippines, and later became head of the zoology department at the university in Manila. He published comprehensive monographs on Philippine amphibians and turtles (1921), snakes (1922), and lizards (1922); all three were reprinted in 1966.

In 1926 Taylor joined the faculty of the University of Kansas. The next year he completed his doctoral thesis (Ph.D. 1927) on Philippine mammals, which was later published as a book (1934). His research interests shifted to American species, especially to lizards of the genus *Eumeces*, on which he published a major monograph (1935 [1936]). At that time the herpetofauna of Mexico was poorly known and Taylor, together with his graduate student, Hobart M. Smith, decided on a major program of field work in that country with the object of producing monographic summaries. During the period 1933-1952 Taylor published 78 titles on the Mexican herpetofauna. Included were three comprehensive treatises: snakes (1945), amphibians (1948), and the remaining reptile groups (1950), each co-authored with Smith and containing keys and checklists; all three titles were reprinted in 1966 in a single volume.

Even before the Mexican program was completed, Taylor shifted his attention to Costa Rica. Emmett R. Dunn had collected extensively there, as part of a project on the herpetology of Lower Middle America. Building on these collections and Dunn's never-published manuscript, Taylor proceeded to make his own collections. His 14 titles on Costa Rican herpetology covered the period 1948-1958 and provided the first major review of the herpetofauna of that small but species-rich country. Once again, however, Taylor's main interest shifted, this time back to Asia where, during World War II, he had been involved in intelligence-gathering activities for the U.S. Government in Southeast Asia, Java, and India (1944-1946). His focal country now was Thailand, where several decades earlier Malcolm A. Smith had begun the study of the Siamese herpetofauna. From 1958 to 1971 Taylor published six titles on Thailand, including four extensive monographs that cover all amphibians and reptiles. In addition to these principal regions of interest, Taylor published papers on the herpetology of China, India, South America, Africa, Australia, and especially Ceylon. He also wrote numerous papers on fossil amphibians, turtles, and lizards.

Taylor's final topic of study was caecilians. Beginning in 1960 he began a series of smaller studies, mostly descriptions of new species, and eventually this work led to his largest book, "The Caecilians of the World" (1968). By this time Taylor was 79 years of age and had retired from his professorship at the university (1959); his phenomenal memory, on which he long had relied for locality records, species characteristics, and bibliographic references, had begun to fail. Nevertheless, as with all of the major research topics of his career—the Philippines, *Eumeces*, Mexico, Costa Rica, Ceylon, Thailand, and finally the Gymnophiona—Taylor provided the all-important foundations on which modern studies have been built.

Above all, Taylor was a field collector. It is calculated that he spent a total of 23 years in the field, logging almost a half million kilometers of which a total of nearly one year was spent at sea travelling by ship. Collecting was his first love and for which he had extraordinary ability, endurance, and a keen sense of where to look. His enthusiasm to collect, however, led him to trust in his memory, and specimens were sometimes not tagged until the end of the day, if then, even when he was conducting altitudinal transects. He named some 500 herpetological species, mostly from specimens he had collected. Once the collections had been studied, they were sold to various American museums, in order to finance the subsequent expeditions.

Taylor's main interest was exploring new areas and unknown taxa, for well-known ones held little interest for him. His keen taxonomic eye not only identified numerous new species, but also succeeded in solving important taxonomic issues. His partitioning of *Eumeces fasciatus* into three species was widely questioned but later accepted. He was the first to properly appreciate the true species diversity of caecilians and Neotropical salamanders, and he established the families Centrolenidae and Anomalepididae, to mention some of his most important contributions.

Throughout his long tenure at the University of Kansas Taylor was one of the university's most colorful teachers and he had several graduate students, including Irwin L. Baird, Claude W. Hibbard, Wayne L. Reeve, Hobart M. Smith, Wilmer W. Tanner, and Robert G. Webb, but his strong personality and presence influenced numerous others who studied at Kansas. Taylor served briefly as Curator of

Herpetology in the Museum of Natural History (1946-1954), which ended after a long, bitter dispute with the museum's director. His research continued to the very end of his long life and his last paper appeared the year he died. All told, he published 202 titles comprising nearly 9500 pages of text and illustrated with more than 2150 figures. He died in Lawrence, Kansas, on 16 June 1978, at the age of 89.

• *References*: "Edward H. Taylor: Recollections of an Herpetologist," essays by E. H. Taylor, A. B. Leonard, H. M. Smith, and G. R. Pisani, Monogr. Mus. Nat. Hist. Univ. Kansas, 4: (vii), 1-159, 1975; "Edward Harrison Taylor 1889-1978," by W. E. Duellman, Copeia, 1978: 737-738, 1978; "Edward Harrison Taylor 1889-1978," by R. G. Webb, Herpetologica, 34: 422-425, 1978; "Edward H. Taylor (1889-1978)," by J. Villa, Brenesia, 16: 1-8, 1979; "Edward Harrison Taylor, 1889-1978," by H. M. Smith, Herpetol. Rev., 10: 47-50, 1979. • *Portrait* (about 1936) *and signature* (1971): Adler collection.

BARBOUR, Thomas (1884-1946).

Thomas Barbour, herpetologist, zoogeographer, and benefactor of American herpetology, was born on the island of Martha's Vineyard, Massachusetts, on 19 August 1884. His father, owner or director of several companies, was a very wealthy man and Thomas Barbour's inheritance was used generously to further the many aims of his life's work. While spending the winter of 1898 in Florida and the Bahamas to recuperate from typhoid fever, he developed interests in reptiles, the tropics, and island faunas that set the course of his career.

The Barbours then lived in New York City, and Thomas soon began to build a collection of preserved reptiles, largely from carcasses of specimens saved for him by the zoo; the donation of this collection was his entrée to the Museum of Comparative Zoology when he entered Harvard College as a student. Barbour graduated from Harvard in 1906 and continued on for graduate study (A.M. 1908, Ph.D. 1911) but in about 1910, while he was still a graduate student, he took over responsibility for the museum's collection of amphibians and reptiles from Samuel Garman whose interests then lay more with fishes.

In 1911, Barbour became Associate Curator of Reptiles and Amphibians at the museum, and thus began the most rapid growth of that great herpetological collection. During his tenure the number of species tripled, as the result of Barbour's own collecting in various parts of the world and an aggressive exchange program with other institutions, but especially through his personal purchases. Money, of course, was no object during those pre-Depression days, and the accumulations piled up, although by the early 1920s Barbour had little time to curate them and the collection fell into disarray. Fortunately, help arrived in 1924 when Arthur Loveridge was hired as curator.

Thus freed from curatorial chores, Barbour's interests shifted. In 1923 he became the executive officer in charge of development of Barro Colorado Island, formerly a forested hill that was made into an island when Gatun Lake was formed during construction of the Panama Canal. Barbour recognized the potential of this site—now the Smithsonian Tropical Research Institute—for research purposes. He personally bought out the banana growers, hired the first superintendent (James Zetek), and under his supervision the first laboratory building was erected. Barbour continued as executive officer until 1945.

Another of Barbour's interests was the development of domestic production of snake antivenin, which at the time did not exist in the United States. Thus, in 1926 the Antivenin Institute of America was founded by Barbour, Raymond L. Ditmars, and other Americans, with Afrânio do Amaral acting as consultant. This organization, involving a venom-collection serpentarium in Honduras and an antivenin-producing laboratory in Pennsylvania, also published the short-lived journal *Bulletin of the Antivenin Institute of America* (1927-1932, reprinted 1973) which Barbour funded and co-edited. His money was also used to support another journal, *Copeia*, and for many years he made up its operating deficits.

In 1927, Barbour succeeded Samuel Henshaw as Director of the Museum of Comparative Zoology, and he continued in that position the rest of his life. He liberally supported museum projects from his personal funds and, in fact, he never drew a salary. During his long tenure at the university he had numerous students. Many herpetologists began their careers under his guidance, including Archie F. Carr, P. E. P. Deraniyagala, Emmett R. Dunn, and G. Kingsley Noble, and there were many non-herpetologists, too, who benefited from his advice and financial support.

For all his influence in the field, Barbour's own research

was rather diffuse and sometimes superficial. His main interests were in systematics and zoogeography, especially of island faunas, on which he made substantive contributions. His first two monographs covered the amphibians and reptiles of the East and West Indies (1912, 1914), and the largest fraction of his more than 200 herpetological titles (over the period 1901-1946) dealt with the West Indies, an area in which Harvard herpetologists have long had an interest beginning with that of Samuel Garman. The study of lizards of the genera *Anolis* and *Sphaerodactylus* was a particular favorite of Barbour. In addition to the West Indies, his other areas of interest included East Asia and the Pacific, Central and South America, and Africa.

North America held little attraction for him, but together with Leonhard Stejneger, the person who most influenced his scientific career, Barbour co-authored five editions (1917-1943) of checklists of North American amphibians and reptiles which were instrumental in stabilizing nomenclature. His other major works included "Herpetology of Cuba" (1919) with Charles T. Ramsden, "*Sphaerodactylus*" (1921), a monograph on the herpetology of a mountainous region in Tanganyika (now Tanzania), with Arthur Loveridge (1928), and "Antillean Terrapins" (1940) with Archie F. Carr. Barbour also published a successful semi-popular book, "Reptiles and Amphibians and Their Adaptations," in 1926 (revised edition 1943).

Despite his many significant accomplishments and the many honors bestowed upon him—including honorary doctorates from four universities and membership in the U.S. National Academy of Sciences—in some respects Barbour was more a wealthy amateur than a serious professional. His book "Naturalist at Large" (1943), one of four autobiographical books completed shortly before his death in Boston on 8 January 1946, gives a glimpse into his complex life.

• *References*: "Naturalist at Large," by T. Barbour, Little, Brown, Boston, xii, 314 pages, 1943; "Thomas Barbour—Herpetologist," by A. Loveridge, Herpetologica, 3: 33-39, 1946; "Thomas Barbour 1884-1946," by E. R. Dunn, Copeia, 1946: 1-3, 1946; "Obituary: Thomas Barbour," by A. Loveridge, Proc. Linn. Soc. London, 158: 63-64, 1947; "Thomas Barbour 1884-1946," by H. B. Bigelow, Biogr. Mem. U.S. Natl. Acad. Sci., 27: 13-45, 1952; Schmidt, 1955 (pp. 611-612); "Thomas Barbour," by A. S. Romer, Syst. Zool., 13: 227-234, 1964; "Thomas Barbour," by A. S. Romer, Dict. Amer. Biogr., suppl. 4: 51-53, 1974. • *Portrait* (1938): By Raymond B. Cowles, courtesy Charles M. Bogert. • *Signature*: Adler collection.

FEJÉRVÁRY, G. J. von (1894-1932).

Géza Gyula Imre Fejérváry von Komlós-Keresztes (or in its German form, Géza Julius Emerich Freiherr, or baron, von Fejérváry), Hungarian herpetologist and evolutionary biologist, was born in Budapest on 25 June 1894, of a noble family. After private schooling in Switzerland, Vienna, and Budapest, in 1913 he became a volunteer assistant to Lajos Méhelÿ at the Hungarian National Museum while he attended the University of Budapest, specializing in zoology and medicine. When Méhelÿ left to become a professor at the university in 1915, Fejérváry unofficially succeeded him as curator of herpetology at the museum, a position that was made permanent in 1916 after he graduated. In 1917, Fejérváry completed his Ph.D. at the University of Budapest, under Méhelÿ's supervision. Most of Fejérváry's short career was spent at the National Museum until 1930 when he moved to the university in the town of Pécs, his boyhood home. He died suddenly on 2 June 1932, following surgery in Budapest.

Fejérváry was a precocious and brilliant student. His first published title, a very respectable herpetological study of a transect in the Rhône Valley of southwestern Switzerland, issued when he was only 15, was prepared with the help of his friend, S. J. Bolkay. Eventually he published 163 titles, of which about 60 were herpetological. Like his teacher, Lajos Méhelÿ, he was interested in systematics and distribution of Hungarian species, especially *Vipera* and *Lacerta*, but in addition, he published extensively on frog osteology and phylogeny and on fossil lizards of the families Varanidae and "Megalanidae," writing several major monographs on both topics. He also published a few papers on exotic species, based partly on his own collections from Switzerland, eastern Europe, Malta, and Morocco. Most of these were published prior to 1928 after which his attention turned increasingly to broader questions of evolutionary biology. Upon his death in 1932, his wife, A. M. Fejérváry-Lángh, who had also assisted at the museum and had taken a Ph.D. in paleontology (Budapest 1919), succeeded him as curator and continued his work on frog osteology.

• *References*: "Baron Géza Gyula Imre Fejérváry von Komlós-Keresztes," by J. Versluys, Verhandl. Zool.-Bot. Gesellsch. Wien, 82: 28-33, 1932; "Die wissenschaftliche und literarische Tätigkeit von Géza Gyula Fejérváry auf dem Gebiete der Zoologie," by O. G. Dely, Vertebr. Hungar., 10: 13-43, 1968, and supplement by I. Boros and O. G. Dely, same volume: 45-142, 1968. • *Portrait*: Courtesy Oliver G. Dely. • *Signature*: From Versluys, 1932.

SMITH, Malcolm A. (1875-1958).

Malcolm Arthur Smith, born at New Malden, Surrey, in 1875, was an amateur herpetologist who, though a trained physician, spent most of his life studying the herpetofauna of Southeast Asia and India and, toward the end of his life, that of his native Britain. He showed a deep interest in amphibians and reptiles as a child and, it is said, he went on to take a medical diploma (M.R.C.S., L.R.C.P., Charing Cross Hospital, London, 1898) as the only practical means then available to ensure continued study of his beloved animals and still earn a decent living.

After practicing medicine in London, in 1902 Smith went to Siam (now Thailand) as medical officer to the British Legation in Bangkok and shortly thereafter he took on the additional duties as doctor to the Royal Court. He spoke Thai fluently and was much loved by the Royal family; indeed, he became a trusted confidant of the king and queen. Smith's unique experiences within the Royal household were written up by him as a book, "A Physician at the Court of Siam" (1947). He co-founded the Natural History Society of Siam, initiated the Society's *Journal* in 1914, and his numerous articles in it, which began that same year, became the backbone of the journal which he also co-edited during the period 1914-1924.

In 1925, Smith retired from medical practice and returned to England where he was given bench space at the British Museum (Nat. Hist.) to facilitate his personal research. George A. Boulenger had retired from curatorship of the museum's herpetological collection by that time, but Smith had long been under his influence through regular correspondence while he was still in Siam, and they had even collaborated on two papers. Smith's interest in Siam continued, but on returning to London the scope of his publications expanded greatly. Prior to leaving Siam he had published papers on the herpetofauna of the Malay Peninsula, Borneo, Hainan Island, and Indochina, based largely on his own collections, but after 1925 he also published regularly on India, Burma, the Indoaustralian region, and Australia, and beginning in 1940 on Afghanistan and Nepal. He also wrote a series of papers on snake venoms and reptilian anatomy. Ironically, he began publishing on the herpetofauna of Britain very late in his career (from 1949), yet it was here that his interest began.

During this long period at the museum, Smith gave guidance and encouragement to younger workers, among them Angus d'A. Bellairs, who co-authored two of his earliest papers with him. Smith was compiler of the herpetological section of the *Zoological Record* (1929-1945) and Editor of the *Record* from 1938 to 1949. While Zoological Secretary of the Linnean Society, he was personally responsible for safeguarding their collections at the outbreak of World War II. In 1947, Smith became founding President of the British Herpetological Society, serving in that capacity until 1954. He even served as Curator of Reptiles at the London Zoo for a short time (about 1937). He died on 22 July 1958 at his home at Ascot, near Windsor.

Smith's first book, "Monograph of the Sea-Snakes (Hydrophiidae)" (1926; reprinted 1964), covered a group of reptiles whose greatest species richness is in Southeast Asia and which had interested him for a long time. In 1930, he published "The Reptilia and Amphibia of the Malay Peninsula," intended to supplement and update George A. Boulenger's book on the same region published in 1912.

Smith then shifted his attention to India, where his father had been a civil engineer. Thus began Smith's most important work, a three-volume series on the reptiles of the Indian and Indochinese regions in the "Fauna of British India" series. At that time, "India" meant Imperial India and included what is today not only India but Pakistan, Nepal, and Bangladesh, plus Sri Lanka and Burma. In addition, Smith extended the limits to cover Indochina as far south as the Isthmus of Kra in the Malay Peninsula and into southern China as far east as Hong Kong. The first volume on crocodiles and turtles appeared in 1931, followed by those on lizards (1935) and snakes (1943); these are still the most definitive works on the region and have been reprinted. The companion volume on amphibians was never completed, but the void was partly filled by René Bourret's book on Southeast Asian amphibians (1942).

In 1951, Smith's "British Amphibians and Reptiles" was published, with a revised edition in 1954. This was a model study, full of life history details, that brought Smith full circle to the interests of his boyhood days in Surrey.

• *References*: "Dr. Malcolm A. Smith," by A. G. C. Grandison, Nature, 182: 697, 1958; "M. A. Smith, M.R.C.S., L.R.C.P.," anonymous, Brit. Med. Jour., 1958(2): 515-516, 1958; "Malcolm Arthur Smith," by A. d'A. Bellairs, Proc. Linnean Soc. London, 170(2): 130-131, 1959; "Dr. Malcolm Arthur Smith, F.L.S., F.Z.S.," testimonials and bibliography by W. P. C. Tenison, J. C. Battersby, A. d'A. Bellairs, R. H. Ahrenfeldt, and H.R.H. Prince Chula of Thailand, Brit. Jour. Herpetol., 2: 136-148, 1959; 2: 186-187, 1960. • *Portrait* (1953): British Museum (Nat. Hist.), courtesy A. F. Stimson. • *Signature* (1944): Adler collection.

LUTZ, Adolpho (1855-1940).

Adolpho Lutz, public health scientist and naturalist, was born in Rio de Janeiro on 18 December 1855, of Swiss parentage. He developed an interest in natural history as a young boy, but for practical reasons decided on a medical career and completed his formal medical education at the University of Berne in 1880; he also had advanced training in several other European universities. He returned to Brazil in 1881 and set up medical practice in São Paulo. Except for the period 1889-1892 when he studied leprosy in Hawaii, Lutz spent the rest of his career in Brazil, where he became one of his country's greatest scientists and laid the foundation for tropical medicine, medical zoology, and parasitology. He was instrumental in identifying numerous tropical diseases and effecting their eradication or control. Despite violent opposition from the medical community, he prescribed methods of control for yellow fever and even had to demonstrate that the disease was mosquito-borne by using himself as a subject, before that opposition was withdrawn.

Lutz became Director of the Bacteriological Institute in São Paulo and his pioneering work there on snake venoms led to the establishment, at his instigation, of the Instituto Butantan. In 1908 he became Chief of the Medical Zoology Division at Instituto de Manguinhos (later called Instituto Oswaldo Cruz), on the condition that he would be free of administrative duties to devote more time to scientific matters and to natural history. Lutz's services to tropical medicine were widely honored in his country and abroad. He won his country's highest honor, the Einstein Prize, and in 1955 Brazil issued a postage stamp with his likeness. He died in Rio de Janeiro on 6 October 1940.

Although he published a few papers on snakes (1922-1923), Lutz is best remembered herpetologically for his work on the taxonomy and life histories of frogs (publications in 1924-1939), an interest that sprung from his research on amphibian parasites. Even at an advanced age he travelled widely in Brazil, including remote regions in the interior, to collect and observe frogs. Lutz was an excellent field naturalist, but his published descriptions of frogs were poor, saved only by the good illustrations which allow modern workers to know which species were involved. In this work he had the enthusiastic collaboration of his children: Bertha Lutz, naturalist at the Museu Nacional in Rio, with whom he co-authored his last papers on frogs and who continued his herpetological research program after his death, and his son Gualter A. Lutz, a professor of medicine in Rio, who was an outstanding nature photographer.

• *References*: "Adolpho Lutz," anonymous, Copeia, 1940: 275-276, 1940; "Adolpho Lutz (1855-1940)," by H. W. Stunkard, Jour. Parasitol., 27: 469-471, 1941; "Adolpho Lutz Zoólogo," by A. do Amaral, Rev. Inst. Adolpho Lutz, 15: 86-99, 1955; "Adolpho Lutz (1855-1955)," anonymous, Comis. Centen. Adolpho Lutz, Consel. Nac. Pesquisas, Rio de Janeiro, 55 pages, 1956. • *Portrait*: Howard K. Gloyd collection, courtesy Kathryn J. Gloyd and Roger Conant. • *Signature* (1930): British Museum (Nat. Hist.), courtesy A. F. Stimson.

MIRANDA RIBEIRO, Alipio de (1874-1939).

Although principally recognized as South America's foremost ichthyologist of his era, Miranda Ribeiro published extensively on all vertebrate groups. He was born on 21 February 1874 in the city of Rio Preto, Minas Gerais, Brazil, and showed an inclination to natural history at the age of 14. After attending Malvina Reis College in Rio de Janeiro, he went to medical school in Rio and concurrently began work at the Museu Nacional as a preparator (in 1894). In 1899, he became Secretary of the museum and later, after various promotions, Professor and Head of the Department of Zoology (beginning 1929), a post he held until his death, in Rio de Janeiro, on 8 January 1939. He participated in several expeditions, including the famous ones with Colonel Rondon to lay the first telegraph lines through Amazonas and the Matto Grosso, as Zoologist of the Telegraph Commission Survey. In 1911 he visited European and American museums, principally to examine Brazilian types, but he also studied fisheries programs. After his return in 1912, he became Inspector General of Fisheries and established the first Brazilian fisheries program.

Miranda Ribeiro's principal book, in several volumes and extending to nearly 2,000 quarto pages, was his "Fauna Brasiliense: Peixes," the most important faunal study on South American fish. Although he wrote a few short herpetological titles previously, it was in 1920, with an invitation to come to the Museu Paulista in São Paulo to arrange their frog collections, that his long series of titles on frogs began. His most monumental herpetological work was "Notas para Servirem ao Estudo dos Gymnobatrachios (Anura) Brasileiros," in quarto format with 22 color plates, published in 1926. Miranda Ribeiro's generic definitions generally have not been accepted, but he monographed the Brazilian species and discovered many new ones. Later, his museum's publication funds exhausted, he had to turn to a Brazilian farm journal, *O Campo*, which had a very limited circulation, as an outlet for his papers; fortunately, the Museu Nacional in Rio issued a reprint of most of them in 1951. Miranda Ribeiro's assistant, Antenor Leitão de Carvalho, succeeded him as curator of the herpetological collections at Museu Nacional, and his son, Paulo de Miranda Ribeiro, carried on his ichthyological research.

• *References*: "Dr. Alipio de Miranda Ribeiro," by G. S. Myers, Copeia, 1939: 184, 1939; "Alípio Miranda Ribeiro," by J. Kretz, Impr. Oficial Estado São Paulo, 24 pages, 1942; "Prof. Alípio de Miranda Ribeiro," by L. Travassos, Arq. Mus. Nac. Rio, 41: xi-xxxvi, 1951. • *Portrait* (1912): George S. Myers collection, California Academy of Sciences, courtesy Alan E. Leviton. • *Signature* (1926): Adler collection.

AMARAL, Afrânio do (1894-1982).

Afrânio Pompílio Bastos do Amaral, Brazilian specialist on snakes, venoms, and antivenin, was born in Belém, Pará, on 1 December 1894. As a boy he collected snakes for Emílio Goeldi in Belém, before moving to Salvador, Bahia, where he completed his bachelor's degree (1911). He received his doctor's degree from the Faculdade de Medicina da Bahia (M.D. 1916); his thesis dealt with parasitic worms. After a brief medical practice he moved to São Paulo in 1917 where he began his lifetime's career at the Instituto Butantan, under the supervision of Vital Brazil, João Florencio Gomes, and Arthur Neiva. On the premature death of Gomes, Amaral was put in charge of the snake section (1919), and published many papers under his name that were taken from Gomes's notes and unfinished manuscripts. In 1921 Amaral succeeded Brazil as director of the institute.

At that time, Brazil was one of the few countries that produced antivenins, including all antivenins used in the United States. In an effort to begin its own domestic production, Thomas Barbour, Raymond L. Ditmars, and other Americans invited Amaral to the United States for discussions, which in 1926 led to the establishment of the Antivenin Institute of America (a division of Mulford

Biological Laboratories, now Merck Sharp & Dohme, at Glenolden, Pennsylvania). Amaral, on leave from Butantan, was its first director; he also had a lecturing position at Harvard University School of Public Health, where he received a D.P.H. degree in 1924. The Antivenin Institute was a cooperative venture between Mulford, Harvard, and the United Fruit Company, which established a serpentarium at Tela, Honduras, under Douglas March; the venoms were extracted in Honduras, and Amaral personally supervised the production of antivenins at Glenolden. In 1927, Amaral initiated and co-edited the *Bulletin of the Antivenin Institute of America*, a quarterly journal devoted to the biology of reptiles (ceased publication in 1932; series reprinted 1973).

In 1928, Amaral was recalled to his directorship at Instituto Butantan. He took on additional duties as Director of the Department of Hygiene, São Paulo School of Medicine, in 1933. In about 1935, political difficulties led to his removal as director of Butantan, but in 1954 he was reinstated by judicial decree. Besides his herpetological work, he also published on serology, venoms, parasitology, and nutrition chemistry; his work in the last-named area was awarded the Brazilian National Prize of Nutrition in 1950. Amaral was a member of numerous national and international organizations, including the International Commission of Zoological Nomenclature. He died in São Paulo on 29 November 1982.

Amaral published over 450 titles, beginning in 1915, of which 207 are on herpetology, including snake venoms. Originally his herpetological papers were exclusively on snakes, but starting in 1932 he also published on lizards. His publications dealt primarily with South America, especially Brazil, and also included occasional ones on North and Central America. The number of titles is deceiving, however, since many of them, beginning in about 1945, were republications or rearrangements, which causes confusion in citation.

All told, Amaral described 15 new genera of snakes and lizards and about 40 species. He also described numerous subspecies, but his subspecies concept was at variance with modern views and many of his names are not used today. His major herpetological works include "A General Consideration of Snake Poisoning and Observations on Neotropical Pit-Vipers" (1925), checklists of Brazilian snakes (1929; second edition, 1937) and lizards (1937), and, in 1965, his "Iconografia Colorida das Serpentes do Brasil," in four volumes (reissued as "Serpentes do Brasil" in 1977, in parallel Portuguese and English texts, with numerous color plates).

• *References*: "Breve Noticia Sobre a Vida Científica de Afrânio do Amaral," by E. C. Falcão, Mem. Inst. Butantan, 39: 3-25, 1975 (1976); "Homenagen a Afrânio do Amaral (1894-1982)," by J. C. de M. Carvalho, Sessão Solene Acad. Brasil. Cienc., Rio de Janeiro, pp. 37-39, 1983; "Edicion Dedicada . . . el Profesor Doctor Afranio do Amaral," by F. Sandner Montilla, Mem. Cientif. Ofidiol., Inst. Venezol. Ofidiol., 7: 1-77, 1985. • *Portrait* (1923): Courtesy Library, Museum of Comparative Zoology, Harvard University. • *Signature* (1926): Adler collection.

NICÉFORO MARÍA, Brother (1888-1980).

Brother Nicéforo María, one of Colombia's leading naturalists, was born Antoine Rouaire Siauzade in Briade (Haute Loire), France, on 29 February of leap-year 1888. Following religious training in Luxembourg and Belgium (1905-1907), he joined the Marist Order of Catholic Brothers, a religious order founded for missionary and educational work in the name of the Virgin Mary, hence his adopted name.

Nicéforo María went to Colombia in 1908 and taught first at the Colegio de San José in Medellín (Antioquia), where he began a museum of natural history, and in 1922 he was transferred to the Instituto de La Salle in Bogotá and spent the remainder of his life there. He eventually became director of the natural history museum at the institute and conducted field work throughout Colombia to collect specimens—especially lower vertebrates—for the museum. Although he had long before given up expeditions, he was actively curating the collection even in his final years when, at the age of 92, he died in Fusagasugá, near Bogotá, on 24 February 1980.

Nicéforo María was interested in all vertebrates and also in crustaceans and arachnids, but he specialized in amphibians and reptiles (publishing 38 titles over the period 1924-1970). His herpetological papers cover all groups and emphasized snakes and turtles; his major work on reptiles, "Los Ofidios de Colombia," was published in 1942. Besides his own research, he was extraordinarily helpful to other scientists, by sending them collections of Colombian material or by facilitating their field work in Colombia. In gratitude, he received many honors, including the naming of no fewer than 21 species of amphibians and reptiles with the epithets *mariarum* or *nicefori,* and a genus of leptodactylid frogs (*Niceforonia*) was also named for him. In 1961 he was awarded an honorary doctorate by the Universidad Católica de La Salle.

• *Reference*: "Brother Nicéforo María," by J. R. Tamsitt, Copeia, 1980: 952-953, 1980. • *Portrait*: Museo de La Salle, courtesy Brother Roque Casallas Lasso. • *Signature* (1953): Smithsonian Institution Archives, courtesy William Cox and Alan E. Leviton.

NOBLE, G. Kingsley (1894-1940).

Gladwyn Kingsley Noble, American herpetologist and pioneer student of animal behavior, was born in Yonkers, New York, on 20 September 1894. His father was a co-founder of the book publishing firm of Barnes and Noble. His earliest interest was in birds, and as a schoolboy he was involved in banding studies with gulls. As a student at Harvard College (A.B. 1917, A.M. 1918) he was a protégé of Thomas Barbour who instilled in him an interest in field studies and trained him in systematics. Noble's first papers, six of them co-authored with Barbour, dealt with taxonomy, geographic distribution, and faunistics. Because of the war in Europe he had attended Officer's Training School while at Harvard, and he enlisted in the navy in 1918. With a commission as Ensign, he served in Washington working with codes in the Office of the Chief of Naval Operations.

After his service in the navy, Noble entered Columbia University as one of William K. Gregory's group of graduate students which at that time included Charles L. Camp and Alfred S. Romer. Noble received his Ph.D. degree in 1922. Previously he had been hired by Mary C. Dickerson as Assistant Curator at the American Museum of Natural History (1917). Upon Miss Dickerson's death in 1923, Noble succeeded her as Curator of Herpetology. As a student of Gregory, Noble's interests naturally shifted to osteology and myology and his thesis (published in 1922 as "The Phylogeny of the Salientia") became his first major work. In it, he expanded the seminal work of Edward D. Cope on osteology and G. E. Nicholls's (1916) system of classification based on articulations of the vertebrae, to which he added his own original work on the arrangement of thigh muscles and tendons. Other papers during this period continued his systematic interests, including a major work, the Amphibia section (1924) of the series "Contributions to the Herpetology of the Belgian Congo," for which Karl P. Schmidt wrote the two parts on reptiles.

In about 1923 Noble's primary interest shifted to studies on life history and its relationship to phylogeny. Among other aspects, he began experiments on the effect of thyroid extracts on metamorphosis, for which research he attended advanced lecture courses on endocrinology at several universities. The first published results concerned the production of cloacal glands in female salamanders by testicular transplants (1926). Thus began an important new phase of experimental work which led, in 1928, to his founding of a new Department of Experimental Biology (later re-named Department of Animal Behavior, but disbanded about 1980) at the museum. Noble's philosophy of a modern museum was that it should not only display objects but also inform the public concerning underlying principles and raise important questions. He designed the museum's Hall of Reptiles and its Hall of Animal Behavior, which were both devoted to this philosophy.

At that time, animal behavior studies in America were the province of psychologists and the evolutionary approach of ethologists was confined almost exclusively to Europe. Noble was practically its sole practitioner in America, where his work was not fully understood or appreciated. His research expanded to cover studies on courtship and mating behavior, including dominance hierarchies, in all vertebrate classes. Bird behavior, which had been his first interest in biology, became a renewed area of interest at this time.

Somewhat later, Noble's interests turned to the nervous system, which required his attendance in more courses, on neurology, and led to his brilliant studies on the sensory physiology of reptiles. Among the most notable of these were the experiments which demonstrated the infrared-sensing ability of the facial pits of snakes, the sensory organs involved in lizard and snake courtship and in aggregation behavior, the sensory basis of orientation in hatchling sea turtles, and the function of Jacobson's organ. Together with his earlier studies in systematics, biogeography, anatomy, life history, and endocrinology, Noble's scope of interest and his insight were staggering, perhaps greater than that of any herpetologist of his era. His mastery of such a broad-based biology of amphibians and reptiles was demonstrated in his classic *magnum opus*, "The Biology of the Amphibia" (1931, reprinted 1954). His knowledge of and ability to integrate such diverse areas of amphibian biology is still unsurpassed.

Despite Noble's preoccupation with experimental studies, the more traditional work of museum-based herpetology was not neglected. He continued to publish taxonomic papers, and by the time of his death the collection of amphibians and reptiles at the American Museum exceeded 110,000 catalogued specimens, most of them acquired during Noble's tenure as Curator of Herpetology. He had many able assistants over those years, including Charles M. Bogert, Carl F. Kauffeld, Clifford H. Pope, and Karl P. Schmidt. It was Bogert who succeeded Noble as curator and continued his tradition of research.

Noble's abrasive personality and tremendous ego are well known, yet he gave credit where it was due and co-authored numerous papers with his assistants. He was prone to controversy and became involved in several public debates, primarily with American colleagues such as Emmett R. Dunn. Because of Noble's interest in amphibian life history and endocrinology, he was naturally drawn into the debate over the alleged inheritance of acquired characteristics in the European midwife toad, *Alytes*. Paul Kammerer claimed to have anatomical evidence that development of the black nuptial spines in the males could be explained by this inheritance mechanism, but Noble examined Kammerer's sole extant specimen in Vienna and exposed the fraud. His report on an examination of the specimen was published in *Nature* in 1926 (two papers) and led to Kammerer's disgrace; Kammerer committed suicide less than two months later, although it was never proven that he personally injected the India ink into the frog's thumb to blacken the spines.

Noble published about 180 titles over the period 1913-1940, some 140 of them on amphibians and reptiles. He received numerous honors during his lifetime, including election to membership in the American Philosophical Society (1933); he was never elected to the U.S. National Academy of Sciences, however, probably due to his premature death. His research had a profound influence on the biological study of not only amphibians and reptiles but all vertebrates, yet his professional career lasted a relatively short period of not quite two and a half decades. He died in Englewood, New Jersey, on 9 December 1940, at the comparatively young age of 47.

• *References*: "G. K. Noble," anonymous, Copeia, 1940: 274-275, 1940; "Gladwyn Kingsley Noble, 1894-1940: A Herpetological Bibliography," by W. L. Necker, Herpetologica, 2: 47-55, 1940; "Gladwyn Kingsley Noble (1894-1940)", by W. K. Gregory, Year Book Amer. Philos. Soc., 1941: 393-397, 1942; "Biographical Note," by R. C. Noble, 3 p. *in* Noble's Biology of the Amphibia, repr. ed, Dover Publ., New York, 1954; Schmidt, 1955 (pp. 613-614); "Gladwyn Kingsley Noble," by C. H. Pope, Dict. Amer. Biogr., 11 (suppl. 2): 489-490, 1958. • *Portrait*: Adler collection. • *Signature* (1921): Library, Academy of Natural Sciences of Philadelphia, courtesy Marsha Gross.

SCHMIDT, Karl Patterson (1890-1957).

For many years the dean of American herpetologists, Schmidt was a broadly-interested naturalist and animal geographer. He was born in the North Chicago suburb of Lake Forest on 19 June 1890 and, influenced by his mother, developed an early interest in nature. Being intellectually gifted, he entered Lake Forest College at the age of 16, in 1906, but withdrew the next year to help on his family's farm where he developed an interest in reptiles. Encouraged by his former biology teacher at Lake Forest who had since moved to Cornell University, Schmidt entered Cornell in 1913 and, because of his age (23), enjoyed the status of a graduate student. Geology was his official major, but he took courses in biology with his eventual mentor, Albert H. Wright. During these years he participated in numerous geological field trips in Chesapeake Bay, Louisiana, and the Dominican Republic, and his first herpetological collection, made on one of these student trips to North Carolina, became his first published paper (1916).

It was Schmidt's brief meeting in 1916 with Mary C. Dickerson, curator at the American Museum of Natural History, that led to his first herpetological position, a temporary job to study the enormous collections brought from the Belgian Congo by the Lang-Chapin expedition, but Dickerson insisted as part of the bargain that he complete his Cornell degree (B.A. 1917). Thus, except for a short period on the family farm and in the army at the very end of the war, Schmidt remained on Dickerson's staff (and later that of G. Kingsley Noble, her successor) until 1922 when he returned to his boyhood town, Chicago, to found a herpetology department at the Field Museum of Natural History, as Assistant Curator (in charge). Although alone in that department until joined by Clifford H. Pope in 1941, Schmidt succeeded in building one of the leading museum research centers in herpetology. The Field Museum sent numerous expeditions to all parts of the world and Schmidt himself participated in those to Central and South America, to the Pacific including New Guinea and the Philippines, to New Zealand, and to Israel.

In addition to his research activities, Schmidt served as herpetological editor for many publications, including *Copeia, Biological Abstracts, American Midland Naturalist, Encyclopedia Britannica*, and for the zoological series at Field Museum. An inveterate bibliophile, he built an extensive herpetological library at the museum. Schmidt was an honored member of the American scientific establishment, being elected president of several societies and, in 1956, to the U.S. National Academy of Sciences. He received an honorary D.Sc. degree from Earlham College (1952) and was elected Eminent Ecologist by the Ecological Society of America just a month before his untimely death.

Schmidt's herpetological interests were cosmopolitan and touched all parts of the world. His research was primarily taxonomic and faunistic, but also emphasized distribution and ecology. Besides his scientific writing, Schmidt wrote many semi-popular articles. He held the position of Chief Curator of Zoology at the Field Museum (1941-1955) and, at the time of his death, he had returned to the topic of his first serious herpetological work, central Africa. While examining a boomslang (*Dispholidus typus*) on 25 September 1957, he was bitten; given that it was such a small specimen and that he was struck by only a single fang, he shunned medical treatment, but died the next day at his home in Homewood just south of Chicago. Schmidt systematically recorded his reaction to the bite during that last day, as published by his colleague, Clifford Pope, in *Copeia* the next year.

Schmidt's major herpetological works include the two-volume "Contributions to the Herpetology of the Belgian Congo" (1919, 1923), the "Amphibians and Land Reptiles of Puerto Rico" (1928), based partly on his own collections there in 1919, "Field Book of Snakes of the United States and Canada" (1941), co-authored with D. Dwight Davis, Schmidt's colleague and anatomist at Field Museum, "Check List of North American Amphibians and Reptiles," (1953, with an addendum published in the 1954 volume of *Copeia*), the sixth edition in a series of checklists previously authored by Leonhard Stejneger and Thomas Barbour, and in 1957, "Living Reptiles of the World," co-authored by Schmidt's former student and successor in charge of herpetology at the Field Museum, Robert F. Inger.

In addition, Schmidt co-authored two other technical books, "Ecological Animal Geography" (1937), a translation of Richard Hesse's original text in German (a revised edition, by Hesse, W. C. Allee, and Schmidt, was published 1951), and "Principles of Animal Ecology" (1949) with Allee, Alfred E. Emerson, Orlando Park, and Thomas Park. Schmidt wrote two semi-popular books as well.

• *References*: "Karl Patterson Schmidt zum Gedächtnis," by R. Mertens, Natur und Volk, 87: 401-402, 1957; "Karl Patterson Schmidt," by C. H. Pope, Biol. Abstr., 32: xii, 1958; "Karl Patterson Schmidt 1890-1957," by D. D. Davis, Copeia, 1959: 189-192, 1959; "Karl P. Schmidt 1890-1957," by J. M. Cei, Investig. Zool. Chilenas, 4: 129-131, 1958; "In the Steps of the Great American Herpetologist Karl Patterson Schmidt," by A. G. Wright, M. Evans Co., assoc. J. B. Lippincott, Philadelphia and New York, 127 pages, 1967; "The Herpetologists—Part II," by R. Pinney, Herp, Bull. New York Herpetol. Soc., 16(1): 23-31, 1980. • *Portrait*: Courtesy Heinz Wermuth. • *Signature* (1946): Adler collection.

DUNN, Emmett Reid (1894-1956).

Emmett Reid Dunn, American herpetologist and specialist on salamanders and the Neotropical herpetofauna, was born in Alexandria, Virginia, now a suburb of Washington, on 21 November 1894. As a boy he had an interest in snakes that was encouraged by Leonhard Stejneger during visits to the nearby National Museum. Dunn's uncle had been a professor at Haverford College which influenced him to attend, and he earned two degrees there (B.A. 1915, M.A. 1916). Immediately after graduation Dunn proceeded to the American Museum of Natural History where Mary C. Dickerson, the Curator of Herpetology, agreed to sponsor him on his first expedition, to the high mountains of western North Carolina. This first serious field experience with salamanders had a lasting influence on Dunn, which culminated in his only book, "The Salamanders of the Family Plethodontidae" (1926, reprinted 1972), to this day one of the most comprehensive and influential treatments of a single family of amphibians or reptiles. Dunn's work revolutionized the biology of salamanders, a group whose systematics had not advanced since the work of Edward D. Cope in the late 1800s and whose taxonomy had been thoroughly confused by George A. Boulenger.

Dunn's first teaching post was at Smith College, an all-women's school in western Massachusetts. Because of the war in Europe, Dunn attended Officer's Training School, but the Army failed him for a commission, reportedly on the grounds that his weekend trips to collect salamanders and snakes represented "conduct unbecoming to an officer candidate." The U.S. Naval Reserve, however, commissioned him an Ensign (1917-1918), and he served on submarine patrol. Dunn returned to Smith College in 1918 where, among his colleagues, were two of the leading specialists on salamanders, Harris H. Wilder and Inez W. Wilder (née Whipple, as used on her earlier papers). Harris was an anatomist who, among other contributions, had discovered lunglessness in plethodontid salamanders (1893 [1894]); Inez is best remembered for her book on amphibian metamorphosis (1925). Within this intellectual environment, Dunn's interest was enlarged from taxonomy, but it had been clear even from his first paper (1915), on variation in a brood of watersnakes, that he was interested in broad evolutionary interpretations.

Dunn had previously published scholarly reviews of major taxa (plethodontids, 1917, with H. W. Fowler; *Desmognathus* and *Leurognathus*, 1917; Harvard University's collection of salamanders, 1918; Hynobiidae, 1923) to which he added important analytical papers (origin of lunglessness, 1920, with Inez Wilder; sound-transmitting apparatus, 1922; breeding habits and phylogeny, 1923; geographical distribution, 1923). The fertility of his mind and his breadth of interests were reminiscent of Edward D. Cope, whose career path Dunn was soon to follow. While on the staff at Smith, Dunn completed his doctorate at Harvard (1921), under the benevolent influence of Thomas Barbour, who encouraged his interests in the tropics and funded Dunn's trips to Jamaica, Cuba, and Central America. In 1926 Dunn participated in the Douglas Burden-American Museum Expedition to Komodo Island in the East Indies, which was his only field experience in the Old World.

Dunn's classic book on plethodontids appeared in 1926 and, although he continued to publish on salamanders, thereafter his interests shifted to Neotropical herpetology. In 1929, Dunn returned to teach at his *alma mater*, Haverford College, where Cope himself briefly had been a professor. In 1937 he became Honorary Curator of Herpetology at the Academy of Natural Sciences in nearby Philadelphia, where Cope had begun his own career and which possessed a large Neotropical collection, including most of Cope's types. Here was an ideal setting in which to pursue his new research interests. Dunn spent the rest of his active career at Haverford, until his death in nearby Bryn Mawr, on 13 February 1956.

Most of Dunn's field work was done in the American tropics. He made more than a dozen collecting trips to Costa Rica and Panama, and later spent over a year in Colombia (1943-1944); he published numerous papers based on his collections and became the leading authority on the amphibians and reptiles of these regions. In addition to numerous new species, his papers often provided important new classifications or discussions of biogeography, phylogeny, and adaptation. His major works of this period were a monograph of the American caecilians (1942) and a long series of 23 papers on the herpetology of Colombia (1943-1946; most of them reprinted in book form, 1957). All told, Dunn published 217 titles (1915-1964), nearly all of them on herpetology.

Dunn had a considerable literary gift, practiced partly as editor of the journal *Copeia* (1923-1936). It was best expressed in his book "Plethodontidae," particularly in his oft-quoted foreword which has set a certain romantic tradition in herpetology. Dunn's eloquent descriptions of his most famous collecting sites—Mount Orizaba in Mexico, Yonahlossee Road and Grandfather Mountain of North Carolina, and Volcan Irazú in Costa Rica, among many others—have transformed them to a classic stature in herpetological lore. Dunn's classical inclination was also manifested in his naming of new taxa, among them the Central American salamanders *Oedipus rex* and *Oedipus complex*, combinations which no longer exist since the generic name turned out to be preoccupied by a genus of fossil grasshoppers (a name itself no longer recognized).

Although Haverford College did not grant Ph.D. degrees, Dunn attracted numerous students for study leading to the bachelor's and master's degrees, including the herpetologists Joseph R. Bailey, Herndon G. Dowling, Douglas C. Robinson, and Henri C. Seibert. One would-be student, W. H. Weller, whom Dunn had advised through voluminous correspondence beginning in 1929, was a precocious naturalist with ten papers to his name by the age of barely 18. Weller had already discovered several new species of salamanders when, less than three months before he was to join Dunn as an undergraduate student, he died on the slopes of Dunn's beloved Grandfather Mountain, from a fall after he had collected additional specimens of a new species later named *Plethodon welleri* by Charles F. Walker in 1931. Dunn was also a mentor for Jay M. Savage, who worked under his supervision at the Academy of Natural Sciences in Philadelphia during 1954.

• *References*: "Dr. Emmett Dunn, Biologist, Was 61," anonymous, New York Times, 14 February 1956, p. 29, 1956; "E. R. Dunn, Herpetologist," by R. Conant, Science, 123: 975, 1956; "Emmett Reid Dunn 1894-1956," by K. P. Schmidt, Copeia, 1957: 75-77, 1957; "Emmett Reid Dunn," anonymous, Natl. Cyclop. Amer. Biogr., 43: 6-7, 1961; "Emmett Reid Dunn and the Development of Scientific Studies of Salamander Biology," by D. B. Wake, p. iii-x. *In* Dunn's Salamanders of the Family Plethodontidae, repr. ed., Soc. Study Amphib. Rept., Athens (Ohio), 1972. • *Portrait*: Library, Academy of Natural Sciences of Philadelphia, courtesy Marsha Gross. • *Signature* (1941): American Museum of Natural History, courtesy Margaret S. Shaw.

POPE, Clifford H. (1899-1974).

Clifford Hillhouse Pope, American herpetologist and writer of semi-popular books on reptiles, was born in Washington (Wilkes Co.), Georgia, on 11 April 1899. He entered the University of Georgia in 1917, but after two years transferred to the University of Virginia where he graduated (B.S. 1921). He then applied for and won a position as "Herpetologist, Chinese Division" in the Roy Chapman Andrews-American Museum of Natural History Central Asiatic Expeditions and spent the years 1921-1926 in China. The main field party worked in northern China and Mongolia where, in the Gobi, they made their most spectacular discovery—the first dinosaur eggs. Pope, on the other hand, employing only local assistants which required that he become fluent in Chinese, spent his time collecting in southern China, particularly in coastal Fukien (now Fujian) and on the island of Hainan. Large collections of vertebrates were made and regularly sent to New York City, where Pope's museum colleague, Karl P. Schmidt, wrote up the first reports.

On Pope's return from China in 1926, he began detailed study of the collections himself and published a dozen technical papers on Chinese amphibians and reptiles (1928-1934), as well as numerous semi-popular articles about his travels (beginning 1924). In 1928 he was promoted to Assistant Curator in the museum's Department of Herpetology, and about 1930 he began in earnest to work on what was to become his only technical book, "The Reptiles of China" (1935). This was the first comprehensive review of Chinese reptiles, based not only on Pope's own collections but on his examination of available museum material elsewhere; even today, despite the active herpetological research and publication taking place in China, his book is still the major synthesis on the topic. Unfortunately, Pope's companion volume on amphibians, virtually complete and with beautiful colored drawings by his native artist, Hao-ting Wang, remains unpublished; only a brief review of the amphibians was published by Pope, jointly with Alice M. Boring, in 1940.

The year 1935 was a turning point in Pope's life. His book had been well received and he was elected President of the American Society of Ichthyologists and Herpetologists, but Pope also lost his curatorship at the museum that same year when he aroused the wrath, and perhaps also the jealousy, of G. Kingsley Noble, his department chairman. Thus Pope was out of a job at the height of the Great Depression and, since he had already had some success at semi-popular writing, he decided to turn to writing as a source of income. His first new book, "Snakes Alive and How They Live," appeared in 1937, followed by "Turtles of the United States and Canada" (1939) and "China's Animal Frontier" (1940). All were well written, technically sound, and successful.

In 1941 Pope became Curator of Amphibians and Reptiles at the Field Museum of Natural History in Chicago, thus succeeding his friend and former colleague, Karl P. Schmidt, who earlier had moved to Chicago and had just become Chief Curator of Zoology at the museum. Pope's research at this time dealt with snakes and, after World War II, salamanders. In fact, salamanders had been a longtime interest since his undergraduate days at the University of Virginia, when he first studied the plethodontid salamanders of North Carolina (first paper in 1924). During the period 1947-1951 he published eight papers on the systematics and biology of plethodontids, together with his wife, Sarah H. Pope, or with Nelson G. Hairston, then a graduate student at nearby Northwestern University, in Evanston just north of Chicago.

In part to devote his time to writing, Pope resigned his position at the museum in 1953. His book "The Reptile World" was published in 1955, with an abridged version two years later ("Reptiles Round the World"). Finally, in 1961, his last book, "The Giant Snakes," was issued. He retired to Escondido, California, and later died there on 3 June 1974.

• *References*: "Clifford H. Pope," by R. F. Inger, Copeia, 1974: 1012, 1974; "Clifford Hillhouse Pope 1899-1974," by C. M. Bogert, Herpetol. Rev., 6: 22-26, 1975; "The Herpetologists—Part II," by R. Pinney, Herp, Bull. New York Herpetol. Soc., 16(1): 23-31, 1980. • *Portrait* (1924): Department of Library Services, American Museum of Natural History, courtesy Barbara Mathe. • *Signature* (1962): Adler collection.

BISHOP, Sherman C. (1887-1951).

Sherman Chauncey Bishop, specialist on American salamanders and spiders, was born on 18 November 1887 in Sloatsburg, near New York City, but soon moved to the small town of Clyde in the Finger Lakes region of western New York State. His first contact with natural history was along the banks of the old Erie Canal which passes through

Clyde. His independent spirit put him in conflict with high school authorities and he withdrew before graduating; fortunately, nearby Cornell University was willing to take him as a special student in 1909. There he had dual interests, in entomology and herpetology, and soon came under the influence of Albert H. Wright. He joined the 1912 Cornell expedition, led by Wright, to the Okefenokee Swamp along the Georgia-Florida border; this resulted in his first paper (1915), jointly with Wright, on the snakes of the Okefenokee.

Bishop graduated from Cornell in 1915 and promptly began graduate work. In 1916 he accepted a post as Zoologist with the New York State Museum in Albany, and concurrently as a zoologist for the State Department of Conservation; his primary duties were entomological. Very soon, however, World War I intervened. He enlisted in the navy in 1917 and worked for Naval Intelligence at the Charleston Naval Base in South Carolina. After returning to his position in Albany, he also continued to work on his Cornell thesis, which was completed for the Ph.D. degree in 1925. In 1928 he moved to the University of Rochester, not far from his boyhood home, where he was promoted to Professor of Vertebrate Zoology in 1933. After more than a decade of prolonged and serious illness, during which his widely-known pixy sense of humor continued unabated, he died in Rochester on 28 May 1951.

In terms of numbers of titles, Bishop was primarily an arachnologist but he also published on all vertebrate classes. His reputation as a herpetologist rests on two well-known books and a series of papers describing new taxa and the life histories of several species. Bishop named 14 taxa of North American salamanders including the distinctive species *Ambystoma mabeei*, *Batrachoseps wrighti*, *Desmognathus aeneus*, *Eurycea neotenes*, *Plethodon dunni*, and *Typhlotriton nereus*, among others, in all representing 12 different genera which indicated the breadth of his knowledge.

Bishop also wrote careful monographs on the life histories of *Hemidactylium* (1920), *Pseudotriton* (1925), and *Necturus* (1926), and this line of work culminated in his book "The Salamanders of New York" (1941) which is still the major source-book for basic natural history of the salamanders of northeastern North America; most of the data were gathered by Bishop himself. His *magnum opus*, "Handbook of Salamanders," followed in 1943, one of the series of handbooks of North American amphibians and reptiles edited by Bishop's former professor, Albert H. Wright.

Bishop had numerous graduate students; among the herpetologists were Arnold B. Grobman and Joseph A. Tihen who both earned doctorates under his supervision. Bishop was primarily a field naturalist who had a broad knowledge of plants and animals. Besides his field work in New York State, he also participated in expeditions throughout North America, including Alaska and the deserts of the Southwest and adjacent Mexico.

• *References*: "Sherman C. Bishop 1887-1951," anonymous, Proc. Rochester Acad. Sci., 9: plate and unnumbered page before p. 327, 1951 (1952); "Sherman Chauncey Bishop," by T. S. Palmer, Auk, 69: 121, 1952; "Sherman C. Bishop 1887-1951," by A. B. Grobman, Copeia, 1952: 127-128, 1952. • *Portrait* (1930s): Courtesy Beth Bishop Flory. • *Signature* (1950): Carnegie Museum, courtesy C. J. McCoy.

BLANCHARD, Frank N. (1888-1937).

Although he died prematurely at the age of 48, Frank Nelson Blanchard became the leading scholar of North American snakes since the work of Edward D. Cope. He was born on 19 December 1888 in Stoneham, Massachusetts, but his family soon afterward moved to Somerville, now a suburb of Boston, where Blanchard spent his youth. As a high school student he was interested in physics and chemistry, but he was converted to biology by his professors at nearby Tufts College, where one of his teachers was the amphibian anatomist J. S. Kingsley. Before graduating in 1913 Blanchard had already discovered two new species of algae (resulting in his first paper, 1914). Unfortunately, his father had died early in 1913 and thus he could not afford to go on directly to graduate school, so he accepted a teaching post at Massachusetts Agricultural College (now University of Massachusetts) in Amherst for three years. Here, in the countryside of western Massachusetts, Blanchard had his first opportunity for field work, which sealed his interest for a career in zoology.

In 1916 he went to the University of Michigan for graduate work and, after a year, settled on a thesis in herpetology and became Alexander G. Ruthven's first doctoral student in what was to become the leading center of graduate training in herpetology, the university's Museum of Zoology. Following in the tradition of his mentor's own thesis on the snakes of the genus *Thamnophis*, Blanchard's thesis, published in 1921, monographed the kingsnakes (*Lampropeltis*). He received his Ph.D. in 1919, but the year before he had joined the staff of the National Museum in Washington as Aide to Leonhard Stejneger.

In 1919, however, Blanchard somewhat reluctantly returned to the University of Michigan as an instructor, where they needed staff in the Department of Zoology to replace those called for war duty; Blanchard was exempted from duty due to his poor eyesight. He became an Assistant Professor in 1926 and, at the time of his death in Ann Arbor on 21 September 1937, he was Associate Professor, a title he had held since 1934. Since 1922 he had also been a member of the teaching staff of the university's Biological Station at Douglas Lake located in the wilds of northern Michigan. It was here that he began his long-term studies on snakes and salamanders, which were to become his major research area.

The first phase of Blanchard's research was on snake systematics and distribution. His treatise on kingsnakes was widely acclaimed, and he also produced careful taxonomic studies on other snake genera (*Arizona, Carphophis, Coluber, Diadophis, Natrix, Pituophis, Tantilla,* and *Virginia*). His "Key to the Snakes of the United States, Canada and Lower California" (1925, reprinted 1939) represented a prodigious amount of research on the specific characters of snakes and their variation. It established Blanchard's reputation as the preeminent student of North American snakes, which resulted in his receipt of numerous aberrant specimens from all over the country which did not fit his key, specimens which have been used in herpetology classes at the University of Michigan for years, to the bewilderment of generations of students there.

Beginning in 1922, concurrent with his appointment at the Biological Station to teach ornithology, Blanchard began the next phase of his research career, although he never totally abandoned systematics. Taking the opportunity of class trips with students, he began what would become the first long-term studies of species of amphibians and reptiles ever carried out. These were conducted on the salamanders *Hemidactylium* (which resulted in a long series of papers, 1922-1936) and *Plethodon* (1928), and also the snakes *Diadophis* (1927-1937, with a final installment in 1979 which contained detailed and still-useful data although it was published more than 40 years after his death), *Opheodrys* (1933), and *Storeria* (1937).

Besides basic life history data, Blanchard also wrote important papers on sex characteristics and ratios, age groups, courtship, and breeding habits. He pioneered the clipping of scales as a method for marking snakes (1933 paper) and performed the first genetic experiments with snakes (published posthumously and jointly with his wife, Frieda C. Blanchard, in 1941 and 1942). Although Blanchard was primarily a field naturalist, he valued the laboratory environment as a proper place to conduct certain kinds of research and was one of the first herpetologists to blend field with laboratory studies of living amphibians and reptiles in a major way.

Blanchard was a very quiet, diffident person, yet he became an outstanding teacher although not a lecturer, and supervised many students, among them the herpetologists William M. Clay, Howard K. Gloyd, and William H. Stickel. Although all of his longitudinal studies were with local species, Blanchard did make two major expeditions outside eastern North America: to Australia, New Zealand, and Tasmania (1927-1928) and to the American Southwest in 1935-1936. The latter trip was with Howard Gloyd and was part of their joint plan, regrettably never fulfilled, to publish a handbook of North American snakes.

• *References*: "Frank Nelson Blanchard, 1888-1937," by K. P. Schmidt, Copeia, 1937: 149-150, 1937; "Frank Nelson Blanchard," by A. F. Shull, Science, 86: 533-534, 1937; "Dr. Frank Nelson Blanchard," by W. H. Stickel, The Biologist, 19: 160-162, 1938; "Frank Nelson Blanchard As He Appeared to His Students," by M. G. Netting, Bios, 10: 131-134, 1939; "Frank Nelson Blanchard, Scholar and Teacher," by H. K. Gloyd, Herpetologica, 1: 197-211, 1940. • *Portrait:* By Howard K. Gloyd, courtesy Kathryn J. Gloyd and Roger Conant. • *Signature* (1937): Adler collection.

KLAUBER, Laurence M. (1883-1968).

Laurence Monroe Klauber, American amateur herpetologist and authority on rattlesnakes, was an electrical engineer by training and profession but made original contributions to the study of reptiles. He was born on 21 December 1883 in San Diego, California, and as a boy collected horned lizards (*Phrynosoma*) and other reptiles in the fields that long ago became the business district of downtown San Diego. He graduated from Stanford University (B.A. 1908) and went on to complete his professional training in the Westinghouse Graduate Apprenticeship Course (graduate, 1910). He returned to San Diego where he became a salesman for the San Diego Gas and Electric Company. Within the first year he was promoted to electrical engineer and eventually he rose to become President (1946-1949) and Chairman (1949-1954) before retiring in 1954. Klauber was an inventor with several electrical devices to his credit; he also was elected to positions of honor in several societies related to his profession.

About 1920, when Klauber was 37, his boyhood interest in reptiles was rekindled when the director of the new San Diego Zoo, which then had no herpetologist on staff, asked him to identify some donated specimens. Eventually Klauber had to appeal to the ranking authority on Western American reptiles, John Van Denburgh, who, in addition to Joseph Grinnell, encouraged Klauber to begin serious field collecting in the Southwest. Aware of the large number of carcasses on the roads each morning, Klauber pioneered the road-running technique at night, using his automobile, and found that many species previously thought to be rare were, in fact, common. He coined the widely-used acronym "DOR," for specimens found dead on the road.

Klauber began a large preserved collection, maintained in the basement laboratory of his home, which numbered about 35,000 specimens at the time of his death. Also in the 1920s he was encouraged by Howard A. Kelly, a physician and a serious amateur herpetologist, to build a personal library which would be necessary for his research. This Klauber did in style, building one of the largest private herpetological libraries ever amassed which, together with his specimens, was eventually donated to the San Diego Natural History Museum. Klauber served as honorary Consulting Curator both for the zoo (from 1922) and for the museum (from 1927), and was a member of their governing boards for several decades.

Klauber's earliest herpetological papers (beginning 1924) dealt with distributional information on reptiles and amphibians, but even in his first article he described a distinctive new local snake, *Trimorphodon vandenburghi*. His interest in rattlesnakes, which was to be his life's major preoccupation, began in earnest in 1927 and thereafter occupied an increasing amount of his time. During this early phase of his herpetological career he proceeded comprehensively to review many of the genera of Southwestern snakes and some lizards, thereby providing the first detailed study of their systematics. These reviews included descriptions of about 50 new taxa in some 15 genera (e.g., the tiny *Leptotyphlops*, the boids *Charina* and *Lichanura*, many colubrid genera, especially the small, inconspicuous burrowing forms, and the lizards *Coleonyx* and *Xantusia*).

In 1930 Klauber named several new subspecies of *Crotalus* and thereafter regularly published papers on crotalid systematics, natural history, and variability. He obtained a series of 804 specimens of *Crotalus viridis* from a hibernaculum in a prairie dog town in northern Colorado, collected by C. B. Perkins, who later became Curator of Reptiles at the San Diego Zoo, which led to a seven-part series (1936-1940) of papers entitled "A Statistical Study of the Rattlesnakes." The summary of such a large sample size required the use of sophisticated statistical techniques and, with his engineer's training in mathematics, Klauber was well suited to develop new methods of statistical and graphic analysis. This interest led to an important booklet entitled "Four Papers on the Application of Statistical Methods to Herpetological Problems" (1941) in which he discussed frequency distributions, the relationship between populations and samples, and the correlation between scutellation and life zones. Biostatisticians have long recognized the important contributions that Klauber made to their discipline.

For many years, Klauber gathered notes for his contemplated *magnum opus* on rattlesnakes, but only about 1946 did he begin to write a book. What resulted was a two-volume monograph, "Rattlesnakes" (1956), comprising more than 1500 pages and virtually everything known about the genera *Crotalus* and *Sistrurus*, those most distinctive of

American snakes. The bulk of the contents was based on Klauber's own work, to which he added every conceivable piece of information gleaned from the literature and from extensive correspondence and conversations with those having contact with rattlesnakes, including farmers, hunters, zoo curators, and amateur herpetologists, all exhaustively documented.

Besides systematics and natural history, the book also included sections on folklore and the annual Hopi Indian "snake dance" ceremony, the latter subject being the topic of papers by him in 1932 and 1947; Klauber himself once observed these dances in northern Arizona. Following publication, Klauber promptly embarked on a revised edition but did not live to see its completion, for he died in San Diego on 8 May 1968. The revision appeared in 1972, but only the first three chapters (of 18) plus the supplemental bibliography were really new. An abridged edition was published in 1982.

Klauber received numerous honors during his lifetime. He served as president of three biological societies and, in 1941, the University of California at Los Angeles conferred upon him an honorary doctorate (LL.D.). Klauber always regarded himself as an amateur and, true to his own origins, encouraged other amateurs to study reptiles; Charles M. Bogert, Bayard H. Brattstrom, Charles H. Lowe, Jr., Findlay E. Russell, and Charles E. Shaw were thus inspired and went on to professional careers. Klauber also fostered amateurs throughout the country, by subscribing to the publications of the rapidly-increasing number of regional societies and by making well-placed donations or providing other encouragement.

• *References*: "Meet Dr. Laurence M. Klauber," anonymous, Herpetologica, 2: 141-150, 1945; Schmidt, 1955 (pp. 616-617); "L. M. Klauber," [by K. Adler], Herpetol. Rev., 1(3): 7, 1968; "Laurence M. Klauber, 1883-1968," by B. H. Brattstrom, Herpetologica, 24: 271-272, 1968; "Laurence Monroe Klauber 1883-1968," by C. E. Shaw, Copeia, 1969: 417-419, 1969. • *Portrait and signature* (1932): Adler collection.

MERTENS, Robert (1894-1975).

Robert Friedrich Wilhelm Mertens, dean of modern European herpetologists, was born in St. Petersburg (now Leningrad) on 1 December 1894, of German parents. His boyhood interest in reptiles was encouraged by his father, a wealthy dealer in furs, during visits to the city's Zoological Museum. He attended the gymnasium in St. Petersburg, but due to the unsettled social conditions that were to give rise to the Russian Revolution five years later, he went to Germany for university training in 1912. He attended the University of Leipzig to study medicine and natural history, and completed his thesis (Dr.Phil. 1915, with Carl Chun) on geographic variation and zoogeography of *Lacerta*.

After brief army duty, Mertens joined the staff of the Senckenberg Museum in Frankfurt (Main) in 1919, as an assistant in mammalogy. At that time the herpetology position was occupied by Richard Sternfeld, known for his herpetofaunal studies of Africa and the Pacific, but he was fired in 1920. Mertens took his place and thus began the growth of the herpetological collection at Senckenberg into one of world class. Following his retirement, curatorship of the collections passed to Konrad Klemmer, who had been Mertens's assistant since 1954.

After Oskar Boettger's death in 1910 the collections had fallen into a chaotic state, but this quickly changed under Mertens who, until 1943, worked alone except for the help of his wife. In view of his enormous productivity, it is scarcely imaginable that he also was in charge of mammals (1919-1953), birds (1923-1947), and fishes (1920-1954); in addition, he was chairman of the zoology section (1934-1955), and from 1947 until his retirement in 1960, Director of the Senckenberg Museum itself, after which time he remained as a research associate. Concurrently, from 1930, he served as a professor at the University of Frankfurt. During this period he published nearly 800 scientific titles, including 13 books, plus over 300 book reviews. His research touched on every part of the world, especially the tropics, and included systematics, zoogeography, behavior, and even extinct forms. He had an encyclopedic knowledge of herpetology and the associated literature, even of the most obscure references.

Under Mertens's curatorship the Senckenberg amphibians and reptiles quadrupled in number and in terms of species richness and type material the collection ranks among European museums with those in London and Paris. This development is due to Mertens's aggressive exchange program, to the numerous collections donated by correspondents, and above all to his own field work. Even during the darkest days of World War II, German soldiers would send him reptiles captured in occupied lands, using the army's field postal system. For safekeeping during that war, Mertens evacuated most of the collections to rural towns; the most important specimens were set up in a dance hall in Oberlais (Hess), where they could still be actively used despite war-time conditions. Fortunately, only a third of the collection remained at Senckenberg, for the museum was badly damaged in the Allied bombing of Frankurt early in 1944. When he became director in 1947, Mertens's main task was to reconstruct the collections and the building from war damage.

Mertens greatly enjoyed field work and, beginning with a trip to Tunisia in 1913, he travelled to about 30 countries in search of specimens. His major expeditions were to the Indoaustralian Archipelago, the Cameroons and Southwest Africa, Hispaniola, El Salvador, Pakistan, and Australia. In most instances major monographs on each country's herpetofauna followed, and for some of them semi-popular books were also published. These collections led to some of his most important works, including "Die Insel-Reptilien" (1934), a study of island biogeography and evolution, his well-known monograph on varanid lizards (in three parts, all 1943), and numerous revisions of genera and descriptions of new species. In addition, Mertens initiated a series of checklists of the European herpetofauna (three editions: 1928 and 1940, both co-authored with Lorenz Müller; 1960, with Heinz Wermuth), designed along the lines of the American lists by Leonhard Stejneger and Thomas Barbour, which served to standardize nomenclature. Mertens also had a special interest in turtles, crocodiles, and the tuatara, and published important checklists of them with Wermuth in 1955, 1961, and 1977.

Besides his interest in the systematics and distribution of amphibians and reptiles, from boyhood Mertens studied living animals, particularly their behavior. He kept large menageries at the museum and at his home in which he made useful observations and published noteworthy studies, the most important of which was "Die Warn- und Droh-Reaktionen der Reptilien" (1946), the first major review of warning and threatening behaviors. He also made several contributions to the coral snake mimicry problem.

Because of his interest in living animals, Mertens contributed regularly to German terrarium journals, and he strongly encouraged amateurs. Accordingly, he also published several semi-popular books on the subject, most notably "Die Lurche und Kriechtiere des Rhein-Main-Gebietes" (1947), "Kriechtiere und Lurche: Welches Tier ist das?" (1952, later editions 1960, 1964, 1968), and the beautifully illustrated "La Vie des Amphibiens et Reptiles" (1959, later editions in English, Italian, and Spanish), all of which demonstrate his extensive knowledge of the biology of these animals and their lives in nature.

Mertens also contributed numerous articles on biography, including a book on Eduard Rüppell (1949),

Senckenberg's first herpetologist. Mertens continued active research into his 81st year. On 5 August 1975 he was bitten by a specimen of *Thelotornis*, an African rear-fanged snake, which had been a longtime pet at his home. Unfortunately, no antivenin existed for this species, and after 18 painful days, Mertens died in Frankfurt. Throughout this ordeal he kept a diary of each day's events, in which he wrote, with black humor, "für einen Herpetologen einzig angemessene Ende," or in translation, "a singularly appropriate end for a herpetologist."

• *References*: Mertens, 1967 (pp. 17-34); "Prof. Dr. Robert Mertens †," by H. Wermuth, Aquar. Terrar. Zeitschr. (DATZ), 28: 430-432, 1975; "Prof. Robert Mertens," by W. D. Haacke, Jour. Herpetol. Assn. Africa, 14: 35-36, 1975; "Robert Mertens (1894-1975)," by K. Klemmer, Natur und Mus., 106: 252-255, 1976; "Robert Mertens, 1894-1975," by C. Gans, Copeia, 1976: 420, 1976; "Die Wissenschaft und das lebende Tier," by H. Wermuth, Natur und Mus., 108: 245-248, 1976; "Robert Mertens, sein Leben und Werk," by K. Klemmer, and other articles. Cour. Forsch.-Inst. Senckenberg, 20: 1-104, 1977; "Erinnerung an Prof. Dr. Robert Mertens (1894-1975)," by O. G. Dely, Vertebr. Hungar., 18: 3-6, 1978. • *Portrait* (1951): By M. Graham Netting, courtesy Carnegie Museum. • *Signatures* (on left, dated 1912): From Klemmer, 1976; (on right, 1972): Adler collection.

KOPSTEIN, Felix (1893-1939).

Felix Kopstein, Austrian physician and naturalist, was born in Vienna on 4 June 1893. Even as a student Kopstein was interested in herpetology, and in 1914 he made a collecting trip to Albania. The specimens thus obtained were donated to the Naturhistorisches Museum in Vienna, where he was associated informally with Otto von Wettstein, and formed the basis for his first paper (1914). He studied medicine and biology in Vienna (1913-1920).

In January 1921, Kopstein moved to the East Indies as a medical doctor in the Dutch government service and during the first three years he worked in Ambon (today called Amboina, in what is now eastern Indonesia), from which base he travelled extensively in the Moluccas and to New Guinea. These trips and others in Indonesia were summarized in his book "Een Zoölogische Reis Door de Tropen" (published in about 1930). Soon after arriving in Ambon, Kopstein contacted the Rijksmuseum van Natuurlijke Historie in Leiden and offered to donate specimens. He sent large collections of vertebrates, including amphibians and reptiles, as well as invertebrates, from Ambon and other parts of Indonesia. His ultimate goal, to obtain a position at the Rijksmuseum, was never achieved, although he worked at the museum during his European furloughs in 1927 and 1929-1930 and from 1938 on.

In 1924 Kopstein was transferred to Java where, among other duties, he was employed by the Institut Pasteur in Bandoeng (now Bandung). At different times he worked in various parts of the island. In the meantime he began to study his own collections of lizards and snakes, which resulted in a large number of papers, including a series entitled "Herpetologische Notizen" (in 18 numbered parts,

1929-1938). In 1930 he published a semi-popular book on the island's poisonous snakes ("De Javaansche Gifslangen en Haar Beteekenis voor de Mensch") which contained much information on natural history. Although some of his papers were taxonomic in content, most dealt with life history and ecology, and among his last papers were titles on the eggs and reproductive biology of Javan reptiles (1938) and, posthumously, one on sexual dimorphism in snakes (1941). In the former paper Kopstein reported delayed fertilization in three species of snakes, which was the first discovery of this phenomenon.

Kopstein died in The Hague on 14 April 1939, at the age of only 45.

• *References*: "Teraardebestelling dr F. Kopstein," anonymous, Het Vaderland, 19 April 1939; "In Memoriam Dr F. Kopstein 1895-1939," by M. A. Lieftinck, Trop. Natuur, 28: 89-90, 1939; "Dr. Felix Kopstein," by M. S. Hoogmoed, unpublished manuscript. • *Portrait* (about 1938) *and signature* (1927): Rijksmuseum van Natuurlijke Historie, courtesy Marinus S. Hoogmoed.

KINGHORN, J. R. (1891-1983).

James Roy Kinghorn, herpetologist and ornithologist at the Australian Museum and author of "Snakes of Australia," was born in Richmond (now a Sydney suburb), New South Wales, on 12 October 1891. He was educated at All Saints College in Bathurst, just west of Sydney, and the North Sydney Church of England Grammar School. In 1907, he became a Scientific Cadet—one of the boy assistants to the scientific staff—at the Australian Museum in Sydney and worked primarily on Crustacea at that time. Following war service in 1915-1918, partly in France, Kinghorn returned to the museum and was put in charge of amphibians, reptiles, and (later) birds, succeeding Allan R. McCulloch. In 1941, he took on the additional duties of Assistant to the Director, but with a change in administration that was not to his liking, he retired in 1956. J. Allen Keast briefly succeeded him in charge of herpetology (1957-1961) and he, in turn, by Harold G. Cogger, who began as a Cadet Preparator at the museum in 1952. Kinghorn died in Sydney on 4 March 1983.

In many respects Kinghorn was as much a museum educator as a researcher. He was the first to involve the Australian Museum in educational programs and conducted nature-oriented radio programs as early as 1924. Kinghorn was an excellent speaker and gave numerous popular lectures in the museum's regular evening series and probably was the first Australian naturalist to be seen on television when it was introduced to Australia in the late 1950s. He regularly published popular articles on amphibians, reptiles, and birds in the museum's news magazine.

Kinghorn's technical papers on herpetology cover the period 1920-1955 and number about 30. Most deal with the Australian fauna but there are also several titles on New Guinea and the Solomon Islands, including a major review of the herpetology of the Solomons (1928). His best known work, however, is his field guide, "The Snakes of

Australia," first published in an elongate, pocket-sized version in 1929 and, later, in a second edition of a more standard size (1956; revised edition 1964, in collaboration with Harold Cogger). In 1943, together with Charles H. Kellaway, then the leading Australian expert on snake venoms, Kinghorn published an important war-time booklet titled "The Dangerous Snakes of the South-west Pacific Area," which was prepared for use by Australian and American troops; it covered the area from Australia through Indonesia and the Philippines.

• *References*: "James Roy Kinghorn," p. 399-400. *In* H. M. Whittell, The Literature of Australian Birds, Paterson Brokensha, Perth, 1954; "Retirement of Mr. J. R. Kinghorn," by E. LeG. Troughton, Austr. Mus. Mag., 12: 125, 1956; Cogger, 1985 (pp. 147-148). • *Portrait* (1937): By Howard K. Gloyd, courtesy Kathryn J. Gloyd and Roger Conant. • *Signature* (1939): Adler collection.

GLAUERT, Ludwig (1879-1963).

Ludwig Glauert, naturalist and Western Australia's first herpetologist, was born in Sheffield, England, on 5 May 1879, and was educated in Sheffield at the Technical College and later at University College. He became a Demonstrator in Geology at the college and university, but finding no permanent employment in his chosen field of geology, in which he had been interested since boyhood, he decided to emigrate to Australia.

Glauert arrived in Freemantle, near Perth, Western Australia, in 1908, and had a temporary appointment as field geologist with the Geological Survey. He soon began his excavations of the famous fossils of monotremes and marsupials at the Margaret River Caves, which resulted also in his first permanent post, at the Western Australian Museum (1910). His paleontological work continued, as well as new research on Australian aborigines, but this was interrupted by service during World War I and, afterward, by study leave in England. Glauert returned to Perth in 1920 and, due to staff changes, took on added duties for natural history. He became Curator of the museum in 1927 (the title was altered to Director in 1954), a position he held until retirement in 1956. He died in Hollywood, Perth, on 1 February 1963.

Financial considerations limited Glauert's field work to the area near Perth, but he did participate in two expeditions to the north, in the Murchison district (1922) and along the Gascoyne River (1928). He also made extensive collections on Rottnest Island and the Swan Coastal Plain, and in the Darling Ranges surrounding Perth. For a long time he was the only herpetologist in the territory, an area comprising one third of Australia. This situation continued until Glauert was joined by A. R. Main at the University of Western Australia and his student, Glen M. Storr, who succeeded Glauert as herpetologist at the museum in 1962.

After 1920, Glauert's interests and publications turned more to zoology, especially herpetology. He published some 33 titles on amphibians and reptiles (1923-1962), 13 of them in a numbered series entitled "Herpetological Miscellanea." His best known works, however, were "Handbook of the Snakes of Western Australia" (first edition 1950; second 1957) and "Handbook of the Lizards of Western Australia" (1961).

• *References*: "Ludwig Glauert—Museum Director and Naturalist," by D. L. Serventy, West. Austr. Nat., 5: 148-165, 1957; "Ludwig Glauert," by R. Mertens, Natur u. Museum, 93: 353-354, 1963; "L. Glauert, M.B.E.," by D. L. Serventy, West. Austr. Nat., 8: 189-193, 1963. • *Portrait* (1956): From West. Austr. Nat., 5: 146, 1957. • *Signature* (1926): Smithsonian Institution Archives, courtesy William Cox and Alan E. Leviton.

HAAS, Georg (1905-1981).

Haas, a reptile morphologist and later a student of the herpetology and paleontology of his adopted Israel, was born in Vienna on 19 January 1905. He studied zoology and paleontology at the University of Vienna, where his teachers included Jan Versluys, Franz Werner, and Otto von Wettstein, and was awarded the Ph.D. degree in 1928 for his thesis on functional cranial anatomy of primitive and venomous snakes. After a short period in Berlin (1931-1932), he emigrated to the Hebrew University of Jerusalem in the then British Mandate of Palestine and remained on its staff until his retirement in 1976.

Besides continuing his anatomical work, Haas began distributional and taxonomic studies on Middle Eastern reptiles and amphibians, and participated in several expeditions to Transjordan, the Negev, and the Sinai. His careful taxonomic studies led to the description of several new taxa and to the distinguishing of long-confused species. Over time his interests turned more to paleontology, and at

KLINGELHÖFFER, Wilhelm (1871-1953).

Wilhelm Karl August Klingelhöffer, terrarist and amateur herpetologist, was born in Gladenbach (Hesse), Germany, on 11 January 1871. He was interested in animals and plants as a young boy, encouraged by his mother, and enrolled in a secondary school in Marburg (1882-1891). He studied medicine at universities in Marburg and Munich and, after passing the state medical examination in 1896, became an intern at the eye clinic in Mannheim. He began medical practice at the Offenburg (Baden) State Hospital in 1899, specializing in ophthalmology. During World War I he was a field doctor and was wounded at the front in 1914.

During Klingelhöffer's years in Offenburg he developed a large vivarium filled with lower vertebrates whose nutrition, growth, and proper care he studied in detail; turtles were his specialty. He became the leading authority on the maintenance of amphibians and reptiles, and championed the idea that they must be kept in naturalistic conditions for reasons of health and reproduction. He published regularly on these topics, beginning in 1900, and wrote an influential chapter in Abderhalden's handbook on biological methods (1928).

Klingelhöffer's best known work, however, was his book "Terrarienkunde" (1931; revised and expanded edition, in four volumes, 1955-1959, edited by Christoph Scherpner) which became the standard for terrarists for many decades. This book was instrumental in popularizing naturalistic methods for keeping amphibians and reptiles and in adapting the

the time of his death, in Jerusalem on 13 September 1981, he had been working for several years on the interesting Lower Cenomanian snake-like fossils collected at 'Ein Yabrud in the Judaean Hills.

Throughout his career Haas continued his longstanding interest in anatomy, especially jaw musculature, research from which he attempted to answer key questions of phylogeny and classification: the Middle American snake *Loxocemus* (1955), the lizards *Xenosaurus* and *Shinisaurus* (1960), and various primitive and venomous snakes, to mention only a few topics. In all he published 77 titles (1929-1982), about 45 of them herpetological. Perhaps his most important legacy was his training of students, among them Elazar Kochva, Eviatar Nevo, and Yehudah L. Werner, and through his inspiring teaching and enthusiasm he broadly influenced Israeli science and society; indeed, one of his former students became President of Israel.

• *References*: "Professor Georg Haas on the Occasion of His Sixtieth Birthday," by Y. L. Werner, Israel Jour. Zool., 14: 5-9, 1965; "Georg Haas 1905-1981," by Y. L. Werner, Copeia, 1982: 491-493, 1982; "Georg Haas 1905-1981," by C. Gans, Amer. Zool., 23: 343-346, 1983; "Now Imagine that I am a Young Amphioxus . . . [biography of Georg Haas]," by Y. L. Werner, Teva' WaArez, 26: 8-9, 1984. • *Portrait* (1964): By M. Graham Netting, courtesy Carnegie Museum. • *Signatures* (Hebrew form): Courtesy Yehudah L. Werner; (romanised form, 1947): Courtesy J. H. Hoofien.

latest scientific discoveries to practical use; in turn, it was highly influential in the development of the several terrarist journals and the terrarist tradition in Germany and in Europe generally.

In 1936, Klingelhöffer moved to Neu-Isenburg, near Offenbach, so that he could be closer to the Senckenberg Museum in Frankfurt (Main), which named him an Honorary Associate in 1943. He retired from medical practice in 1952 and died on 17 September 1953. Only a month before his death he willed his house to the museum to support the further development of its herpetology collection.

• *References*: "Dr. Wilhelm Klingelhöffer†," by W. B. Sachs, H. Weise, Die Aquar. Terrar. Zeitschr. (DATZ), 6: 299-300, 1953; "Wilhelm Klingelhöffer," by R. Mertens, Natur u. Mus., 83: 363-364, 1953; "In Memoriam Dr. med. Wilhelm Klingelhöffer," p. 9-10. *In* C. Scherpner, Klingelhöffer's Terrarienkunde, ed. 2, vol. 1, A. Kernen, Stuttgart, 167 pages, 1955. • *Portrait* (1935) *and signature* (1928): Senckenbergischen Naturforschenden Gesellschaft, courtesy Konrad Klemmer.

CHERNOV, S. A. (1903-1964).

Sergius Alexandrovich Chernov was head of the Soviet Union's leading center for herpetology for over 30 years. He was born in Kharkov, in the Ukraine, on 28 July 1903, and finished his studies at the University of Kharkov in 1926 where he had been a student of A. M. Nikolsky, then the leading Soviet herpetologist. Chernov was appointed Curator of the Department of Herpetology at the Academy of Sciences in Leningrad in 1930, succeeding Sergius F. Tsarevsky. This post was once held by Nikolsky himself (1896-1903), and Chernov in turn was succeeded by his own student, Ilya S. Darevsky, in 1961. During his tenure in Leningrad, Chernov did extensive field work in Transcaspia (1932), the Caucasus (1937-1939), and in Tadjikistan (1942-1944).

Chernov's major studies were on the systematics and distribution of amphibians and reptiles of the arid and semi-arid regions of the southern USSR, and he published many papers and books on this subject over the period 1926-1962. These included major monographs on Tadjikistan (1935, 1959) and Armenia (1939), as well as shorter works on Turkmenia (1934), the Gissar Valley of Tadjikistan (1945), and the Kazakhs Highlands and Lake Balkhash (1947). Chernov also was responsible for revising the snakes which A. M. Nikolsky and others had hopelessly lumped into the genus *Contia*. His most widely-known work, however, was the "Synopsis of the Reptiles and Amphibians of the USSR," in collaboration with P. V. Terentjev; this was issued in three editions (1936, 1940, 1949), the last of which was published in English translation in 1965. Unfortunately, a major project on the karyology of Palaearctic snakes was left incomplete at the time of his death, in Leningrad, on 2 January 1964.

• *References*: "Sergius Alexandrovich Chernov," by A. E. Leviton, Copeia, 1964: 466, 1964; Scarlato, 1982 (pp. 70-71). • *Portrait*: Academy of Sciences, Leningrad, courtesy Ilya S. Darevsky and Natalia B. Ananjeva. • *Signature* (Russian form, 1958): Courtesy Carl Gans.

TERENTJEV, P. V. (1903-1970).

Paul Victorovich Terentjev, a leading Soviet herpetologist, authored the first textbook on herpetology. He was born in Sevastopol on 23 December 1903 and completed his academic training at Moscow State University. During his postgraduate studies there (1922-1926), he headed the terrarium section at the Moscow Zoo, but then moved to Leningrad where he held a series of posts from 1934 to 1970, including the Chair of Zoology at Leningrad State University. He also had a joint appointment at the Zoological Institute of the nearby Academy of Sciences and played a crucial role in the protection of the herpetological collections during the German siege of Leningrad (1941-1944). He attained the D.Sc. degree in zoology in 1944 and was widely regarded as an inspiring teacher. Terentjev also organized the first two national conferences of Soviet herpetologists (1964, 1967), both held in Leningrad under his chairmanship and summarized in *Herpetological Review*, by Hobart M. Smith et al. (1967), and by Terentjev (1968), respectively. Terentjev died in Leningrad on 30 December 1970.

Terentjev's first herpetological title was published in 1921 and thereafter he produced numerous books and papers,

SCORTECCI, Giuseppe (1898-1973).

Giuseppe Scortecci, authority on African and desert herpetofaunas, was born in Florence, on 2 November 1898. After service in World War I, he obtained a doctorate in natural sciences at the University of Florence (1921) and joined the staff of the Institute of Comparative Anatomy there. He later moved to Milan as Curator of Lower Vertebrates at the Museum of Natural History and, finally, in 1942, to the University of Genoa, where he became Professor of Zoology and, later, dean. Scortecci's special interest was deserts and, thus, also lizards. He conducted numerous explorations in the Sahara, north to Libya and east as far as Somalia and Arabia, often into previously unexplored regions. His interests coincided both geographically and chronologically with the expanding Italian colonial empire. Besides his taxonomic work, Scortecci also published on zoogeography, ecology, anatomy, and physiology, and he was widely honored for his work. He died in Milan, on 18 October 1973, from injuries suffered in a street accident.

Beginning in 1928, Scortecci published about 50 herpetological titles. His major contributions, besides sizeable papers on collections from various regions in northern Africa, include small books on the amphibians (1933) and poisonous snakes of Italian Somalia (1934), another on the amphibians of the Tripolitania region in Libya (1936), a major work on the venomous snakes of Italian Africa (1939), and the herpetological section of the

principally on taxonomy and biogeography, especially of Anura, biometrical studies of variation, faunal studies on Central Russia, the northern Caucasus, the Far East, and Sakhalin and Kuril Islands, as well as evolutionary and ecological studies. Terentjev was an early proponent of the use of statistics in herpetology. He is perhaps best known outside the Soviet Union for his book "Synopsis of the Reptiles and Amphibians of the USSR," co-authored with S. A. Chernov, and issued in three editions, 1936, 1940, and 1949, the last of which was published in English translation in 1965. He published a college manual on the frog in 1950 and the first university-level text of herpetology in 1961 (English edition, 1965).

• *References*: "Paul V. Terentjev," by I. S. Darevsky and L. I. Khosatzky, Copeia, 1971: 382-384, 1971; "Paul V. Terentjev zum Gedenken," by R. Mertens, Salamandra, 7: 1-4, 1971; "In Memoriam Prof. Pavel Viktorovich Terentyev, 1903-1970," by L. J. Khosatsky and I. S. Darevsky, Herpetol. Rev., 3: 68-70, 1971; ["In Memory of P. V. Terentjev"], by A. S. Malchevsky, J. I. Polyansky, and L. J. Khosatsky, Vest. Leningrad Univ., Biol., 2: 156-158, 1971.
• *Portrait*: Academy of Sciences, Leningrad, courtesy Ilya S. Darevsky and Natalia B. Ananjeva. • *Signatures* (Russian form, 1946): Courtesy Natalia B. Ananjeva; (romanised form, 1960): Courtesy Alan E. Leviton.

Sagan-Omo mission to Ethiopia (1943). He also authored the enormous herpetological volume (1967), with its lavish photographs, in the nine-volume, 5,000-page Italian series "Animals," which was entirely written by him.

Scortecci published a book on the Sahara ("Biologia Sahariana," 1941 [1940]) which, because it was issued during World War II, was not widely distributed and is little known; it emphasized reptiles and was illustrated with many photographs by the author. Another more general book on the Sahara was published by him in 1945. Scortecci also published four major papers (1937-1941) on the sensory organs of the skin of agamid and iguanid lizards, following up on research by the German anatomist Franz Leydig in 1868, and Scortecci was the first to distinguish the two principal types of skin receptors.

• *References*: "Giuseppe Scortecci (1898-1973)," by M. Sarà, Ann. Mus. Civ. Stor. Nat. Genova, 79: 402-403, 1972-1973 (1974?); "Prof. Giuseppe Scortecci. 1898-1973," by E. Tortonese, Copeia, 1974: 294, 1974; "In Memoria di Giuseppe Scortecci," by M. Sarà, Boll. Zool., 41: 141-143, 1974; "Giuseppe Scortecci (Firenze, 1898-Milano, 1973)," by M. Sarà, Boll. Mus. Ist. Biol. Univ. Genova, 42: 5-9, 1974.
• *Portrait* (about 1946): Courtesy Charles M. Bogert. • *Signature* (1940): Adler collection.

PITMAN, Charles R. S. (1890-1975).

Charles Robert Senhouse Pitman, game warden of Uganda and field naturalist, was author of the most important guide to East African snakes. Born in Bombay, 19 March 1890, he graduated as Second Lieutenant from the Royal Military College at Sandhurst (1909) and returned to India where he saw service throughout the country. During 1914-1920 he performed military duty in Egypt, France, Mesopotamia, and Palestine, and won much-coveted military decorations (D.S.O., M.C.). He rose to the war-time rank of Lieutenant Colonel, but preferred to be addressed by his regular army rank of Captain. He resigned his military commission in 1921 to farm in western Kenya. From 1925 to 1950 he was Game Warden of the Uganda Protectorate, except in 1931-1932 when he served as Acting Game Warden in Northern Rhodesia (now Zambia), and during 1941-1946 when he was Director, Security Intelligence (Uganda). He was awarded the C.B.E. in 1950 and in about 1951 retired to London, where he became an active member of local scientific and conservation organizations. He died in London on 22 September 1975.

Pitman's first publications on reptiles, beginning in 1913, were miscellaneous notes on Indian reptiles, a subject on which he continued to write even after moving to Africa. His major book on reptiles, and the one upon which his scientific reputation rests, was "A Guide to the Snakes of Uganda," published originally in serial form in the *Uganda Journal* (1936-1937) and containing 23 beautiful colored plates. Five hundred extra copies of each issue of the journal were set aside as published and later reassembled into a book with added preliminaries and indexes (with each copy individually numbered), which was published late in 1938. Sales dried up when World War II broke out in 1939, so it was decided to ship the remaining 236 copies to London to increase sales, but an enemy bomb dropped on the warehouse destroyed the lot (and the printing blocks for all the color plates), thus causing the book's scarcity. However, in 1974 a revised edition was published, covering an additional 15 species with five more color plates but in 70 less pages due to Pitman's peculiar mode of writing, as defended in his preface ("'Staccato' style adopted throughout achieved enormous reduction superfluous words").

Pitman published numerous articles on African reptiles, perhaps the most important being a description of the unusual biting mechanism of the viper *Atractaspis* (1959), co-authored with N. L. Corkill and C. J. P. Ionides, fellow British expatriates working in eastern Africa. He also wrote several popular books on his experiences as game warden (1931, 1942, 1956), but his most influential book, in two volumes, "A Report on a Faunal Survey of Northern Rhodesia" (1934), led to the establishment of the Zambian wildlife conservation department.

• *References*: "Retirement of Game Warden—Captain C. R. S. Pitman," p. 60. *In* Ann. Rep. Game Fish. Dept., Uganda Protect., Entebbe, 1951; "The Shamba Raiders, Memories of a Game Warden," by B. Kinlock, Collins and Harvill, London, 384 pages, 1972 (see pp. 201-210); "Charles R. S. Pitman," by D. G. Broadley, Jour. Herpetol. Assoc. Africa, 14: 36-38, 1975; "Charles Robert Senhouse Pitman," by C. W. Benson, Ibis, 118: 427-428, 1976; "Captain Charles Robert Senhouse Pitman," p. 628-629, *in* Who Was Who, 1971-1980, vol. 7, A. and C. Black, London, 1981. • *Portrait* (about 1940): Uganda Society, Kampala, courtesy Derek E. Pomeroy. • *Signatures* (script and printed versions): Courtesy Donald G. Broadley.

PARKER, H. W. (1897-1968).

Hampton Wildman Parker, British herpetologist and systematist, was born on 5 July 1897 in Giggleswick, Yorkshire. He attended the University of Cambridge (Selwyn College, B.A. 1923, M.A. 1935) and in 1923 succeeded Joan B. Procter, then temporarily in charge of reptiles and amphibians at the British Museum (Nat. Hist.) in succession to George A. Boulenger. During World War II Parker was seconded for work at the Admiralty, but on returning to the museum in 1945 he became involved in reconstruction from war damage. His administrative responsibilities increased in 1947 when he was promoted to Keeper, or head, of the Department of Zoology, a post once held by both John E. Gray and Albert Günther.

At the museum, Parker is credited with the infusion of the so-called New Systematics, especially a concern for variability and the treatment of large samples. His *alma mater*, however, was not so broad minded. Cambridge was unwilling to grant a doctorate in pure systematics, so Parker submitted his thesis on the snakes of Somalia to the University of Leiden in the Netherlands and was granted a D.Sc. degree in 1949. He retired from the keepership and the museum in 1957, was awarded the C.B.E. that same year, and retired to the Kent countryside. He died on 2 September 1968.

Parker's primary research became systematics, although his university training was broadly-based in biology and chemistry; indeed, he had never heard of Boulenger before he was hired by the museum! Like his great predecessor, his geographic interests were cosmopolitan. His nearly 100 scientific papers, beginning in 1924, cover all continents, especially South America and Africa, and include fossil forms as well. His particular interests were frogs, about which he wrote monographs of the Microhylidae (1934) and the Australasian Leptodactylidae (1940), and the lizards and snakes of Somalia (1942, 1949). These works, like all his papers, were characterized by careful systematics within a broader biological context.

After retirement, Parker's interests turned to semi-popular writing, for which he had great aptitude, and he published two books: "Snakes" (1963) and "Natural History of Snakes" (1965; second edition 1977, co-authored with Alice G. C. Grandison, his successor in herpetology at the museum). Regrettably, a book manuscript on the lives of amphibians, completed just before his death, was never published.

• *References*: "Dr Hampton Parker," [by A. C. G. Grandison], The Times, London, 6 September 1968; "H. W. Parker," by A. G. C. Grandison, Copeia, 1969: 416-417, 1969; "Hampton Wildman Parker," by J. C. Battersby, Herpetol. Rev., 1(6): 5-6, 1969; "Hampton Wildman Parker," by A. Leutscher, Brit. Jour. Herpetol., 4: 105-106, 1969. • *Portrait* (1963): By M. Graham Netting, courtesy Carnegie Museum. • *Signature* (1967): Adler collection.

ANGEL, Fernand (1881-1950).

Fernand Angel, French herpetologist and ichthyologist, was born in Douzy (Ardennes), on 2 February 1881. His work at the Muséum National d'Histoire Naturelle in Paris began in 1905 as an Assistant Preparator under Léon Vaillant, professor in charge of the laboratory of reptiles and fishes. In his first years, Angel was trained by François Mocquard, Vaillant's chief assistant from 1884 to 1908, and later by George A. Boulenger in London. One of his first tasks was to draw the last plates for the volume on reptiles in the series "Mission Scientifique au Mexique et dans l'Amérique Centrale" (1870-1909), complementing the other drawings by M.-F. Bocourt.

Following Vaillant's retirement in 1910, the three subsequent holders of the professorship were ichthyologists, thus Angel was practically the only qualified herpetologist at the museum for several decades and provided the sole link between the herpetologists of the 19th Century and the modern generation at the museum. Angel's service in curating one of the world's premier herpetological collections extended without interruption for 45 years, except for duty in World War I, until his death on 13 July 1950.

Angel's primary interests were the herpetofaunas of Africa and Asia, particularly the former French colonies in West Africa, Madagascar, and Indochina. Most of his papers were on systematics, but he also published a series on the biology of an African live-bearing toad (*Nectophrynoides occidentalis*) in collaboration with Maxime Lamotte. He wrote several major works, among them "Les Serpents de l'Afrique Occidentale Française," (1932 [1933]), "Les Lézards de Madagascar" (1942), and the herpetological section (1946) in the series "Fauna de France."

Besides these technical books, Angel published numerous semi-popular works—La Vie des Caméléons et Autres Lézards," "Vie et Mœurs des Serpents," and the two-volume "Petit Atlas des Amphibiens et Reptiles"—all of which were illustrated by him; there were several editions and translations. His herpetological publications span the period 1907-1954, including several posthumous titles co-authored with Lamotte and Jean Guibé.

• *References*: "Allocution Prononceé aux Obsèques de M. F. Angel" by L. Bertin, Bull. Mus. Natl. Hist. Nat., ser. 2, 22: 541-542, 1950; "Fernand Angel, 1881-1950," by J. Guibé, Copeia, 1951: 1-2, 1951. • *Portrait*: George S. Myers Collection, California Academy of Sciences, courtesy Alan E. Leviton. • *Signature* (1952): Adler collection.

BOURRET, René (1884-1957).

René Léon Bourret, specialist on Indochinese herpetology, was born on 28 January 1884 in Nérac (Lot-and-Garonne), France. He arrived in French Indochina in 1900, for military purposes in Tonkin. Later, in 1907, he became a surveyor at the Cadastral Survey. After the First World War he was an assistant at the Mining and Geographical Survey (1919-1925), during which time he published important papers on the geology of Tonkin and Laos; he had previously earned a doctorate in geology at the Université Toulouse. In 1925, he became a professor at the École Supérieure des Sciences, Université Indochinoise, in Hanoi.

Bourret's first zoological publication was a general review of the vertebrates of Indochina (Hanoi 1927). His scientific titles on herpetology began in 1933, with a report on new snakes collected by him which was co-authored with his colleague Fernand Angel in Paris. Thereafter, Bourret's technical papers on amphibians and reptiles appeared in a series of 25 numbered papers (1934-1944), published in Hanoi, in which he reported new distributional records and numerous new species. In addition, he authored several identification manuals, including a major one on sea snakes (1934), and also on mammals and other vertebrate groups.

During the Japanese occupation of French Indochina, beginning in the fall of 1940, Bourret remained in Hanoi and continued to publish regularly, but his last Hanoi papers were issued in 1944, about a year before nationalist forces resumed control of Hanoi. By this time Bourret had moved south and in May, 1947, he returned to France and lived in Toulouse. Most of his specimens, including types, were presented to the museums in Paris and Toulouse. He never resumed his scientific work and died, alone, on 28 July 1957, in Toulouse.

Bourret's reputation largely rests on three major monographs of the Indochinese herpetofauna, all nicely illustrated: a two-volume treatise on snakes (1936), one on turtles (1941), and the last on amphibians (1942). Regrettably, his manuscript on lizards, completed in 1943, was never published. These were the first herpetological reviews of French Indochina since those of his compatriots Gilbert Tirant (1885) and François Mocquard (1907 [1906]), but the geographic coverage is far greater: from Burma east through southern China, including Hainan, and south to Singapore. Bourret's amphibian book thus filled a special gap because Malcolm Smith's amphibian volume in the series "Fauna of British India" never appeared.

Working in a region far from reference collections and at a time when communication was often disrupted by war conditions, Bourret could not make direct comparisons of his new species with existing types. As a result some of his conclusions have not been accepted by other workers and a number of his names have been synonymized, yet after half a century his great works remain the most definitive reviews of Indochinese herpetology.

• *Reference*: "René Léon Bourret," by C. P. Blanc, unpublished manuscript. • *Signature* (1941): Adler collection.

BORING, Alice M. (1883-1955).

Alice Middleton Boring, an American biologist in China, trained an entire generation of Chinese herpetologists. She was born on 22 February 1883 in Philadelphia, and attended nearby Bryn Mawr College (B.A. 1904). Her first paper, on the cleavage of frog embryos (published 1903), was co-authored with her teacher, Thomas Hunt Morgan, later a years in China. She was deeply devoted to the advancement of science in China, to her many students, and to the welfare of the Chinese people, causes she continued to espouse for the rest of her life. After an appointment as part-time Professor of Zoology at Smith College (1951-1953), she retired to Cambridge, Massachusetts, where she died on 18 September 1955.

Boring's earliest research interests were in embryology, regeneration, and chromosome behavior, but for practical purposes she shifted to herpetology in 1929, particularly to the distribution and life histories of frogs. Her major publications were a "Handbook of North China Amphibia and Reptiles," co-authored with Cheng-chao Liu and Shu-ch'un Chou (1932), a three-part series on the amphibians of Hong Kong (1934-1936), "Survey of Chinese Amphibia" with Clifford H. Pope (1940), and a book, "Chinese Amphibians: Living and Fossil Forms," published in Peking (now Beijing) in 1945.

• *References*: "Alice Middleton Boring—[class of] 1904," by C. W. Hull, S.-C. Shen, Bryn Mawr Alumni Bull., Winter 1956, p. 17; "Alice Middleton Boring," p. 43-45. *In* M. B. Ogilvie, Women in Science: Antiquity Through the Nineteenth Century. Mass. Inst. Technol. Press, Cambridge, 1986.
• *Portrait*: United Board for Christian Higher Education in Asia, courtesy Yale Divinity School Archives and Clifford Choquette. • *Signature* (1936): Adler collection.

OKADA, Yaichirō (1892-1976).

Okada was one of Japan's leading zoologists of this century. He was born on 24 June 1892 in Daishoji-cho (now Daishoji-nakamachi, Kaga City), Ishikawa Prefecture,

geneticist and Nobel laureate. Indeed, Boring had the good fortune to work with some of the greatest biologists of her day, including Edwin G. Conklin at the University of Pennsylvania, Theodor Boveri at Würzburg and, later, after completing her Ph.D. at Bryn Mawr (1910), Raymond Pearl at the University of Maine, where Boring held her first position (1910-1918). In 1918 she went to China for two years as Assistant Professor in the premedical Department of Biology of Peking Union Medical College, an appointment that changed the course of her life.

Although Boring returned to America for a brief period (1920-1923) at Wellesley College, she soon accepted a position at Peking (later called Yenching, now a part of Beijing) University. Here she stayed until after the attack on Pearl Harbor when the Japanese closed the university and removed her and many others to an internment camp for non-combatants. In 1943 Boring was repatriated. Up until then she had trained large numbers of Chinese biologists, including the herpetologists Tso-kan Chang, Cheng-chao Liu, and Han-po Ting (now Ding).

During 1944-1945 Boring was a lecturer at Columbia University Medical School and later a Visiting Professor at Mount Holyoke College, but in 1946 she returned to her beloved Yenching University, where she served faithfully even through the difficult years of civil war. She finally returned to the United States in 1950, after a total of 26 educated at the Imperial Fisheries Institute, and later received advanced training in America and Europe. His zoological interests were exceptionally broad, and he published more than 400 titles. Early in his career herpetology was his principal fascination, especially distributional and faunal studies, but his research later shifted more to the study of fishes. His special interest in biogeography led to his becoming founding President of the Biogeographical Society of Japan in 1928.

Okada published many books, several of them herpetological: "Monograph of Japanese Frogs," in Japanese (1930; English edition, "Tailless Batrachians of the Japanese Empire," 1931) and covering the Japanese Empire including Formosa (now Taiwan) and Korea, "Ecology and Evolution of Reptiles" (1932, co-authored with Yosiyuki Takakuwa), and "Fauna Japonica, Anura (Amphibia)" (1966), the last in English. In addition, Okada wrote the herpetological sections ("Amphibia and Reptilia of Jehol," 1935) of a report on the expedition to Manchoukuo in 1933, led by Shigeyasu Tokunaga, covering that portion of northeastern China then under Japanese control.

Okada received many honors during his lifetime, including The Order of the Sacred, conferred upon him by Emperor Hirohito. Okada died at Yokosuka, Kanagawa Prefecture, on 28 April 1976.

• *References*: "Yaichiro Okada, 1892-1976," by C. L. Hubbs and K. Kuronuma, Copeia, 1977: 206-207, 1977; "Dr. Yaichiro Okada, 1892-1976," by T. Matsui, Japan. Jour. Herpetol., 7: 1-3, 1977; "Yaichiro Okada D.Sci. (1892-1976); by M. Watanabe, Bull. Biogeogr. Soc. Japan, 36: i-xiv, 1981. • *Portrait*: Mrs. Yaichirō Okada, courtesy Toshijiro Kawamura. • *Signatures* (Japanese form, 1963): Courtesy Hajime Fukada; (romanised form, 1938): Adler collection.

MAKI, Moichirō (1886-1959).

Maki's book, "A Monograph of the Snakes of Japan," is one of the classics on snakes. Although he also published eight short herpetological papers, his reputation rests on this great book which was published in several sections. In 1931, the English text, selected accounts from the Japanese text, and the first 39 plates were issued, followed in 1933 with the full Japanese text and the remaining plates (numbers 40-86). All of the plates are beautifully colored. The geographic coverage of the book is Imperial Japan, including Formosa (now Taiwan) and Korea. The book was reprinted in 1978.

Maki was born in Matsuyama, Ehime Prefecture, on 20 February 1886, and was educated at Hiroshima Higher Normal School, graduating in 1909, and from the school's research division in 1911. From then until 1926 he held various positions with the Japanese colonial government in Formosa, during which time he studied local amphibians and reptiles, but his main occupation was the study of pest insects on which he published extensively. From 1927 to 1946 he was a lecturer at Doshisha University in Kyoto. He earned his D.Sc. degree at Kyoto Imperial University in 1932, under the supervision of Taku Komai. His "Monograph" served as his thesis.

Ironically, Maki was deathly afraid of snakes—there were so many venomous ones in Formosa where he did much of his field work—and would always take someone along on his field trips, usually his wife, who would catch them once they were found. Beginning in 1947 he became Professor of Biology at Hanazono Buddhist College, a position he held until his death, in Kyoto, on 19 April 1959.

• *References*: "Moichiro Maki," by H. Fukada, Copeia, 1959: 272, 1959; ["Dr. Moichiro Maki, 1886-1959,"] by H. Fukada, Acta Herpetol. Japon., 1: 5-7, 1964; Hasegawa, 1967 (pp. 75-76). • *Portrait*: Courtesy Toshijiro Kawamura. • *Signature*: Courtesy Hajime Fukada.

LOGIER, E. B. S. (1893-1979).

Eugene Bernard Shelley Logier, longtime dean of Canadian herpetologists, wrote the first comprehensive reviews of the amphibians and reptiles of his adopted country. Born on 27 February 1893 in Clontarf, near Dublin, where, living near the Irish Sea, Logier developed a childhood interest in animals. His family emigrated to Toronto in 1906, where he came under the influence of C. W. Nash, of the Ontario Department of Agriculture, who was writing the first checklists of Ontario vertebrates. Logier had artistic skills, which he developed commercially, and in 1915 he accepted a volunteer position at the Royal Ontario Museum as artist to paint displays.

A decade later Logier assumed responsibility for developing the museum's herpetological collection, finally becoming curator in title in 1947. During his custodianship the collections grew from virtually zero to about 10,000 specimens, many collected by Logier (especially during 1919-1933) from throughout Canada: Atlantic and Pacific coasts, Great Lakes Region, and especially Ontario. For ten

COCHRAN, Doris M. (1898-1968).

Doris Mable Cochran, born in North Girard, Pennsylvania, on 18 May 1898, was a specialist on Neotropical herpetology and for many years custodian of the American national collection in Washington. Her interest in natural history began as a child in a small town near the shores of Lake Erie. She attended George Washington University in Washington, D.C. (A.B. 1920, M.S. 1921) but previously, in 1919, while still an undergraduate student, she became Aide in the Division of Herpetology at the United States National Museum, succeeding Frank N. Blanchard. In this post Cochran assumed day-to-day responsibility for the herpetological collections which were formally under the curatorship of Leonhard Stejneger who had heavy administrative duties.

Although Stejneger was on the museum's staff until 1943, when he died at the age of 92, Cochran had been effectively in charge of the herpetological collections for many years. In 1927 she was finally given a proper title (Assistant Curator) and was promoted to Associate Curator the year before Stejneger's death, after which she formally assumed charge of the collections. In the meantime, she completed her doctorate at the nearby University of Maryland (Ph.D. 1933) with a thesis on crab myology. Eventually she was promoted to Curator (1956), but retired in 1968 when her successor, James A. Peters, assumed responsibility for the collections. She died less than a month later, on 22 May 1968, in Hyattsville, Maryland, a Washington suburb.

years, beginning in 1948, he taught the herpetology course at the University of Toronto. He retired in 1961, after 46 years on museum staff, but remained active through involvement with the Canadian Amphibian and Reptile Conservation Society. He died on 16 March 1979, in Toronto.

Logier's primary interests were in systematics and distribution. His task was enormous: to survey almost singlehandedly the distribution of the amphibians and reptiles of the third largest country in the world. The results were published in three books, "Frogs, Toads and Salamanders of Eastern Canada" (1952), and two editions, co-authored with G. C. Toner, of "Check-list of the Amphibians and Reptiles of Canada and Alaska" (1955, 1961). These contain detailed spot maps for each species, showing for the first time the correct distribution for Canadian taxa. In addition, Logier published several shorter books on Ontario amphibians (1937), reptiles (1939), and snakes (1958). His books are beautifully illustrated by himself. All told he published 46 titles (1925-1976), including papers on melanism in snakes, the life histories of frogs, and the impact of human activities on habitats.

• *References*: "Eugene Bernard Shelley Logier, 1893-1979," by E. J. Crossman, Canadian Field-Nat., 94: 469-473, 1980; "Eugene Bernard Shelley Logier," by E. J. Crossman, Copeia, 1980: 572-574, 1980. • *Portrait* (1936): By Howard K. Gloyd, courtesy Kathryn J. Gloyd and Roger Conant. • *Signature* (1942): Carnegie Museum, courtesy C. J. McCoy.

Cochran was a kind and gracious lady, always willing to assist others, be they distinguished foreign visitors or young amateurs. She always put her own research interests second and recognized her service function for the national collection first. Her gentle ways, however, led to chronic underfunding of the collection, as a result of which for many years she did all the curating, cross-cataloging, literature indexing, and caring for the division library. The only help she had were untrained workers who shelved or withdrew jars of specimens from the collection rooms. This resulted in a perplexing system of filing specimens, in which they were shelved according to the size of the bottle and not according to taxonomic group, but this was the only system that Cochran's unskilled assistants understood. Thus, until the arrival of James A. Peters in 1964, the collection was organized in an outmoded manner that antedated Cochran herself who simply never had enough assistance to modify it.

Cochran's research interests primarily emphasized the herpetofauna of Southeast Asia, the West Indies, and South America (particularly Brazil and Colombia). She published nearly 90 titles (over the period 1922-1970), most of them short taxonomic papers describing numerous new genera (8) and species and subspecies (125). Her interest in the West Indies spanned a 20-year period and culminated in a book, "Herpetology of Hispaniola" (1941). Thereafter her major interest was South American frogs, about which she published many papers and two books, "Frogs of Southeastern Brazil" (1954 [1955]), based in part on five months of field work there in 1935 in the company of Adolpho and Bertha Lutz, and "Frogs of Colombia" (1970), published posthumously with her longtime collaborator, Coleman J. Goin. She also wrote numerous semi-popular articles as well as wartime booklets for use by military personnel. Her most widely-known book, however, was the profusely-illustrated "Living Amphibians of the World" (1961), which was translated into about ten languages. Another book co-authored with Goin, "The New Field Book of Reptiles and Amphibians" (1970), was not a success and was never a serious competitor to the existing field guides by Roger Conant and Robert C. Stebbins.

• *References*: "The Doris Mable Cochran Number," by C. J. Goin, Herpetologica, 8: 109-110, 115-120, 1953; "Doris Mable Cochran," by C. J. Goin, Copeia, 1968: 661-662, 1968; "Doris M. Cochran, 1898-1968," by H. M. Smith, Herpetologica, 24: 268-270, 1968. • *Portrait*: Howard K. Gloyd collection, courtesy Kathryn J. Gloyd and Roger Conant. • *Signature* (1961): Adler collection.

LOVERIDGE, Arthur (1891-1980).

Arthur Loveridge, naturalist-collector and specialist on African and Austro-New Guinean herpetology, was born on 28 May 1891 in Penarth (Glamorganshire), Wales. He developed an interest in natural history as a boy and decided at the age of ten to be a museum curator. After Loveridge finished secondary school, however, his family insisted on a two-year commercial apprenticeship before he could attend college. After one year at the University College of South Wales in Cardiff, Loveridge held posts at the Manchester University Museum and the National Museum of Wales when a propitious opportunity took him to British East Africa in 1914 as the first Curator (or director) of a newly-founded museum in Nairobi. War had broken out in Europe that summer and its effects were soon felt in the far-flung empires of the belligerents, including their colonies in East Africa, where hostilities began early in August 1914.

Loveridge joined the East African Mounted Rifles in 1915, and during three years of service, most of it in German East Africa (later Tanganyika, now Tanzania), he managed to collect amphibians and reptiles, sometimes under the most difficult of circumstances. As recounted in his later autobiographical book covering this period ("Many Happy Days I've Squandered," 1944), Loveridge sometimes had to dodge sniper fire to capture specimens and one time held up his regiment's departure in order to photograph a chameleon. His sergeant once indignantly asked him: "Is this a war or a blooming museum expedition?" Toward the end of his service he was stationed in northern Mozambique, but soon he returned to the museum in Nairobi. In 1921 he became Assistant Game Warden of Tanganyika, a position that put him where he wanted to be most—in the field—but it did not last long, for in 1924 he moved to the United States.

Beginning in 1919, Loveridge regularly sent herpetological specimens to the Museum of Comparative

Zoology at Harvard College. The museum collections were then in the charge of Thomas Barbour, who in 1924, using his own funds, purchased all of Loveridge's collections for the museum with the stipulation that he come with them. The Harvard collections had been ill curated and were desperately in need of the reorganization and meticulous care that Loveridge gave them for the next 33 years. His new title was Associate in Zoology, but in fact he was assistant to Barbour and, by default, curator, although he did not become curator in title until after 1927 when Barbour was named director of the museum. From 1931, Loveridge had the aid of Benjamin Shreve, an unpaid volunteer whose interests in the Neotropics complemented his own.

Loveridge made five year-long expeditions to East Africa: the Uluguru and Usambara mountains (1926-1927), the highlands of southwestern Tanganyika (1929-1930), Kenya and eastern Uganda (1933-1934), Kenya, Tanganyika, and Uganda (1938-1939), and Nyasaland (now Malawi) and Tete, in Mozambique (1948-1949). He was one of the last broad field collectors and brought back specimens of all groups of vertebrates as well as many invertebrates and ethnographic material. Parts of these enormous collections were used for exchange with other institutions and, together with the purchases made possible by Barbour's wealth, Loveridge was enabled to continue the rapid growth of Harvard's herpetological collection. Loveridge was the consummate curator: a stickler for neatness and order, attributes that extended to his concepts of species and classification. His passion for simplicity sometimes led him to grossly underestimate species diversity and he treated species more as objects to be catalogued than as biological entities. Nevertheless, among his some 200 titles (over the period 1913-1979), he published a dozen major revisions of African genera and families of lizards, snakes, and turtles (1936-1958) which provide a valuable basis for today's research. Loveridge retired in 1957 and moved to the British island of St. Helena in the South Atlantic Ocean where he continued some zoological work. On 16 February 1980, he died at the age of almost 89.

Loveridge's major works include a monograph, with Thomas Barbour, on Tanganyikan herpetology (1928), "Reptiles of the Pacific World" (in two editions, the first a pocket-sized version intended for the use of military forces during World War II issued in 1945, the second a trade edition in 1946 that was reprinted in 1974), "Revision of the African Tortoises and Turtles of the Suborder Cryptodira" (1957), co-authored by Ernest E. Williams, his eventual successor as curator at Harvard, and later, in 1957, his "Check List of the Reptiles and Amphibians of East Africa."

The book on Pacific reptiles (and amphibians), which was very widely distributed, led to the receipt by Loveridge of large collections made by military and other personnel in that region, and no wonder; at the end of the chapter on proper techniques for preservation and shipment, there was a helpful statement showing how the box should be addressed to Harvard! Loveridge had long had an interest in the herpetofauna of Australia and New Guinea, and he published about 20 papers on the subject from 1927 to 1956. His major research interest, of course, was African herpetology, but he also managed to publish papers concerning Tropical America, North America, Borneo, and the Philippines.

Besides his technical works, Loveridge was a skilled popular writer and published several books, most of which went through numerous editions and translations: "Tomorrow's a Holiday" (1947), "I Drank the Zambesi" (1953), and "Forest Safari" (1956), in addition to his 1944 book noted earlier. These were necessitated in part by the meagre salary of a curator who was not concurrently a professor. The stock of one other book, "Snake Hunting in East Africa" (1938), was destroyed in the bombing of London, except for two copies.

• *References*: "Obituary: Arthur Loveridge (1891-1980)," by D. G. Broadley, Jour. Herpetol. Assoc. Africa, 23: 2, 1980; "Arthur Loveridge 1891-1980," by R. Conant, Herpetol. Rev., 11: 87-88, 1980; "In Memoriam: Arthur Loveridge," by C. Gans, Herpetologica, 37: 117-121, 1981; "Arthur Loveridge—A Life in Retrospect," by E. E. Williams, Breviora, Mus. Comp. Zool., Harvard Univ., 471: 1-12, 1982. • *Portrait* (1932): Library, Museum of Comparative Zoology, Harvard University, courtesy Roxane Coombs. • *Signature* (1973): Adler collection.

GLOYD, Howard K. (1902-1978).

The American herpetologist Howard Kay Gloyd, a leading specialist on pit vipers, was born in De Soto (Johnson Co.), in the farm country of eastern Kansas, on 12 February 1902. His mother encouraged his interest in natural history, especially ornithology, when he was a young boy, and by the age of 12 he was already an accomplished taxidermist.

He attended nearby Ottawa University (B.S. 1924) and at Kansas State College completed his Master's thesis (1929) on reproduction in the copperhead (*Agkistrodon contortrix*) while serving on the teaching staff. One of Gloyd's undergraduate students at that time was Hobart M. Smith, whom he inspired to switch from an intended career in entomology.

During the summer of 1925, Gloyd had attended the University of Michigan's Biological Station where he met Frank N. Blanchard, then the leading student of North American snakes. By 1927, when he again returned to the station, an agreement was reached whereby, after finishing his Master's degree, Gloyd would go to the University of Michigan as Blanchard's student to write a doctoral thesis on rattlesnakes. That was completed in 1937 and later published as a book, "The Rattlesnakes, Genera *Sistrurus* and *Crotalus*: A Study in Zoogeography and Evolution" (1940, reprinted 1978), which firmly established Gloyd's reputation.

Shortly before receiving his Ph.D. degree, Gloyd went to the Chicago Academy of Sciences as its Director. This is a small museum by comparison to its downtown rival, the Field Museum of Natural History, and Gloyd quickly made herpetology its major research focus. Despite a limited budget, the academy was able to issue numerous publications, but only because Gloyd took full advantage of the academy's primitive printing press. This was operated by a journeyman printer assigned by the Works Progress Administration, or WPA, the government agency organized to provide employment during the Great Depression of the 1930s. For a number of years (1936-1941) the journal *Herpetologica* was also printed on this same press by its co-editor, Walter L. Necker, who was a staff member at the academy. In 1958 Gloyd left the academy and became Lecturer and Research Associate in the University of Arizona's zoology department, until becoming Professor Emeritus in 1974. He died in Tucson, Arizona, on 7 August 1978.

Gloyd's interest in rattlesnakes and other pit vipers began when he was a boy in Kansas and became the main theme of his life's research, although he also published on other reptiles, amphibians, and birds. However, his study of rattlesnakes was not without competition, for Afrânio do Amaral and, particularly, Laurence M. Klauber also studied them. Early in their work Gloyd and Klauber agreed to an amicable division of taxa for separate study—Gloyd to do *Sistrurus* and the eastern *Crotalus*—and they even matched the format of their data sheets in order to share information. Both wrote many technical papers on rattlesnakes in which numerous new taxa were described (Gloyd himself named eight rattlesnakes), and both wrote books summarizing the two genera, but these were quite different and largely complementary in content. Whereas Klauber's book was more wide-ranging and popular in tone, Gloyd's was the more technical.

Gloyd's other major interest was *Agkistrodon*, on which he published throughout his life and which became his major area of research after moving to the University of Arizona in 1958. In the early 1930s, he and Roger Conant undertook the task of monographing the American species, but that project soon expanded by plans to include all members of the genus worldwide. They co-authored three papers (1934-1943), but the project fell into abeyance until rejuvenated by Gloyd and then rejoined by Conant in 1976. This undertaking, which included Asian species now allocated to four genera, after a gestation period in excess of half a century, is finally to be published this year under the title "Snakes of the *Agkistrodon* Complex: A Monographic Study."

Gloyd authored nearly 100 titles, including about 75 on herpetology over the period 1926-1989. An active field naturalist who participated in numerous expeditions to the American South and Southwest, he was also an accomplished photographer who developed several new photographic techniques.

• *References*: "Howard Kay Gloyd: Biographical Notes and the Genesis of His 'Rattlesnakes'," by H. M. Smith, p. vii-xvi. *In* Gloyd's Rattlesnakes, repr. ed., Soc. Study Amphib. Rept., Athens (Ohio), 1978; "Howard Kay Gloyd 1902-1978," by R. Conant, Herpetol. Rev., 9: 127-129, 1978; "Howard Kay Gloyd (1902-1978)," by W. M. Clay, Copeia, 1979: 187-189, 1979. • *Portrait* (1939): Courtesy Kathryn J. Gloyd and Roger Conant. • *Signature* (1977): Adler collection.

MYERS, George S. (1905-1985).

George Sprague Myers, American ichthyologist, herpetologist, and biogeographer, was born in Jersey City, New Jersey, on 2 February 1905. He had an early interest in aquarium fishes about which he published his first paper at the age of 15. At that time he began regular visits across the Hudson River into New York City and the American Museum of Natural History. There he came under the influence of several staff ichthyologists and herpetologists, including John T. Nichols, Charles M. Breder, Karl P. Schmidt and, especially, G. Kingsley Noble who sometimes took Myers along on his field trips to collect frogs in the New Jersey Pine Barrens.

Myers spent so much time as volunteer assistant at the museum (1922-1924) that his school work was neglected and, as a result, he never graduated. Fortunately, the ichthyologist Carl H. Eigenmann learned about Myers's plight and invited him to Indiana University as a special student and his assistant. During this period Myers published several papers on the herpetology of Indiana. Soon, he had an invitation to join Eigenmann's own teacher, David Starr Jordan, then the preeminent American ichthyologist, at Stanford University in California, where he arrived in the fall of 1926.

Before Jordan died in 1931, Myers had completed his A.B. (1930) and A.M. degrees (1931). His doctoral thesis (Ph.D. 1933) was on the African cyprinodont fishes and their geographic distribution, which gave an early indication of his lifelong interest in biogeography. He was then hired by Leonhard Stejneger to become Assistant Curator in charge of fishes at the National Museum, and in 1936 Myers returned to Stanford University as Associate Professor and

Head Curator of the zoological collections. He actively built the collections, especially those of lower vertebrates, as well as the associated library.

During World War II, Myers was posted to Brazil by the U.S. Government, as part of an overall effort to foster inter-American cooperation. He was associated with the Museu Nacional in Rio de Janeiro in 1942-1944, during which time he made extensive collections of fishes, amphibians, and reptiles and later published several papers, some of them with his museum colleague, Antenor Leitão de Carvalho. Among other results, they settled the perplexing status of the plethodontid salamander "*Ensatina platensis*," described by M. Jiménez de la Espada as from Uruguay which, in fact, turned out to be from California (published 1945). Later, Myers conducted expeditions to other parts of Central and South America, but also to various Pacific Islands and western North America.

Myers published more than 600 titles, about half of them on ichthyology. Herpetological papers (1922-1962) total 51 in number and cover such diverse topics as South American lizards and frogs, Californian salamanders, Solomon Islands frogs, Chinese newts, and Indian frogs and snakes, but he never published any major *opus*.

Myers had his most significant impact as a teacher of graduate students. After the war, Myers built at Stanford University a dynasty of students in ichthyology and herpetological; at times the number of doctoral students exceeded a dozen and this program continued until his retirement in 1970. Most of his students specialized in fishes, but among his herpetological doctoral students were Angel C. Alcala, Steven C. Anderson, Benjamin H. Banta, Walter C. Brown, Frank S. Cliff, Alan E. Leviton, T. Paul Maslin, and Jay M. Savage. His systematics courses in ichthyology and in herpetology were renowned for his virtuoso performances: two-hour lectures, often without notes or a stop, having an historical bent complete with displays of the masterworks from his personal library. Surely his deepest disappointment came when, shortly after his retirement, Stanford University abandoned teaching in systematics and donated the collections to the California Academy of Sciences.

After retirement, Myers lived in Scotts Valley, near Santa Cruz, California, where he died on 4 November 1985.

• *References*: "On the Natural History of George Sprague Myers," by L. A. Walford [and partly autobiographical], Proc. Calif. Acad. Sci., ser. 4, 38: 1-17, 1970; "George Sprague Myers 1905-1985," by A. E. Leviton, D. C. Regnery, and J. H. Thomas, Campus Report (Stanford Univ.), 9 April 1986, p. 19, 1986; "George Sprague Myers 1905-1985," by D. M. Cohen and S. H. Weitzman, Copeia, 1986: 851-853, 1986; "Guide to the Papers of George Sprague Myers," by W. E. Cox, Guide Colls. Smithsonian Inst. Arch., 8: v, 1-86, 1988.
• *Portrait* (1936): By Howard K. Gloyd, courtesy Kathryn J. Gloyd and Roger Conant. • *Signature* (1965): Adler collection.

CARR, Archie (1909-1987).

Archibald Fairly Carr, Jr., conservationist, writer, and authority on sea turtles, was born in Mobile, Alabama, on 16 June 1909. His father, a minister, instilled in him a love of nature and for good writing which provided the pillars for his life's work. He began at the age of three with a turtle, and kept many reptiles as pets. After family moves to Fort Worth, Texas, and later to Savannah, Georgia, he entered Davidson College, near Charlotte, North Carolina, as an English major and with writing as a career goal, then transferred to Florida's Rollins College near Orlando, and finally to the University of Florida where he graduated with a B.S. in English (1932). At Florida his interest in natural history was rekindled and he switched to biology for his graduate work (M.S. 1934, Ph.D. 1937), under the supervision of J. Speed Rogers, a limnologist. His doctoral thesis, which was the first comprehensive account of the herpetology of Florida, was later published as a book-length monograph in 1940.

Carr was appointed to the biology staff at the University of Florida in 1937 and for the next seven years spent each summer at Harvard University where he was associated with Thomas Barbour, who was clearly the dominant figure during Carr's formative years. His research with turtles increased and, in collaboration with Barbour, he published a beautifully-illustrated monograph, "Antillean Terrapins" (1940). Carr's main interest during the 1930s and 1940s was systematics, and he published numerous papers beginning in 1934, including descriptions of several new turtles (*Graptemys* and *Pseudemys*) from North America and the West Indies, as well as the spectacular blind cave salamander (*Haideotriton wallacei*) of Georgia and Florida.

Carr's first paper on sea turtles appeared in 1942, but not until the 1950s did this topic predominate. During this period he published his "Handbook of Turtles" (1952), an outstanding technical review of all species north of Mexico. This book also gave the first example of his superb literary skills and won for Carr the Elliott Medal of the U.S. National Academy of Sciences (1952). In 1955, together with his University of Florida colleague, Coleman J. Goin, Carr published a guide to the lower vertebrates of Florida, but this was the last of his major works on groups other than sea turtles, except for a semi-popular book, "The Reptiles," issued in the *Life* "Nature Library" series (1963, later edition 1970).

In the early 1950s Carr began his research on the life histories of sea turtles and soon recognized their plight worldwide and especially in the Caribbean. Alarmed by the rapid decline in populations and the decimation of the remaining nesting beaches, he wrote "The Windward Road" (1956), which has been recognized as one of the classics of nature writing and earned him the O. Henry Memorial Award for short stories (1956) and the Burroughs Medal for nature writing (1957). From this point Carr's interests included both basic research and conservation of sea turtles. In 1954 Carr began a tagging program in northeastern Costa Rica, near the village of Tortuguero, which is the principal nesting area for green turtles (*Chelonia mydas*) in the Caribbean and from which animals regularly fan out as far as Florida, Mexico, and Surinam. Up to 1987, more than 35,000 adult females had been tagged at Tortuguero by Carr and his collaborators in order to study migration routes and orientation mechanisms, nesting behavior and physiology, nutrition, and other aspects of their behavior, as summarized in their numerous papers. Through Carr's efforts the area became Tortuguero National Park and the tagging program continues today, making it the longest continuous population study of any reptilian or amphibian species.

In 1967 Carr summarized the work on green turtles to date in his well-received book "So Excellent a Fishe," but his conservation interests extended to other subjects as well. His concern for the problems of game management in Africa resulted in a book, "Ulendo" (1964), and in 1969, together with his wife, Marjorie H. Carr, he founded Florida Defenders of the Environment. For his conservation efforts Carr was awarded the Gold Medal of the World Wildlife Fund (1973) and the Borland Award of the National Audubon Society (1984). His life's work was recognized by Tulane University (honorary D.Sc. 1985) and finally, less than two weeks before his death—at his home in Micanopy, just south of Gainesville, on 21 May 1987—by his election as Eminent Ecologist of the Ecological Society of America.

Besides his seminal contributions to turtle biology and conservation, Carr is remembered as one of the most literate of biologists. His love of the sounds and rhythms of English made him a superb writer and his ability to generalize the simplest observation of nature into something of general interest earned him a wider readership than any other herpetologist.

• *References*: "Archie Carr: In Memoriam," by D. Ehrenfeld, Conserv. Biol., 1: 169-172, 1987 (reprinted in Copeia, 1987: 1087-1090, 1987); "Archie Carr: To the Edge of Hope 1909-1987," by J. R. Spotila, Herpetologica, 44: 128-132, 1988. • *Portrait*: University of Florida Archives, courtesy Joyce Dewsbury. • *Signature* (1976): Adler collection.

GOIN, Coleman J. (1911-1986).

The American herpetologist Coleman Jett Goin was senior author of the most widely-used textbook of herpetology and the first in English. He was born on 25 February 1911 in Gainesville, Florida, where he learned about natural history by wandering the fields as a boy. Due to family circumstances, he had to work and could not finish high school, but he won a scholarship to the University of Pittsburgh in 1935. After a year, he transferred to the University of Florida where he completed all of his academic work (B.S. 1939, M.S. 1941, Ph.D. 1946). His herpetological interest began during his student days in Pittsburgh when he did volunteer work at the Carnegie Museum; during the period 1936-1946 he was employed at the museum each summer. After finishing his doctoral degree, under the limnologist J. Speed Rogers, Goin became an instructor in the zoology department at the University of Florida and was promoted to Professor in 1956. He retired in 1971 and moved to Flagstaff, Arizona, where he was associated with the Museum of Northern Arizona. Goin died in Flagstaff on 12 May 1986.

Goin's primary research interest was amphibians, and he published diverse studies on these animals. He was an excellent field collector, in spite of an arm crippled by polio during childhood. Among some 120 herpetological titles (1938-1978), he described numerous new frogs from the West Indies and North and South America, several North American salamanders including fossil taxa, and a few reptiles. He published many papers on Floridian species, including detailed studies on the eggs and larvae of salamanders and on frog behavior. Beginning with his doctoral thesis (published 1947), he had a longtime interest in color pattern inheritance in frogs (*Eleutherodactylus*). A topic of special concern was the family Sirenidae, which Goin steadfastly championed as a separate order of amphibians. Other areas of interest and publication included dentition of frogs, adaptation of amphibians to land environments, and the nuclear DNA content of amphibians and its relationship to life history and phylogeny.

Goin is best remembered for his numerous books. The first, co-authored with his University of Florida colleague, Archie F. Carr, was a guide to the lower vertebrates of Florida (1955, reissued 1959). All of Goin's other books were written either with his wife, Olive B. Goin, or with Doris M. Cochran. With the former he published a vertebrate comparative anatomy manual (1965) and a human-oriented introductory biology text (1970, second edition 1975), as well as a little book about amphibians entitled "Journey Onto Land" (1974). With Cochran he published "Frogs of Colombia" (1970), which covers 196 species, and a field guide to North American amphibians and reptiles (1970) which did not receive wide use.

Goin's most important herpetological book, however, was his textbook, "Introduction to Herpetology" (1962), co-authored with Olive B. Goin. This was the first university-level herpetology text in English (P. V. Terentjev had published the first in 1961, in Russian), and it was adopted in undergraduate herpetology courses at about fifty colleges and universities in the United States, Canada, and elsewhere. Further editions were issued in 1971 and 1978 (the latter co-authored also by George R. Zug).

• *References*: "Coleman J. Goin, 1911-1986," by C. J. McCoy, Herpetol. Rev., 17: 55, 1986; "Coleman J. Goin 1911-1986," by E. H. Colbert, Copeia, 1986: 1041-1043, 1986. • *Portrait*: Courtesy Olive B. Goin. • *Signature* (1947): Courtesy M. Graham Netting.

COWLES, Raymond B. (1896-1975).

Raymond Bridgman Cowles, American herpetologist and conservationist, was one of the founders of physiological ecology. He was born at the Adams Mission Station in Amanzimtoti, Natal, South Africa, on 1 December 1896, where his grandfather Bridgman, originally from Massachusetts, had been the first Christian missionary in the Umzumbi Valley. As a boy Cowles was interested in birds and reptiles and also in the native Zulus and their relationship to the environment, subjects that were to become important themes of his professional career. Cowles went to southern California to complete high school and for his undergraduate training (Pomona College, B.A. 1921), during which time he published his first article on herpetology (1920), and then went to Cornell University where be became a student of Albert H. Wright.

Cowles's doctoral thesis concerned the life history of the Nile monitor lizard (*Varanus niloticus*), which he studied in Natal in 1925-1927 (published 1930). In particular, he investigated an observation he had made as a boy, that Nile monitors often lay their eggs in termite mounds which act as incubators. Before receiving his Ph.D. in 1929, he had joined the faculty of the University of California at Los Angeles (UCLA), and ten years later established a field station near Indian Wells, in the Coachella Valley just southeast of Palm Springs, where he began his first detailed studies of thermoregulation in reptiles. He remained at UCLA until his retirement in 1962 when he moved to the University of California at Santa Barbara. He died on 7 December 1975 in Santa Barbara.

Long interested in the adaptation of reptiles to hot environments, Cowles was particularly intrigued by work done by two friends in southern California: Laurence M. Klauber, an amateur in San Diego, who had discovered the importance of air temperature in predicting the activity of particular reptilian species at night, and Walter Mosauer, his UCLA faculty colleague, who performed the first laboratory experiments on tolerance of reptiles to high temperatures and pioneered work on the functional morphology and locomotion of snakes. When Mosauer died tragically in 1937, at the age of 32, Cowles extended his physiological research and expanded it to include field studies at his Coachella Valley station.

Cowles's main interest at that time was the extinction of the dinosaurs at the end of the Mesozoic. He had developed the unorthodox view that these mass extinctions were due to *increased* temperatures, and he saw in the local lizards and snakes an opportunity to study living models. During the 1940s Cowles and his students investigated the thermophysiology of reptiles and demonstrated that reptiles can and do regulate body temperature by behavioral means, thus rendering obsolete the classic division of animals into homeotherms and poikilotherms. It was Cowles (1940), in fact, who coined the new terms endotherm and ectotherm, to denote the source of the energy that warms the body.

Cowles's research on this topic was scattered in numerous publications (over the period 1939-1965), of which the best known was that co-authored with Charles M. Bogert ("A Preliminary Study of the Thermal Requirements of Desert Reptiles," 1944, reprinted 1974). This classic paper, in turn, led to studies on alligators, jointly with Edwin H. Colbert and Bogert (1946, 1947), which were more relevant to the original question of dinosaur extinction; these experiments provided the first real evidence about the importance of body size in thermoregulation. In his studies on temperature regulation Cowles had the euthusiastic collaboration of his several students, among them Bogert, Bayard H. Brattstrom, Charles H. Lowe, Kenneth S. Norris, Robert C. Stebbins, and Richard G. Zweifel.

In the 1950s Cowles returned twice to the Umzumbi Valley of his youth to find it vastly different: the land was overgrazed and barren and thus susceptible to serious erosion during flooding. The modern agricultural and medical practices introduced partly by his own family had unwittingly led to human overpopulation with disastrous results. He recounted these experiences in his book "Zulu Journal" (1959), and thus was one of the first writers to warn of the danger of over-use of natural resources by a burgeoning human population. Cowles's interest in his later years was the environment and its preservation, which became the subjects of several scientific and popular papers.

• *References*: "Preface," by F. H. Pough, p. i-iv. *In* Cowles and Bogert's Preliminary Study of the Thermal Requirements of Desert Reptiles, repr. ed., Soc. Study Amphib. Rept., Athens (Ohio), 1974; "Desert Journal" [partly autobiographical], by R. B. Cowles, collab. with E. S. Bakker, Univ. Calif. Press, Berkeley, xiii, 257 pages, 1977; "Raymond Bridgman Cowles 1896-1975," by B. H. Brattstrom, Copeia, 1977: 611-612, 1977; "Raymond B. Cowles and the Biology of Temperature in Reptiles," by J. S. Turner, Jour. Herpetol., 18: 421-436, 1984. • *Portrait*: Courtesy Archives, University of California at Los Angeles. • *Signature* (1930): Courtesy Cornell University Graduate School.

KAUFFELD, Carl (1911-1974).

Carl Frederick Kauffeld followed in the tradition of Raymond L. Ditmars to become one of America's leading popularizers of reptiles. He was born in Philadelphia on 17 April 1911 and became intrigued with reptiles at the age of nine when he read Ditmars's "Reptile Book." As a high school student he was a volunteer assistant in entomology at the Academy of Natural Sciences and also volunteered in the reptile house at the Philadelphia Zoo. He graduated from high school (1929) and a college preparatory school (1930) and later briefly took some courses at Columbia University in New York City. During 1930-1936 he was on the staff of the Department of Herpetology, headed by the autocratic G. Kingsley Noble, at the American Museum of Natural History. At first he was apprenticed to Clifford H. Pope but when Pope began work on his book on Chinese reptiles, Kauffeld had day-to-day responsibility for the herpetological collections.

Despite the opportunities at the museum, Kauffeld's real interest was in living reptiles, so when the post of Curator of Reptiles at the newly-formed Staten Island Zoo, in New York City, was offered in the summer of 1936, he accepted. Fully half the exhibition space at the small zoo was devoted to reptiles, so the curatorship was all the more attractive.

Kauffeld was able to develop one of the largest collections of living reptiles in the United States, one that once boasted having on exhibition all taxa of North American rattlesnakes, which were animals of lifelong interest to him.

Kauffeld pioneered techniques for maintaining captive reptiles, including cage sanitation, odor manipulation for feeding recalcitrant snakes, the use of ultraviolet light, and humidity regimes. His development of new techniques to combat disease doubtless benefited from his service during World War II as a technician for the Army Medical Corps (1943-1946). His longevity records for many species of snakes are still unsurpassed. Kauffeld was a popular lecturer and made numerous appearances on radio and television, and regularly lectured to students at Staten Island's Wagner College, all primarily to popularize reptiles.

Kauffeld published more than 200 titles, most of them of a popular or semi-popular nature intended for the general public or the amateur. He published three books which had a profound influence on amateurs, although not always a beneficial one. In 1937, with C. Howard Curran, he wrote "Snakes and Their Ways," followed by his own "Snakes and Snake Hunting" (1957) and "Snakes: The Keeper and the Kept" (1969). These books were based upon Kauffeld's zoo experiences and his field work. His expeditions to the New Jersey Pine Barrens, Lake Okeechobee in Florida, the canyons of Arizona's Huachuca Mountains, Dutchess County in New York State, and especially to Okeetee in the flat pine country of southernmost South Carolina are now part of the folklore of American herpetology. Conservationists, however, bitterly criticized Kauffeld for pinpointing favorable localities, some of which were later destroyed by overzealous collectors. Caravans of cars regularly descended upon Okeetee each spring and on one occasion a group of eager collectors arrived in a bus especially chartered for the purpose!

Kauffeld's influence at the zoo was distinctly positive. Many of his former assistants, all trained by him, went on to curate collections of reptiles at major American zoos, including Louis Pistoia, William H. Summerville, John E. Werler, Robert T. Zappalorti, and David Zucconi. Kauffeld guided a local group, called the Staten Island Herpetological Society, which provided a stimulus for many New York City amateurs to study herpetology, among them William G. Degenhardt, Carl Gans, and Richard Highton who later became professionals. In 1948 Kauffeld turned down an opportunity to direct the Milwaukee Zoo and elected to stay at the Staten Island Zoo which named him its Director in 1963. He held that post, in addition to his curatorship, until retirement in 1973. He died on Staten Island on 10 July 1974.

• *References*: "Carl Frederick Kauffeld," by R. Conant, Herpetol. Rev., 6: 27-28, 1975; "The Herpetologists—Part I," by R. Pinney, Herp, Bull. New York Herpetol. Soc., 15(2): 27-43, 1980; "Carl F. Kauffeld Memorial Issue," Herp, Bull. New York Herpetol. Soc., 18(1): 1-64, 1984 [reprints of 13 autobiographical essays by Kauffeld]. • *Portrait*: Staten Island Zoo, courtesy William H. Summerville. • *Signature* (1933): Carnegie Museum, courtesy C. J. McCoy.

VOGEL, Zdeněk (1913-1986).

Zdeněk Vogel, Czech amateur herpetologist and popularizer of reptiles, was born in Kutná Hora, Middle Bohemia, on 10 November 1913. He attended primary and secondary schools in Prague, but had no formal university training. By profession Vogel was a writer and supported himself primarily through income from his many books, but also through trade in live animals. Because of his long interest in amphibians and reptiles, in 1947 he founded the Herpetological Station for the Study of Amphibians and Reptiles in the village of Suchdol near Prague. The Station, which Vogel supported privately, was the center of operation for his studies on the behavior and ecology of amphibians and reptiles until his death in Prague, on 9 December 1986.

Vogel played an important role in the popularization of amphibians and reptiles, through his lectures in Czechoslovakia and other countries and particularly through his several books. Beginning in 1930 he travelled widely in Europe, Asia, Africa, and South America. During a visit to Cuba he was awarded an honorary title of Professor by the Cuban Academy of Sciences (1962). His first major work, "Reptile Life," appeared in 1946, followed by "Aus dem Leben der Reptilien" (1954), "Wunderwelt Terrarium" (1962 and 1963 editions; various English editions, 1964-1966), "Reptiles and Amphibians" (1966), and "Tierfang am Orinoko" (1973). All of these were extensively illustrated, and for Western audiences they provided the first glimpse of terrarium studies of amphibians and reptiles in the socialist countries. Vogel's special interest was snakes, particularly boas and pythons, about which he wrote numerous articles and also a book, in 1968 (second edition, 1973).

• *References*: "Zdenek Vogel," anonymous, Herpetol. Rev., 18: 26, 1987; "Prof. Zdeněk Vogel Zemřel," by J. Čihař, Akvar. Terar., Prague, 1987(7): 23-24, 1987. • *Portrait*: Courtesy Zdeněk V. Špinar. • *Signature* (1955): Courtesy Oliver G. Dely.

MARTÍN DEL CAMPO, Rafael (1910-1987).

Rafael Martín del Campo y Sánchez, long-time dean of Mexican herpetologists, was born in Guadalajara, Jalisco, on 3 January 1910. His entire professional career was spent in Mexico City at the Universidad Nacional Autónoma de México (UNAM), where he began as a student at this then newly-created university, graduating in biology in 1931 and earning a Master's degree in 1937 under Isaac Ochoterena, who published on the anatomy of amphibians and reptiles. While still a student, he was an assistant in the Instituto de Biología at the university, then became curator and head of the old Museo de Historia Natural. In 1964 he took a professorial position on the science faculty at UNAM and later became its dean. During his long tenure at the university he supervised many herpetological students, including Gustavo Casas Andreu, Oscar Flores Villela, and Zeferino Uribe Peña. He became Professor Emeritus in 1985 and died in Mexico City on 25 December 1987.

Martín del Campo's contributions to Mexican herpetology and biology were diverse, beginning with his first paper, in 1934, on the Bidder's organ of *Bufo*. His herpetological titles span the period 1934-1983, with several now in press, but he published on all vertebrate groups. Besides numerous publications concerning distribution, zoogeography, and descriptions of new taxa, he had a special interest in ethnobiology, particularly the zoology of the Aztecs and other native Mexican peoples. His most extensive herpetological work was the volume "The Gila Monster and Its Allies" (1955), written in collaboration with Charles M. Bogert.

• *References*: "Semblanza del Maestro Rafael Martín del Campo," by C. Juárez López, Ciencias Informa, Ciudad Univ. [UNAM], 36(February): 8-9, 1988; "Rafael Martin del Campo 1910-87," by G. Casas Andreu and C. J. McCoy, Copeia, 1989: 546-547, 1989. • *Portrait* (1982): Courtesy Gustavo Casas Andreu. • *Signature* (1940): Courtesy C. M. Bogert.

LUTZ, Bertha (1894-1976).

Bertha Maria Julia Lutz, Brazilian naturalist and feminist, was born in São Paulo on 2 August 1894, the daughter of one of Brazil's leading biologists, Adolpho Lutz. Her interest in frogs began in childhood, when she went on collecting trips with her father, but her formal work in herpetology did not begin until after she was forty. She studied science at the Sorbonne (University of Paris) and, later, law at the Federal University of Rio de Janeiro. The latter was part of her preparation to provide legal assistance to the women's movement in Brazil, a cause that she espoused even as a youth. In 1922 she co-founded the first women's rights movement in Brazil, which in turn led to her membership on a committee to draft the new Brazilian Constitution (1932) that finally resulted in equal suffrage for women (1933).

Previously, Lutz had been appointed Secretary of the Museu Nacional in Rio de Janeiro and, by 1931, she had become head of the department of natural history. Despite her formal employment as a naturalist, she continued to hold important political posts, among them delegate to the San Francisco meeting at which the United Nations was founded (1945) and, at the age of 80, Brazilian representative to the Inter-American Commission for Women held in Washington (1974).

Bertha Lutz's herpetological work began as an assistant to her father, who had studied frogs as an avocation. She published her first strictly herpetological papers jointly with him (1938-1939) but after his death in 1940 she carried on his plan to monograph the frogs of Brazil. This resulted in a long series of papers (through 1977), particularly about species in the family Hylidae, that emphasized systematics, life histories, development, and behavior. One numbered series of eight titles (1949-1952) reviewed the frogs in her father's collection at the Instituto Oswaldo Cruz. Her *magnum opus*, "Brazilian Species of *Hyla*" (1973), was drawn from her father's and her own notes, together with photographs taken by her brother, Gualter A. Lutz. Unfortunately, the promised additional volumes, which were to cover other hylid genera and include the extensive series of watercolors of frogs painted for her father, never materialized. She died in Rio de Janeiro on 16 September 1976.

• *References*: "Bertha Lutz, Naturalist-Feminist (1894-1976)," by J. P. Kennedy, Herpetol. Rev., 8: 8, 1977; "Bertha Lutz 1894-1976," by J. P. Kennedy, Copeia, 1977: 208-209, 1977. • *Portrait and signature* (1974): Courtesy J. P. Kennedy.

DERANIYAGALA, P. E. P. (1900-1973).

Paulus Edward Pieris Deraniyagala, noted Sinhalese anthropologist, artist, and zoologist, was born in Colombo on 8 May 1900, the son of one of Ceylon's most renowned lawyers. His interest in natural history began as a boy, when he kept animals, even including a baby cobra, as pets in his bedroom. At the age of eight, when the family lived in the coastal town of Kalutara where his father was District Judge, he saw his first live sea turtles and crocodiles, animals that later became his chief herpetological interest. He attended various schools in Ceylon including, briefly, medical school, where he studied zoology. In 1919 he went to the University of Cambridge in England where he graduated (B.A. Trinity College 1922); he studied zoology, although he spent more time in the gym (he was a championship boxer). He stayed a fourth year (M.A. 1923), before going to the marine station at Wood's Hole in the United States and then a year at Harvard College (A.M.

1924). There he worked on frog physiology in the laboratory of George H. Parker, but he was also influenced by Thomas Barbour at the Museum of Comparative Zoology.

Up to this point in time he was known as Paul Edward Pieris, following prevailing fashion in Ceylon to adopt Portuguese Christian names at baptism, hence the family name Pieris. On returning home, however, he resumed the ancient native name of his family, Deraniyagala, and took a position in the Museums and Fisheries Department. In 1939, he became Director of the National Museums in Ceylon and concurrently held various positions at Vidyodaya University (1956-1964), eventually becoming a professor and Dean of the Faculty. Besides his special concern for reptiles, he was broadly interested in many other subjects, on which he wrote several books: elephants, the Pleistocene and fossils of Ceylon, and the weapons, combat, and sports of native tribes. To honor his career, he received honorary D.Sc. degrees from Vidyodaya (1960) and Ceylon (1963) universities. He retired from the museum directorship in 1963 and died in Colombo on 1 December 1973.

Deraniyagala published about 75 titles on the reptiles of Ceylon (now Sri Lanka), covering the period 1927 to 1975. Among his special interests were crocodilians and marine turtles and he wrote extensively on their nesting habits and embryology. His first herpetological book—and his most important scientifically—was "Tetrapod Reptiles of Ceylon. Volume I, Testudinates and Crocodilians" (1939); the volume on lizards was not published and a volume on snakes was never planned because of the existence of Frank Wall's book on Ceylon snakes (1921). In addition to systematics and natural history, Deraniyagala's book included extensive data and illustrations on embryological development. In 1949 he issued "Some Vertebrate Animals of Ceylon," a pictorial atlas. His most lavishly illustrated reptile book, "A Colored Atlas of Some Vertebrates from Ceylon," was published in two volumes (1953, 1955; volume 1 in the series was on fish) and contained a total of 49 color plates, all drawn by the author, as were most of the drawings he published during his career.

• *References*: "P. E. P. Deraniyagala," p. 446. *In* Marquis Who's Who in Science, Marquis, Chicago, 1968; "Paul Edward Pieris Deraniyagala," p. 89. *In* Fifth International Directory of Anthropologists, Univ. Chicago Press, Chicago, 1975; "Paul Deraniyagala Scholar and Man," by R. W. M. Dias, p. 1-59. *In* T. T. P. Gunawardana et al. (eds.), P. E. P. Deraniyagala Commemoration Volume, Lake House Invest. Ltd., Colombo, 1980 [1982?]. • *Portrait*: Courtesy Siran Deraniyagala. • *Signatures* (romanised form, 1931): Adler collection; (Sinhalese form): From Deraniyagala, 1955.

FITZSIMONS, Vivian F. M. (1901-1975).

Vivian Frederick Maynard FitzSimons, a member of the noted family of South African herpetologists, wrote the first handbooks of his country's reptiles. He was born in Pietermaritzburg on 7 February 1901 and educated in Port Elizabeth. After receiving his M.Sc. degree at Rhodes University, he joined the staff at the Transvaal Museum in Pretoria (1924) as head of the Department of Lower Vertebrates and Invertebrates. He travelled throughout the country and participated in several major expeditions, notably those to the high rainfall parts of the Kalahari (1930) and to South West Africa (1940), adding some 20,000 herpetological specimens to the museum's collection. He completed a D.Sc. degree at the University of the Witwatersrand (1942), his book "The Lizards of South Africa" (1943, reprinted 1970) serving as his thesis. FitzSimons became director of the museum in 1946 and, despite great financial difficulties, he directed the establishment of the Namib Desert Research Station in 1962. He retired as director in 1966, and was succeeded in that position by W. J. Steyn, a specialist on reptiles. FitzSimons died in Pretoria on 1 August 1975.

FitzSimons's father and younger brother were both herpetologists. The father (Frederick W. FitzSimons), an Irish immigrant, established and directed Africa's first snake park, in Port Elizabeth, and published four popular books on snakes; the brother (Desmond C. FitzSimons) founded Durban snake park and also directed the Clinsearch Serum Laboratories in Cape Town. With such a family it comes as no surprise that Vivian FitzSimons became interested in reptiles. Although an accident blinded him in one eye as a youth, he overcame this handicap and became both a prize-winning long-distance runner in college and an excellent collector of lizards.

Vivian FitzSimons published nearly 50 herpetological papers (1929-1966), but is best known for three books.

Besides his lizard book already noted, he published "Snakes of Southern Africa" (1962), an imposing volume covering 137 taxa (species and subspecies) then known from the area. The book was illustrated by 75 outstanding color plates drawn by P. J. Smit, who long before had done the illustrations for John Anderson's book on Egyptian reptiles. A revised version of FitzSimons's book, under the title "FitzSimons' Snakes of Southern Africa" (1983), covering 160 taxa, was prepared by Donald G. Broadley. Lastly, in 1970 FitzSimons published "A Field Guide to the Snakes of Southern Africa," with a second edition in 1974 revised by his herpetological successor at the museum, W. D. Haacke.

• *References*: "V. F. FitzSimons, A Biographical Appreciation," by R. F. Lawrence, Sci. Pap. Namib Desert Res. Station, 37: 1-4, 1969; "Dr. Vivian FitzSimons," anonymous, S. Afr. Jour. Sci., 71: 288, 1975; "Obituary Dr. V. F. M. FitzSimons," by W. D. Haacke, Jour. Herpetol. Assoc. Africa, 14: 33-35, 1975; "Dr. V. F. M. FitzSimons," anonymous, Transvaal Mus. Bull., 16: 14, 1978. • *Portrait* (1966): British Museum (Nat. Hist.), courtesy A. F. Stimson. • *Signature* (1964): Adler collection.

WITTE, Gaston-François de (1897-1980).

Gaston-François de Witte, Belgian-born specialist on the herpetology of Central Africa, was born in Anvers on 12 June 1897. He had an interest in reptiles as a boy and, at the age of 15, had a fortuitous encounter with George A. Boulenger, a Belgian by birth and then Europe's leading herpetologist, who became the major influence in his life.

Boulenger had come to examine some fossil fish collected at Denée by the Benedictine monks of Maredsous College, where de Witte attended secondary school. Boulenger returned regularly to examine other specimens and, incidentally, tutored de Witte. After Boulenger retired from the British Museum, he settled in Brussels and continued to advise his protégé who never had any formal training in zoology. After service in the First World War, de Witte was assigned to the Musée du Congo Belge in Tervuren (1920) and eventually rose to become head of zoology and entomology there before moving in 1937 to the Musée Royal d'Histoire Naturelle (after 1950 called Institut Royal des Sciences Naturelles) in Brussels, as Curator of Recent Vertebrates. He retired in 1951 and died on 1 June 1980 in Brussels.

De Witte's major research interest was the herpetology of the Belgian Congo (now Zaire) and he carried out seven major expeditions there over the period 1924-1957, primarily to the various national parks. He was a tireless field naturalist and he and his native assistants made truly enormous collections, some 173,000 specimens of amphibians and reptiles alone! These were carefully studied and written up in a long series of papers and books. De Witte's major herpetological monographs deal with the several parks—Albert (1941), Upemba (1953), and Garamba (1966)—but there are also reports on groups of particular interest: colubrid snakes (1947, co-authored with his colleague and compatriot Raymond Laurent), snake genera (1962), and chameleons (1965). He also co-authored a book on the protected animals of the Belgian Congo (1941, revised edition 1953) and wrote a herpetology of Belgium (1942, second edition 1948). All told, he published about 90 titles (1918-1980), most of them herpetological.

• *Reference*: "Gaston de Witte (1897-1980). Notice Biographique et Liste Bibliographique," by X. Misonne, Bull. Inst. Roy. Sci. Nat. Belgique, 52(20): 1-8, 1980. • *Portrait* (1961): By M. Graham Netting, courtesy Carnegie Museum. • *Signature* (1972): Adler collection.

SATO, Ikio (1902-1945).

Born in Okuakegata-mura, Gujo-gun, Ghifu Prefecture, on 22 November 1902, Sato became Japan's leading expert on salamanders despite dying tragically at a young age. After graduation from Hiroshima Higher Normal School, he taught natural history at the Prefectural Nagano Middle School in Nagano Prefecture before going to Hiroshima University where he spent the rest of his career. He graduated in 1932, attained the D.Sc. degree in 1941, and became assistant professor in 1943. On 6 August 1945, while waiting outdoors for a city streetcar to his laboratory, he suffered extensive radiation exposure during the atomic bombing of Hiroshima by the Americans and died five days later at Furuta-cho, Hiroshima City, at the age of 42 by Western accounting, having been just promoted to professor on 8 August.

In addition to his research on arachnids, Sato was primarily interested in the systematics and cytology of

Asiatic salamanders. His major work, "Tailed Amphibians of Japan," published in 1943, covers Imperial Japan, including Formosa (now Taiwan) and Korea. This is a masterful treatise, with beautiful color plates, and includes systematics, life histories, embryology, and chromosome morphology; its continuing value prompted a reprint edition in 1978. Unfortunately, the manuscript for the English edition of Sato's book, which he was working on at the time of his death, was burned when his home was destroyed in the bombing.

• *References*: ["Dr. Ikio Sato"], by S. Makino, Seibutu, 1(3): 192, 1946; ["Ikio Sato (1902-1945)"], p. 340. *In* Y. Ikoma (ed.), Collected Works on the Giant Salamander, Tsuyama Mus. Sci. Education, Tsuyama City, 1973. • *Portrait and signature*: Courtesy Toshijiro Kawamura.

LIU, Cheng-chao (1900-1976).

Liu, China's most prominent herpetologist, was born in the birthplace of Confucius—Tai'an, Shandong Province—on 12 August 1900. He graduated from Yenching University (now part of Beijing University) in 1929, where he was a student of the American biologist Alice M. Boring. He then taught at Northeastern University in Muckden (now Shenyang), where he lost his library and collections during the conflict between China and Japan in 1931. During 1932-1934 Liu attended Cornell University in the United States and completed his Ph.D. thesis there with Albert H. Wright before returning to China and a teaching position at Soochow (now Suzhou) University in Jiangsu Province. In 1939, he moved from Suzhou to Chengdu in western China just ahead of the advancing Japanese troops, to West China Union University (now West China University of Medical Science), where he later was promoted to Professor of Biology. In 1950-1951, he was head of the biology department of Yenching University before returning to Chengdu, in 1951, as President of Sichuan Medical College. Later he was elected Academic Member of Academia Sinica, the Chinese academy of sciences. He died in Chengdu on 9 April 1976.

Liu published 55 papers and two books on herpetological topics, beginning in 1929. Following in the tradition of his teachers Alice M. Boring and Albert H. Wright, his primary interest was to work out the life histories of Chinese amphibians. He was an indefatigable field collector, even well into his seventies. During 1946-1947 he visited America again and while headquartered at the Field Museum in Chicago he completed his *magnum opus*, "Amphibians of Western China," issued in 1950, a lavishly illustrated work with a wealth of natural history information based on his careful study. In 1961, together with his wife and lifelong colleague, Shu-qing Hu, Liu published the book "Chinese Tailless Amphibians," with 34 plates (including 28 outstanding color plates of tadpoles and adult frogs). Besides these major contributions, Liu had many prominent students, among them S.-q. Hu, N.-z. Huang, W.-s. Tian, F.-h. Yang, E.-m. Zhao, and C.-g. Zhu, and founded what is now the largest herpetological collection in China, originally at West China Union University but later transferred to the Chengdu Institute of Biology.

• *References*: "A Catalogue of the Works and Publications of Professor Liu Cheng-chao," anonymous, Acta Herpetol. Sinica, [old ser.], 3: 10-15, 1980; "In the Memory of My

Teacher—Liu Chengzhao," by E.-M. Zhao, Sichuan Jour. Zool., 2: 1-3, 1983. • *Portrait* (1960): Courtesy Shu-qing Hu. • *Signatures* (Chinese form and script romanised form): Chengdu Institute of Biology, courtesy Ermi Zhao; (printed romanised form, 1947): Courtesy M. Graham Netting.

BANNIKOV, A. G. (1915-1985).

Andrei Grigoryevich Bannikov, zoologist and conservationist and senior author of the best known field guide to the herpetofauna of the Soviet Union, was born in Moscow on 24 April 1915. He was long associated with the Moscow Pedagogical Institute and later became Chair of Zoology at the Moscow Veterinary Academy, the position he held at the time of his death on 17 October 1985, in Moscow. During 1942-1945 he was in Mongolia, during which time he helped organize the first Mongolian university, and his field work there later led to his 1958 paper, which was the first significant review of Mongolian herpetology. Besides his zoological work, Bannikov was an active conservationist. He began the Red Data Book project in the Soviet Union and won awards from the World Wildlife Fund (1972) and the International Union for the Conservation of Nature (1973) for his conservation activities, which included the establishment of several nature reserves in the Soviet Union.

Most of Bannikov's more than 400 titles deal with mammals and birds, but he made many importan contributions to herpetology, especially to the popularization of amphibians and reptiles in his native land. His special research interest was amphibian ecology. Bannikov's herpetological books included "Contributions to the Biology of Amphibians" (1956, with his wife and longtime co-worker, M. N. Denisova), "Amphibians and Reptiles of the USSR" (1971, with I. S. Darevsky and A. K. Rustamov), "Animal Life: Amphibians and Reptiles" (1969, with I. S. Darevsky, M. N. Denisova, N. N. Drosdov, and N. N. Iordansky; second edition 1985), "Field Guide to Amphibians and Reptiles of the USSR" (1977, with I. S. Darevsky, V. G. Ischenko, A. K. Rustamov, and N. N. Szczerbak), with excellent color illustrations and also detailed distribution maps for each species, and "Problems of Conservation and Natural Uses of Reptiles" (1978, with V. E. Flint and V. M. Makeev).

• *References*: ["In Memoriam Andrei Grigorevich Bannikov (1915-1985)"], by I. A. Abdusalyamov, G. S. Davydov, and A. I. Sokov, Izvest. Akad. Nauk. Tadzhik. SSR Otdel. Biol. Nauk., 1986: 92, 1986; ["Andrei Grigoryevich Bannikov"], by S. M. Uspensky, Ekologiia [Sverdlovsk], 1986(2): 95-96, 1986; "Andrei Grigoryevich Bannikov (1915-1985)," by I. S. Darevsky and K. Adler, Herpetol. Rev., 18: 25-26, 1987. • *Portrait*: Courtesy M. N. Denisova. • *Signature*: Academy of Sciences, Leningrad, courtesy Natalia B. Ananjeva.

VILLIERS, André (1915-1983).

André Villiers, French entomologist and herpetologist, was born on 9 April 1915. As a boy he was interested in natural history and at the age of 13 began to bring insects and reptiles that he had collected in the Parisian region, especially the forests at Fontainebleau, to the laboratories of the Muséum National d'Histoire Naturelle for identification. After military service in Algeria, he joined the staff of the museum as a technical assistant in entomology in 1937 and was promoted to Attaché in 1944, the year after he received his doctorate at the University of Paris.

In 1945, Villiers moved to the Institut Français d'Afrique Noire in Dakar, Senegal, where he founded the Section of Entomology and became its first Chief. Until he returned to Paris, he and his assistants made large collections of insects—and coincidentally reptiles and amphibians—on their expeditions throughout what was then French West Africa. On Villiers's return to the museum in 1956 he had an appointment in entomology and became a professor in 1976. While based in Paris, he continued to make expeditions to West Africa, as well as to Iran, Madagascar, and the French Antilles. He died on 8 June 1983 in Le Perreux, an eastern suburb of Paris.

Most of Villiers's 661 titles (over the period 1938-1982) were on entomology, but shortly after his arrival in Senegal his boyhood interest in reptiles was reawakened. He published about 35 papers on reptiles, principally snakes and turtles, and five books. The first of his books, "Les Serpents de l'Ouest Africain," was issued in 1950. This was an extensive illustrated key to the 124 species and subspecies of snakes found in a vast region of West Africa, extending from southern Morocco and Algeria south to the countries bordering the Gulf of Guinea and from there east to Chad and Cameroon. Second and third editions of this book were published in 1962 and 1975 (the number of taxa in the latter was 135); a fourth edition, co-authored with Villiers's former colleague in Dakar, Michel Condamin, is now in press.

In 1958 Villiers published "Tortues et Crocodiles de l'Afrique Noire Française," an excellent review of the biology and systematics of these animals which was far more comprehensive than his books on snakes. Unfortunately his promised final volume on lizards was never published. As far as herpetology is concerned, Villiers was primarily a field-naturalist interested in distribution and systematics; he also published several papers on the relationship of reptiles to humans, a topic that also formed the subject of a chapter in each of his books.

• *References*: "Notice sur les Titres et Travaux Scientifiques de A. Villiers," by A. Villiers, Mâcon, Protat, 35 pages, 1961; "André Villiers," by R. Paulian, A. Descarpentries, R. M. Quentin, L'Entomologiste, 39: 155-207, 1983; "In Memoriam: André Villiers (1915-1983)," by J. Carayon, Ann. Soc. Entomol. France, new ser., 19: 355-356, 1983. • *Portrait* (about 1965): L'Entomologiste, Paris, courtesy R. M. Quentin. • *Signature* (1956): Muséum National d'Histoire Naturelle, Paris, courtesy V. Van de Ponseele.

MCCANN, Charles (1899-1980).

Yule Mervyn Charles McCann, author of the first scientific monograph on New Zealand lizards, was born 14 December 1899 at Castle Rock in Portuguese Goa (now India). Educated at St. Xavier's College in Bombay (1916-1920), in 1921 he joined the Bombay Natural History Society (as mammal survey collector, later becoming Assistant Curator) where he remained until 1947 when, after Indian independence, he moved to New Zealand and the Dominion Museum (now National Museum of New Zealand) in Wellington. He retired in 1964 but then took a position at the New Zealand Oceanographic Institute, 1965-1969. Financially destitute, he later worked as a garage mechanic. He died at his home in Lower Hutt, across the harbor from Wellington, on 29 November 1980.

McCann's work in India covered botany as well as zoology, including herpetology. His manual of Indian grasses, co-authored with his former teacher at St. Xavier, Father E. Blatter, is still a standard work. In New Zealand, herpetology was again only one of his many interests, on which he published numerous papers as well as his major work, "The Lizards of New Zealand" (1955). He was an excellent field naturalist and prided himself on his considerable ability at observation and specimen preparation.

• *References*: "Charles McCann," by C. McCann, Notornis, 20: 190-191, 1973; "Charles McCann, 1899-1980," by H. Abdulali, Jour. Bombay Nat. Hist. Soc., 77: 494-495, 1980 (1981); "Dedication [to Charles McCann]," anonymous, p. 3-4. *In* D. G. Newman (ed.), New Zealand Herpetology, Wellington, 1982. • *Portrait* (1964): National Museum of New Zealand, courtesy John Yaldwyn. • *Signature* (1949): British Museum (Nat. Hist.), courtesy A. F. Stimson.

MITCHELL, Francis J. (1929-1970).

Francis John Mitchell, Australian herpetologist, was born on 8 August 1929 in Adelaide, South Australia. As a schoolboy he became interested in the identification of reptiles which led to his joining the staff of the South Australian Museum in 1946 as a Junior Cadet, while still attending the Adelaide Technical High School. He never attended a university, but spent his entire career at the museum, being promoted to Assistant Curator of Reptiles in 1955, Curator in 1956, and Senior Curator of Vertebrates a decade later. He was active in many other organizations, including the Underwater Research Group which he founded and served as President (1958-1964), and the Royal Society of South Australia (he held various offices, including the presidency in 1968-1969). After a long illness, he died at the early age of 40, on 23 February 1970, in Belair, a suburb of Adelaide.

Mitchell's primary research interests were the systematics and distribution of amphibians and reptiles, particularly squamates. He published only about 15 papers (1948-1965) during his short career, but these included important studies on lizards of the families Agamidae, Gekkonidae, and Scincidae. In 1964 he commenced a behavioral and ecological study of *Amphibolurus maculosus*, which inhabits the salt crust of Lake Eyre in northern South Australia, and he designed special chambers at the museum for their observation. Unfortunately this work could not be completed due to his premature death.

• *References*: "Francis John Mitchell, 1929-1970," by M. J. Tyler, Trans. Royal Soc. South Austr., 94: 249, 1970; "F. J. Mitchell," anonymous, Copeia, 1970: 598, 1970. • *Portrait and signature* (1965): South Australian Museum, courtesy Adrienne Edwards.

DONOSO-BARROS, Roberto (1922-1975).

Donoso-Barros, author of "Reptiles de Chile" and a leading specialist on South American amphibians and reptiles, was born on 5 October 1922 in Santiago, Chile. He attended the Universidad de Chile (M.D. 1947) and later joined the staff of the Instituto de Biología Juan Noé. In 1954, he became Professor of Biology at the Universidad de Chile, thus joining a group of professors at that university who led the study of all branches of natural history in Chile both by their own research and training of many students, and in 1965 he moved to the University of Concepción. Beginning in the 1940s and for many years thereafter, he also did voluntary work at the Museo Nacional de Historia Natural in Santiago. Although most of his work was centered in Chile, Donoso-Barros also held temporary appointments at the Universidad de Oriente in Venezuela (1963-1964) and in Argentina at the Universidad de Buenos Aires (1973) before his untimely death.

Donoso-Barros's herpetological publications spanned the period 1947-1975 and were concerned with systematics, life history, and behavior, primarily of Chilean species but also those of Venezuela, Bolivia, and other South American countries. His first major work, "Reptiles de Chile" (1966), was the first competent review of his country's reptiles. During 1967-1968, he collaborated with James A. Peters and Braulio Orejas-Miranda at the Smithsonian Institution in Washington to produce a monumental work, "Catalogue of the Neotropical Squamata" (1970; reprinted with addenda, 1986), which contains a checklist and key to all Latin American lizards and snakes (south of Mexico), in parallel

Spanish and English texts. In 1970 he also issued "Catálogo Herpetológico Chileno," a checklist with illustrated keys that was widely used in Chile. Donoso-Barros died in an automobile accident at the age of 52, in Concepción, on 2 August 1975.

• *References*: "Roberto Donoso-Barros (1922-1975) (In Memoriam)," by J. C. Ortiz Z., Anal. Mus. Hist. Nat. Valparaiso, 8: 1-3, 1975; "Roberto Donoso-Barros (1922-1975)," by M. Codoceo R., Notic. Mens. Mus. Nac. Hist. Nat., 230-231: 3-5, 1975; "Roberto Donoso-Barros," by J. E. Péfaur, Copeia, 1976: 219-220, 1976. • *Portrait*: From Péfaur, 1976. • *Signature* (1974): Adler collection.

HOGE, Alphonse Richard (1912-1982).

A specialist on the snakes and lizards of South America, Hoge was born on 5 September 1912 in Cacequi, Rio Grande do Sul, Brazil. His father, a Belgian engineer, was temporarily in Brazil to supervise the building of bridges. From 1913, his family lived in Gent, Belgium, where he attended the Rijksuniversiteit during 1929-1934 and studied science and medicine. He became an assistant to Georges Bobeau in Paris, who was investigating the use of snake venoms in arresting the growth of cancer cells. Hoge went to Brazil in about 1939 to work for the pharmaceutical firm Ciba, Roche and Torres. He joined the staff of the Instituto Butantan in 1946, was promoted to Director of its Division of Biology in 1969, and retired from the institute's staff just a few months before his tragic death, as the result of a medical accident, in São Paulo on 25 December 1982.

Hoge published 112 papers (1946-1981), nearly all of them herpetological and dealing with systematics and distribution, with descriptions of many new taxa. His major works were a distributional study of Neotropical viperids (1965), a review of the poisonous snakes of the Neotropics (1971), a synopsis of Brazil's poisonous snakes (1972 [1973]; second edition 1978/1979 [1981]) and part one (pit vipers) of a checklist of the poisonous snakes of the world (1978/1979 [1981]). All but the first title were co-authored with his wife, Alma Romano Hoge.

Hoge's most interesting work from a biological point of view was that on the pit viper *Bothrops insularis*, a species that is restricted to rugged Queimada Grande Island off the Brazilian coast near São Paulo. This species, discovered by Hoge's Butantan colleague, Afrânio do Amaral, has a high proportion of intersexuals, with the proportion of intersexes increasing markedly from the time of Amaral's original paper (1921) to the 1950s when Hoge and his colleagues studied the phenomenon in detail (major paper, 1959 [1960]).

• *Reference*: "Alphonse Richard Hoge (1912-1982)," by J. C. Machado, Mem. Inst. Butantan, 46: 1-12, 1982 (1983). • *Portrait*: Courtesy Alma Romano Hoge. • *Signature* (1965): Adler collection.

FUHN, Ion E. (1916-1987).

Ion Eduard Fuhn, author of the major reviews of Rumanian herpetology, was born in Bucharest on 23 January 1916. After an early interest in birds, he shifted to amphibians and reptiles, and the German herpetologist Willy Wolterstorff, with whom Fuhn corresponded, arranged for publication of his first paper in a terrarium journal (1932). Despite his amateur interest, Fuhn developed a very different career. He attended the University of Bucharest and graduated with degrees in law (1937) and philosophy (1938). From 1939 to 1943 he taught philosophy at a secondary school, then practiced law for two years while completing a doctorate in law (1946). In 1946 he was appointed as Attaché in the government's Ministry of Foreign Affairs but in the fall of 1947 he lost his position and all of his family's property in the political turmoil that soon led to creation of a socialist state. He was then assigned to the Academy of Sciences, as editor of their journals for nearly eight years, during which period he began his scientific career. In 1954 he was placed in charge of the academy's sections of herpetology and arachnology, from which posts he retired in 1976. He died in Bucharest on 31 August 1987.

Fuhn's scientific publications in herpetology span the period 1950-1986. His earliest titles were on amphibians, but in the 1960s lizards, particularly skinks, were his preoccupation, not only those of southeastern Europe but also of Asia, Australia, and Africa. During the same period he published several papers on the local herpetofauna, including studies on ecology and natural hybridization, and more recently on *Vipera* systematics and conservation. His most widely-known works, however, are his two volumes entitled "Amphibia" (1960) and "Reptilia" (1961, co-authored with Ştefan Vancea) in the series "Fauna Republicii Populare Romîne." These were the first major reviews of Rumanian herpetology since the monographs of Constantin Kiritzescu (1930) and Raul I. Călinescu (1931), and contain extensive information on systematics, distribution, and natural history. In 1969, Fuhn published "Broaşte, Şerpi, Şopîrle," a popular account of the Rumanian herpetofauna, in a pocket-sized format.

• *References*: "Ion Eduard Fuhn," p. 618. *In* A. G. Debus (ed.), World Who's Who in Science. Marquis-Who's Who, Chicago, 1968; "Ion E. Fuhn 1916-1987," by K. Klemmer, Natur und Museum, 118: 319, 1988; "Dr. Ion E. Fuhn 1916-1987," by G. H. Thommen, unpublished manuscript. • *Portrait* (1971): Courtesy G. Heinrich Thommen[-Fuhn]. • *Signature* (1969): Adler collection.

ROMER, Alfred Sherwood (1894-1973).

The American paleontologist Alfred Sherwood Romer, whose special interest was the osteology of fossil amphibians and reptiles, was born in White Plains, just north of New York City, on 28 December 1894. After high school, Romer worked as a railroad clerk, but in 1913 he won a scholarship to Amherst College in western Massachusetts. His major subjects were history and German literature, but a lone biology course—on evolution—reawakened in him his boyhood visits to the American Museum of Natural History and its displays of dinosaurs. So in a curious way, this one course, the required "science" course in the Amherst curriculum and typical of the so-called distribution courses of American universities, decided his career. He graduated from Amherst (A.B. 1917), then World War I intervened and he served as a junior officer in France. On returning, he became William K. Gregory's student at Columbia University and counted among his fellow students Charles L. Camp and G. Kingsley Noble. His thesis, on the limb muscles of extinct reptiles, was finished in two years (Ph.D. 1921), a phenomenal feat for someone with virtually no formal background in biology.

After a two-year job at Bellevue Medical School in New York City, Romer accepted a position as Associate Professor of Vertebrate Paleontology at the University of Chicago, a distinguished post previously held by the reptilian osteologists Georg Baur and Samuel W. Williston. It was during this period, in the mid-1920s, that Romer began his field work in the famous Permian "red beds" near Wichita Falls, Texas, where he usually worked alone and lived with local farmers and ranchers.

In 1934, Romer became Professor of Zoology at Harvard College and, concurrently, Curator of Vertebrate Paleontology in the Museum of Comparative Zoology. As time went on, he became closely identified with the museum and was named its Director in 1946, succeeding the herpetologist Thomas Barbour. Romer inherited an underfunded program and a grossly underpaid staff, the legacy of the two Agassizs, Louis and Alexander, and of Barbour who ran it as a fiefdom somewhat independent of the university and who, because of their own independent wealth, had no understanding of staff financial needs. Romer changed all that. He increased the museum's endowment tenfold and made the museum a regular department of the university and thus subject to its higher standards.

Although Romer retired as museum director in 1961, he continued his very active research program to the very end of his life. Among his students at Chicago and at Harvard were many paleontologists and anatomists of herpetological interest, including Donald Baird, Robert L. Carroll, David H. Dunkle, Everett C. Olson, Thomas S. Parsons, and during the early part of his doctoral work, Carl Gans. Romer died prematurely and tragically, however, in Cambridge, Massachusetts, as the result of an accident. He choked on some food on 20 October 1973 and lingered in a coma until his death, on 5 November.

Romer is best known to herpetologists for his massive source book, "Osteology of the Reptiles" (1956). Originally intended as an update of Samuel W. Williston's book of the same title (1925, reprinted 1971), Romer's was completely new and contains the most thorough account of the skeleton of both fossil and living forms that exists, as well as a classification of reptiles that lists all valid genera with their synonyms. He also published numerous other books on vertebrates, including "Vertebrate Paleontology" (1933, later editions in 1945 and 1966, with a separate volume of notes and comments in 1968), "Man and the Vertebrates" (1933, 1937, 1941 editions) and its replacement, "The Vertebrate Story" (1959). His most widely used book, however, was "The Vertebrate Body" (1949, with numerous editions and foreign language translations through 1972). This book became a popular mainstay of medical school students and university comparative anatomy courses.

Romer's skill as a teacher was only partly reflected in these clear, concise, well-illustrated books, for he was a flamboyant lecturer with an impish sense humor. He was famous for singing songs to make a point and for taking off his suit coat, rolling up his sleeves, and getting down on all fours to demonstrate to students some particular characteristic of limb movement. This was the same eminent scientist who held the presidencies of several of America's leading professional societies, was president of the 1963 International Zoological Congress, a member of the U.S. National Academy of Sciences (elected 1944) and the Royal Society (1969), and holder of numerous prizes and honorary doctorates, but it was quite in keeping with Romer's humor and self-effacing personality.

Amphibians and reptiles were the groups of particular interest to him. His most important research included "Review of the Pelycosauria" (1940, with L. I. Price), "Review of the Labyrinthodontia" (1947), smaller works on the Branchiosauria (1939), the Microsauria (1950), whose true relationships are still uncertain, and the mammal-like Therapsida (1956, with D. M. S. Watson). To support these investigations of fossils Romer also performed developmental studies on tetrapod limb musculature, using *Lacerta* as a model system (papers in 1942 and 1944), work that also gave important clues to the interpretation of the crossopterygian fin and its relation to the origin of tetrapods. He was an early proponent of continental drift, based on his studies of the fossil reptiles of the southern continents. Romer was also interested in the environments of early vertebrates and long championed the idea that the prototetrapod limb developed not as a means of escape from the aquatic environment, but to find a new pool when the old one dried up.

• *References*: "Alfred Romer, Evolution Authority, Dies," by V. K. McElheny, New York Times, 7 November 1973, p. 50, 1973; "Alfred Sherwood Romer 1884-1973 [sic]," by M. M. Dick, Copeia, 1974: 293-294, 1974; "Alfred Sherwood Romer," by T. S. Westoll and F. R. Parrington, Biogr. Mem. Fellows Royal Soc., 21: 497-516, 1975; "Alfred Sherwood Romer," by E. H. Colbert, Biogr. Mem. Natl. Acad. Sci. USA, 53: 265-294, 1982. • *Portrait* (1961): By M. Graham Netting, courtesy Carnegie Museum. • *Signature* (1969): Adler collection.

PETERS, James A. (1922-1972).

James Arthur Peters, American specialist on the Neotropical herpetofauna, was born in Durant, Iowa, just west of Davenport, on 13 July 1922. His father was a small-town doctor and, after the family moved to Greenup, a town in southern Illinois, Peters's interest in snakes began as the result of a boyhood friendship with a neighbor, Philip W. Smith, who also later became a professional zoologist. Peters's specialty was snakes and he became so knowledgeable about them that one summer he travelled the lecture circuit from one county fair to another with his snake show. Despite these activities, his interest in herpetology was quite serious and he attended his first meeting of the American Society of Ichthyologists and Herpetologists (ASIH) in 1939, at the age of barely 17. After graduation from high school he briefly attended college in Illinois, but it was wartime and in 1942 he joined the U.S. Army Air Force. He served as a radio operator, both in the India-Burma theater and in North Africa, where he always managed to sample the local herpetofauna.

On his release from the army in 1945, Peters entered the University of Michigan, from which he quickly earned three degrees (B.S. 1948, M.S. 1950, Ph.D. 1952). His graduate work was done in the Museum of Zoology, under the supervision of Norman E. Hartweg. During this time his interest in the Neotropics was initiated when he joined museum expeditions to Mexico in 1949 and 1950, and his publications on the Mexican herpetofauna began in 1950. In 1952-1958 Peters was on the faculty of Brown University, in Providence, Rhode Island, and during 1958-1959 he was a Fulbright Lecturer at the Universidad Central in Quito, which was the first of four trips to Ecuador (others in 1962, 1966, and 1969). About 1954 Ecuador had become a major focus of his research, and he continued to publish on its herpetofauna the rest of his life. In 1959 he joined the faculty of San Fernando State College, at the northern edge of Los Angeles, and worked there until 1964 when he became Associate Curator of Reptiles and Amphibians at the National Museum. Peters's life's goal had been to work at a museum, so he accepted the new post eagerly even though it meant a decrease in salary.

The collection, which at that time was the responsibility of Doris M. Cochran, was soon to be moved into a newly-constructed section of the museum. Fortunately so, for not only were the quarters cramped but also the arrangement of the specimens was antiquated and but little advanced since the early days of Leonhard Stejneger, Cochran's predecessor. Cochran retired in 1968, two years after Peters had become Curator-in-Charge, a position he held until his premature death in Washington on 18 December 1972, at the age of 50. He was succeeded by George R. Zug, who had been his associate since 1969.

Peters made numerous contributions to herpetology. As in his youth, snakes were his specialty. His doctoral thesis monographed the colubrid snakes of the subfamily Dipsadinae and this finally was published in 1960. His interest in order and precision of terminology led to his widely-used compendium, "Dictionary of Herpetology" (1964), which gives origins and concise definitions for about 3000 terms. This handy directory of the working language of herpetology is still in print a quarter century later, testimony to its continuing usefulness. Peters's last book was the two-volume monograph "Catalogue of the Neotropical Squamata" (1970; reprinted with addenda, 1986), co-edited with Braulio Orejas-Miranda and Roberto Donoso-Barros. This massive work, covering all Latin American lizards and snakes south of Mexico, contained keys and checklists, in parallel Spanish and English texts.

Peters's influence was also manifested in his activities as Secretary (1961-1966) and President (1970) of ASIH. During his secretaryship, the ASIH Business Office was established at the National Museum. Peters was a pioneer in the applications of computer technology to questions in systematics, biogeography, and museum curation. Beginning in the late 1940s, he was one of the first to use this then-new technology in herpetology and later published numerous papers describing new programs and techniques.

• *References*: "James A. Peters," by B. B. Collette, Copeia, 1973: 388-390, 1973; "James A. Peters 1922-1972 An Appreciation," by H. G. Dowling, HISS News-Jour., 1: 187-188, 1973; "James A. Peters: A Biographical Sketch (1922-1972)," by G. R. Zug, p. 1-4. *In* F. J. Irish and G. R. Zug, Biography and Bibliography of James A. Peters. Smithsonian Herpetol. Inform. Serv., 51, 1982. • *Portrait*: Courtesy Heinz Wermuth. • *Signature* (1958): Adler collection.

BLAIR, W. Frank (1912-1984).

William Franklin Blair, American mammalogist and herpetologist, was born on 25 June 1912 in Dayton, Texas, just east of Houston. His family lived in Westville and, later, Tulsa, Oklahoma, during 1916-1930, where he and his younger brother, Albert P. Blair, who later also published on herpetology, together developed an interest in natural history while they were still boys. Frank Blair attended the University of Tulsa (B.S. 1934), then the University of Florida (M.S. 1935), and completed a doctoral thesis at the University of Michigan (Ph.D. 1938) under the mammalogist and ecologist Lee R. Dice. He stayed at Michigan as Research Scientist, but later was drafted into the U.S. Army Air Force during World War II, after which service he briefly returned to the University of Michigan before taking a position as Assistant Professor of Zoology at the University of Texas at Austin in 1946. He became Professor in 1955, retired in 1982, and died on 9 February 1984 in Austin.

During 1935-1954 nearly all of Blair's nearly seventy papers dealt with mammals, their systematics, biogeography, behavior, and genetics. His first paper in herpetology was published in 1936, but beginning in 1955 his research abruptly shifted to that subject, principally amphibians. In the early 1950s Blair began work on the function of frog calls in mate selection, which led to important discoveries concerning isolating mechanisms and speciation. His study of the relative differences between the breeding calls of two species when in sympatry and when in allopatry—a phenomenon later termed character displacement—was of particular importance. Much of the work of this period was published in an influential book, "Vertebrate Speciation" (1961), edited by Blair. During the 1960s his research turned to the genetics and evolution of toads which was later summarized in another Blair-edited book, "Evolution in the Genus *Bufo*" (1972).

Blair published two other books concerning amphibians and reptiles. The first, an extensive identification handbook, "Vertebrates of the United States" (1957, second edition 1968), became a popular university-level text; it was co-authored with Albert P. Blair, Pierce Brodkorb, Fred R. Cagle, and George A. Moore. His other book, "The Rusty Lizard, A Population Study" (1960), was the summary of a six-year study of a population of *Sceloporus olivaceus* on a ten-acre tract of land surrounding Blair's home.

Besides his numerous publications, Blair's influence in biology resulted from his legion of students and his service activities for various international, national, and state organizations. He supervised 49 Ph.D. and 51 Master's students, as well as many postdoctoral fellows, most of them herpetologists. He held important positions in many professional societies, including the presidencies of the American Institute of Biological Sciences, Ecological Society of America, Society for the Study of Evolution, and the Southwestern Association of Naturalists. He served as Chairman of the U.S. National Committee for the International Biological Program and held several governmental advisory posts.

• *References*: "William Franklin Blair 1912-1984," by Clark Hubbs, Copeia, 1985: 529-531, 1985 (preprint and condensation from "In Memoriam William Franklin Blair," by Clark Hubbs, E. L. Lundelius, Jr., and M. J. Ryan, Doc. Resolut. Gen. Faculty, Univ. Texas at Austin, pp. 16928-16946, [1987]). • *Portrait* (about 1970): Courtesy J. P. Kennedy. • *Signature* (1977): Courtesy Clark Hubbs.

TINKLE, Donald W. (1930-1980).

The American ecologist Donald Ward Tinkle was born in Dallas, Texas, on 3 December 1930. After undergraduate training at Southern Methodist University in Dallas (B.S. 1952), he went to study with Fred R. Cagle at Tulane University in New Orleans, where he completed his graduate degrees (M.S. 1955, Ph.D. 1956). He held faculty positions at West Texas State University and Texas Tech University and, in 1965, became Curator of Amphibians and Reptiles in the University of Michigan Museum of Zoology and, concurrently, Professor of Zoology. In 1975 Tinkle was appointed Director of the Museum of Zoology, a post he held until his early death, at the age of 49, at his home in Saline, Michigan, just south of Ann Arbor, on 21 February 1980.

Tinkle's first publications, beginning in 1951, were on systematics and distribution, but his training with Fred Cagle, who was a major proponent of detailed studies of the life histories of amphibians and reptiles, shifted his main interest to population ecology in which he made brilliant contributions. Tinkle approached the study of life histories from a modern perspective and addressed critical questions of ecology and evolutionary biology. Although he was also talented in developing theoretical arguments leading to testable hypotheses, he was at his best in the field gathering empirical data. Indeed, his major contribution to the study of amphibian and reptilian populations was to emphasize the critical need for intensive, long-term field studies. The great value of this approach was demonstrated in his classic monograph, "The Life and Demography of the Side-blotched Lizard, *Uta stansburiana*" (1967).

During the last two decades of Tinkle's career, he and his associates studied the evolution and ecology of numerous life history parameters—natality, breeding structure, reproductive effort, densities, and the meaning of geographic variation in these parameters—as well as studies on home range, sex ratios, and energy expenditure. His broad interests were further reflected in his co-editing, with his University of Michigan colleague, Carl Gans, of a volume on ecology and behavior (1977) in the series "Biology of the Reptilia."

Tinkle's enthusiasm and approach to research had a profound influence on many persons at the University of Michigan, especially graduate students, but his influence often extended beyond his own official students. For example, even when he was a graduate student himself at Tulane, Tinkle inspired J. Whitfield Gibbons, then a junior high school student, to study reptiles and over two decades later they collaborated on several studies, including an important synthesis entitled "The Distribution and Evolution of Viviparity in Reptiles" (1977). Tinkle supervised many doctoral students at Michigan, among them James P. Collins, Arthur E. Dunham, Gary W. Ferguson, and Stephen G. Tilley, but he also influenced Royce E. Ballinger, Justin D. Congdon, Laurie J. Vitt, and Henry M. Wilbur. Tinkle received numerous honors for his research, perhaps most significantly being named Eminent Ecologist by the Ecological Study of America in the year of his death.

• *References*: "Donald W. Tinkle (1930-1980)," by A. Kluge, Copeia, 1980: 572, 1980; "Dedication to Donald W. Tinkle," by J. W. Gibbons, Herpetologica, 38: 3-4, 1982. • *Portrait* (about 1968) and *signature* (1969): Adler collection.

BAŞOĞLU, Muhtar (1913-1981).

The first Turkish scientific herpetologist, Başoğlu was born on 6 April 1913 in Ödemiş, near İzmir (formerly Smyrna) along the Aegean Coast. He was educated at Istanbul University, graduating in 1936, and, after military service, became an assistant on its Science Faculty. He received his D.Sc. degree in 1942, under the supervision of Curt Kosswig, a German then teaching at the university. Beginning in 1958, Başoğlu had productive collaborations with Walter Hellmich in their joint herpetological explorations of Anatolia. In 1961, Başoğlu moved to Ege (or Aegean) University in Bornova-İzmir as chair of the Section of Systematic Zoology and he founded there the Turkish Herpetological Center which still flourishes under his former students Neclâ Özeti and Mehmet K. Atatür. Başoğlu died on 21 February 1981, at his home in Karşıyaka, İzmir.

Başoğlu's main interests are reflected in his papers on the salamandrid *Mertensiella* and in his numerous studies on distribution and systematics of the Turkish herpetofauna, which culminated in three books written with his former students: "Amphibians of Turkey," with Neclâ Özeti (1973), and "Reptiles of Turkey," parts I and II, both co-authored with İbrahim Baran (1977, 1980). All are in Turkish and have extensive English summaries and keys, as well as illustrations and distribution maps. These are university-level texts which double as guides to the fauna, and represent an achievement unparalleled in the Middle East.

• *Reference*: "Muhtar Başoğlu," by J. Eiselt and İ. Baran, Amphibia-Reptilia, 2: 291-294, 1981. • *Portrait* (1980) *and signature*: Courtesy Mehmet K. Atatür.

MEDEM, Federico (1912-1984).

An immigrant to Colombia, Federico Medem became South America's leading student of turtles and crocodiles. Born on 29 August 1912 in Riga, Latvia, Friedrich Johann Graf (or count) von Medem (see *Note*) was of German Baltic nobility although he considered himself part Latvian. After the Bolshevik Revolution he moved to Berlin where he was a student at Humboldt University and, later, at the University of Tübingen. He conducted his doctoral research (completed in Berlin, 1942) at the famous marine station in Naples, under Gustav Kramer with whom Medem collaborated in studies on *Lacerta*. After obligatory service in the German army (Soviet front), he held positions in Germany and Switzerland before emigrating to Colombia in 1950. Soon afterward he became associated with the Estación de Biología Tropical "Roberto Franco" in Villavicencio, at the foot of the Andes along the Rio Meta, where he was to remain, eventually becoming director in 1966, except for a period in Cartagena on the Caribbean coast (1962-1966). He died in Bogotá on 1 May 1984.

Medem's primary interest in the systematics and life histories of turtles and crocodilians is reflected in more than 90 titles, including his master work, the two-volume "Los Crocodylia de Sur America" (1981, 1983). A manuscript, "Turtles of Colombia," was left unfinished at his death. Besides his scientific work, Medem was an active conservationist, which took him, as a consultant, to many parts of the world including Botswana, Papua, and northern South America.

• *References*: "In Memoriam—Federico Medem," by M. Williamson, Herpetol. Rev., 15: 62-63, 1984; "Federico Medem (29 August 1912–1 May 1984)," by W. W. Lamar, Herpetologica, 40: 468-472, 1984; "In Memoriam Federico Medem Medem (1912-1984)," anonymous, Caldasia, 14: 363-370, 1986. • *Portrait* (1965): By Kraig Adler. • *Signature* (1963): Adler collection. • *Note*: His full Spanish name was Federico Medem Medem, since his mother was also a Medem and a distant cousin of his father.

BARRIO, Avelino (1920-1979).

Avelino Barrio, specialist on frog biology and snake venoms, was born in La Coruña, a port city in northwestern Spain, on 10 August 1920. After emigrating with his family to Argentina, he attended the Colegio Nacional in Buenos Aires, completed his medical degree at the Universidad de Buenos Aires in 1948, and also earned a doctorate in medicine at the latter institution in 1954. Except for a brief period at the Universidad de El Salvador (1955-1957), Barrio spent most of his career at the Universidad de Buenos Aires and the Instituto Nacional de Microbiologia, until 1969 when he founded the Centro Nacional de Investigaciones Iologícas, a research institute devoted to the study of venomous animals. Barrio died in Buenos Aires on 30 June 1979.

Barrio's broadly-based research covered two main areas: herpetology and animal venoms. His doctoral thesis, published in 1954, concerned the neuromuscular and enzymatic action of snake venoms. Barrio was the co-discoverer of gyroxin, a rattlesnake venom neurotoxin, and he authored important papers on its properties and purification as well as on the geographic variation of gyroxin and crotamin, another neurotoxin, in South American rattlesnakes.

Most of Barrio's nearly 100 titles (published 1942-1980), however, dealt with the biology and systematics of amphibians, lizards, and snakes, and he named many new taxa. He authored numerous papers on the reproduction, behavior, cytogenetics, and renal physiology of frogs and caecilians. In the early 1960s, Barrio established the first bioacoustics laboratory in South America and, beginning in 1962, many of his papers included spectrographic analyses of frog vocalizations. This allowed him to distinguish numerous sibling species of frogs. One of his most interesting discoveries was that the green coloration in frogs of the families Centrolenidae, Hylidae, and Pseudidae is due partly to the bile pigment biliverdin, which is deposited in the bones, soft tissues, and even in the eggs (published in three papers, 1965 and 1968).

• *References*: "Dr. Avelino Barrio (1920-1979)," by M. E. Miranda, Physis, 38: 1-6, 1978 (1979); "Avelino Barrio," anonymous, Cabildo, 4(26): 11, 1979; "Avelino Barrio en los Recuerdos de un Amigo," by O. R. Vidal, Publ. Mus. Argent. Cienc. Nat., Buenos Aries, 3 pages, 1981; "Dr. Avelino Barrio," anonymous, Quid, 2: 186, 1983. • *Portrait* (1977) *and signature* (1965): Courtesy Marta E. Miranda.

LITERATURE CITED AND FURTHER READING

ALL OF THESE TITLES include biographical information about herpetologists. In many instances, herpetology was only one of a person's several interests, especially for those who lived before 1900, and for others it was an area of secondary emphasis.

ADLER, K. 1979. A Brief History of Herpetology in North America Before 1900. Soc. Study Amphib. Rept., Athens (Ohio), Herpetol. Circ., 8: 1-40.

ADLER, K. 1986. Collecting Antiquarian Books on Amphibians and Reptiles. AB Bookman's Weekly, 78: 321-332.

ANONYMOUS. 1983. Society for the Study of Amphibians and Reptiles: A Brief History, 1958-1982, p. 40-46. *In* Soc. Study Amphib. Rept., Athens (Ohio), Herpetol. Circ. 13.

BANTA, B. H. and W. W. Tanner. 1964. A Brief Historical Résumé of Herpetological Studies in the Great Basin of the Western United States. Part I. The Reptiles [all publ.]. Great Basin Nat., 24: 37-57.

BELL, T. 1832-1842. A Monograph of the Testudinata. S. Highley, London, xxiv, [80] pages (history, pp. i-xii).

BELLAIRS, A. 1969. The Life of Reptiles. Vol. I. Weidenfeld and Nicholson, London, xii, 282 pages (history, pp. 1-16).

BELTZ, E. Unpublished. Translations of the Scientific Names of the Reptiles and Amphibians of North America. In manuscript, 59 pages.

BERRA, T. M. 1984. A Chronology of the American Society of Ichthyologists and Herpetologists Through 1982. Amer. Soc. Ichthyol. Herpetol., Gainesville (Florida), Spec. Publ. 2: 1-21.

BOGERT, C. M. 1958. The Role of the Zoological Curator as a Writer (With Some Impertinent Digressions). Curator, 1: 48-63.

BOULENGER, G. A. 1906. Reptiles and Batrachians, p. 517-531. *In* History of the Collections Contained in the Natural History Departments of the British Museum, vol. 2. Brit. Mus. (Nat. Hist.), London.

CLOQUET, H. 1819. Considérations Générales sur l'Erpétologie. Dict. Sci. Nat., 15: 13-46.

COGGER, H. G. 1985. Australian Proteroglyphous Snakes—An Historical Overview, p. 143-154. *In* G. Grigg, R. Shine and H. Ehmann (eds.), The Biology of Australasian Frogs and Reptiles. Surrey Beatty and Sons, Chipping Norton (New South Wales).

COLE, F. J. 1944. A History of Comparative Anatomy. Macmillan Co., London, viii, 524 pages.

CONANT, R. 1980. The Reproductive Biology of Reptiles: An Historical Perspective, p. 3-18. *In* J. B. Murphy and J. T. Collins (eds.), Reproductive Biology and Diseases of Captive Reptiles. Soc. Study Amphib. Rept., Contr. Herpetol. 1, Athens (Ohio).

CONANT, R. 1982. Herpetology in Ohio—Fifty Years Ago. Spec. Publ. Toledo Herpetol. Soc., Toledo (Ohio), 64 pages.

DERANIYAGALA, P. E. P. 1939. The Tetrapod Reptiles of Ceylon. Vol. I. Testudinates and Crocodilians. Colombo Mus., Colombo, xxxii, 412 pages (history, pp. 1-3).

DONOSO-BARROS, R. 1966. Reptiles de Chile. Ed. Univ. Chile, Santiago, 458, cxlvi pages (history, pp. 21-25).

DOWLING, H. G. and I. Gilboa. 1975. Landmarks in Herpetology, p. 3-7. *In* H. G. Dowling (ed.), 1974 Yearbook of Herpetology. Herpetol. Inform. Search Syst. (HISS), Amer. Mus. Nat. Hist., New York.

DUELLMAN, W. E. and L. Trueb. 1986 (1985). Biology of Amphibians. McGraw-Hill Book Co., New York, xix, 670 pages (history, pp. 1-9).

DUMÉRIL, A. M. C. and G. Bibron. 1834-1854. Erpétologie Générale on Histoire Naturelle Complète des Reptiles. Libr. Encyclop. de Roret, Paris (history: vol. 1, pp. 225-344, 414-439; vol. 2, pp. 659-673; vol. 4, pp. 6-34; vol. 5, pp. 6-15; vol. 6, pp. 12-65, 212-232; vol. 8, pp. 11-47, 245-258; vol. 9, pp. 18-33).

GANS, C. and T. S. Parsons. 1970. Taxonomic Literature on Reptiles, p. 315-333. *In* C. Gans and T. S. Parsons (eds.), Biology of the Reptilia, vol. 2. Academic Press, London and New York.

GOIN, C. J., O. B. Goin, and G. R. Zug. 1978. Introduction to Herpetology. Third Edition. W. H. Freeman, San Francisco, xiii, 378 pages (history, pp. 9-14).

GOMEZ, R. M. and B. Sanchíz. 1987. Notes on the Spanish XIX Century Herpetology, p. 151-154. *In* J. J. van Gelder, H. Strijbosch, and P. J. M. Bergers (eds.), Proc. 4th Ordin. Gen. Meet Soc. Europ. Herpetol., Nijmegen (Netherlands).

GOODE, G. B. 1901. A Memorial of George Brown Goode, Together With a Selection of His Papers on Museums and on the History of Science in America. Ann. Rept. Board Regents Smithsonian Inst. for 1897 (Rept. U.S. Natl. Mus., 2): xii, 1-515.

GÜNTHER, A. C. L. G. and A. S. Woodward. 1911. History of Herpetology, p. 136-141. *In* A. Günther, H. F. Gadow, and A. S. Woodward, Reptiles. Encyclop. Britannica, ed. 11, 23: 136-176.

GUNTHER, A. E. 1980. The Founders of Science at the British Museum 1753-1900. Halesworth Press, Halesworth (Suffolk), ix, 219 pages.

HARPER, F. 1940. Some Works of Bartram, Daudin, Latreille, and Sonnini, and Their Bearing Upon North American Herpetological Nomenclature. Amer. Midl. Nat., 23: 692-723.

HASEGAWA, H. 1967. Materials on the Lives and Contributions of the Deceased Japanese Entomologists Since the Meiji Era. Kontyû, 35 (suppl.): 1-98 (in Japanese).

HUBBS, C. L. 1964. History of Ichthyology in the United States After 1850. Copeia, 1964: 42-60.

JORDAN, D. S. 1905. A Guide to the Study of Fishes. H. Holt, New York, vol. 1 (history of ichthyology, p. 387-428).

JOURNAL OF THE SOCIETY FOR THE BIBLIOGRAPHY OF NATURAL HISTORY. Volume 1 (1936) and following (continued as *Archives of Natural History*, beginning with volume 10, 1981). London.

KLAUBER, L. M. 1945. Some Herpetological Book Prices Then and Now. Herpetologica, 2: 151-174 (supplement, same journal, 3: 19, 1945).

LOVERIDGE, A. 1936. The Reptile-Amphibian Collection, p. 53-56. In Notes Concerning the History and Contents of the Museum of Comparative Zoölogy. Publ. Tercentennial Foundation Harvard College, Cambridge (Massachusetts).

MALM, T. 1979-1980. [Swedish Herpetological and Paleontological Books, Papers and Theses.] Snoken (Swedish Herpetol. Soc.), 9: 121-129 (in Swedish).

MATSUI, M. 1988. [History of Herpetology], p. 4-20. In Systematic Zoology, vol. 9, no. 2B, Reptilia A. Nakayama Shoten, Tokyo (in Japanese).

MERTENS, R. 1967. Die herpetologische Sektion des Natur-Museums und Forschungs-Institutes Senckenberg in Frankfurt a. M. nebst einem Verzeichniss ihrer Typen. Senckenberg. Biol., 48A: 1-106 (English translation by R. B. Smith and H. M. Smith, 1968, Bull. Maryland Herpetol. Soc., 5: 1-52).

MIALL, L. C. 1912. The Early Naturalists Their Lives and Work (1530-1789). Macmillan, London, xi, 396 pages.

MURTHY, T. S. N. 1983. A Historical Resume and Bibliography of the Snakes of India. The Snake, 15: 113-135.

MYERS, G. S. 1964. A Brief Sketch of the History of Ichthyology in America to the Year 1850. Copeia, 1964: 33-41.

NISSEN, C. 1969-1978. Die zoologische Buchillustration. Ihre Bibliographie und Geschichte. A. Hiersemann, Stuttgart, 2 volumes (vol. 1, bibliography, viii, 666 pages; vol. 2, text, xvi, 604 pages).

NORDENSKIÖLD, E. 1928 (1949). The History of Biology. Tudor Publ. Co., New York, xii, 629, xv pages (translated from Swedish edition, entitled "Biologins Historia," Björck and Börjesson, Stockholm, 3 vols., 1920-1924).

OBST, F. J., K. Richter, and U. Jacob. 1984. Lexikon der Terraristik und Herpetologie. Edition Leipzig, Leipzig, 466 pages (English edition, TFH Publ., Neptune City, New Jersey, 1988).

PALMER, T. S. and others. 1954. Biographies of Members of the American Ornithologists' Union [reprinted from *The Auk*, 1884-1954]. Privately printed by the Editor, Paul H. Oehser, Washington, 630 pages.

PAPAVERO, N. 1971-1973. Essays on the History of Neotropical Dipterology, With Special Reference to Collectors (1750-1905). Mus. Zool., Univ. São Paulo, São Paulo, 2 volumes (vol. I, vii, 216 pages; vol. II, iii, 217-446 pages).

PECHMAN, H. H., H. T. Gaige, and C. H. Hubbs. 1946. Ichthyologia et Herpetologia Americana. Bull. William L. Clements Libr., Univ. Michigan, 25: 1-22.

PORTER, K. R. 1972. Herpetology. W. B. Saunders, Philadelphia, xi, 254 pages (history, pp. 1-16).

SCARLATO, O. A. (ed.). 1982. Zoological Institute of the Academy of Sciences of the USSR—150 Years. Acad. Sci. USSR, Leningrad, 243 pages (in Russian; history of herpetology, pp. 68-71).

SCHLEGEL, H. 1837. Essai sur la Physionomie des Serpens. Van Stockum, The Hague, and Schonekat, Amsterdam, part 1, xxviii, 251 pages (history, pp. 109-126; English translation, by T. S. Traill, 1843, Maclachlan, Stewart and Co., Edinburgh; history, pp. 110-125).

SCHMIDT, K. P. 1955. Herpetology, p. 591-627. In E. L. Kessel (ed.), A Century of Progress in the Natural Sciences—1853-1953. Calif. Acad. Sci., San Francisco. Reprinted Arno Press, 1974.

SCHMIDT, K. P. and D. D. Davis, 1941. Field Book of Snakes of the United States and Canada. G. P. Putnam's Sons, New York, xiii, 365 pages (history, pp. 11-16).

SIEBENROCK, F. 1901. Amphibien und Reptilien, p. 444-462. In A. Handlirsch and R. von Wettstein (eds.), Botanik und Zoologie in Österreich in den Jahren 1850 bis 1909. Festschr., Zool.-Bot. Gesellsch., Vienna.

SMITH, H. M. 1986. Chapman Grant, *Herpetologica*, and the Herpetologists' League. Herpetologica, 42: 1-32.

SMITH, H. M. and R. B. Smith. 1973. Synopsis of the Herpetofauna of Mexico, Volume II. E. Lundberg, Augusta (West Virginia), xxxiii, 367 pages (history, pp. x-xix).

SMITH, M. A. 1931. The Fauna of British India, including Ceylon and Burma. Reptilia and Amphibia. Vol. I.—Loricata, Testudines. Taylor and Francis, London, xxviii, 185 pages (history, pp. 2-13). Reprinted R. Curtis Books, 1973.

SMITH, M. A. 1951. The British Amphibians and Reptiles. Collins, London, xiv, 318 pages (history, pp. 1-13).

SMITH, M. A. 1952. The History of Herpetology in India. Jour. Bombay Nat. Hist. Soc., 50: 907-909.

STRESEMANN, E. 1975. Ornithology from Aristotle to the Present. Harvard Univ. Press, Cambridge (Massachusetts), xii, 432 pages (translated from German edition, entitled "Die Entwicklung der Ornithologie von Aristoteles bis zur Gegenwart," F. W. Peters, Berlin, xv, 431 pages, 1951).

SWAINSON, W. 1840. Taxidermy; With the Biographies of Zoologists, and Notices of Their Works. Longman, Orme, Brown, Green & Longman's, London, (2), 392 pages (biographies, pp. 98-386).

TERENTJEV, P. V. 1957. [Material for a History of Russian Herpetology.] Trudy Inst. Istor. Estestvoz. Akad. Nauk SSSR, 16: 99-122 (in Russian).

TERENTJEV, P. V. 1961. [Herpetology. A Manual on Amphibians and Reptiles.] Moscow, 336 pages (in Russian; history, pp. 5-10) (English translation, Israel Progr. Sci. Transl., Jerusalem, 1965).

VANZOLINI, P. E. 1977-1978. An Annotated Bibliography of the Land and Fresh-water Reptiles of South America (1758-1975). Mus. Zool., Univ. São Paulo, São Paulo, 2 volumes (vol. 1, iv, 186 pages; vol. 2, 316 pages).

WOOD, C. A. 1931. An Introduction to the Literature of Vertebrate Zoology. Oxford Univ. Press, London, xix, 643 pages. Reprinted Arno Press, 1974.

WRIGHT, A. H. 1949. Scientific and Popular Writers on American Snakes (1517-1944), a Check List and Short Biography. Herpetologica, 5 (suppl. 1): 1-55.

INDEX TO BIOGRAPHIES

PAGE NUMBERS in **boldface** refer to the biographies of the 152 individuals featured in this book. Where the same page number is repeated for a single entry, the individual is mentioned in both accounts on that page. Alternate or more complete names are given in parentheses for some individuals. Full names have been provided so far as known.

- A -

Adams, Charles Christopher, 74
Agassiz, Alexander Emmanuel Rodolphe, 40, 61, 128
Agassiz, Jean Louis Rodolphe, 24, 36, 37, **39-40**, 41, 42, 42, 61, 62, 74, 75, 128
Alcala, Angel Chua, 114
Aldrovandi, Ulisse, 7, 12
Allee, Warder Clyde, 92
Alton, Johann Samuel Eduard d', 45
Amaral, Afrânio Pompílio Bastos do, 84, **88-89**, 113, 127
Anderson, John, 39, **54-55**, 78, 122
Anderson, Steven Clement, 114
Andersson, Lars Gabriel, 12, 68, **68**
Andreu, see Casas Andreu
Andrews, Roy Chapman, 94
Angel, Fernand, 44, **106-107**, 107
Annandale, Thomas Nelson, **78-79**
Aristotle, 7, 13
Atatür, Mehmet Kutsay, 132
Audubon, John James, 26, 27, 41
Auguste Dorothea, Princess of Arnstadt-Schwarzburg, 10

- B -

Baer, Karl Ernst von (Karl Maksimovich), 32
Bailey, Joseph Randle, 74, 93
Baird, Donald, 128
Baird, Irwin Lewis, 83
Baird, Spencer Fullerton, **40-42**, 42, 43, 47, 48, 62
Baird, William McFunn, 41
Ballinger, Royce Eugene, 132
Bannikov, Andrei Grigoryevich, **124**
Banta, Benjamin Harrison, 114
Baran, İbrahim, 132
Barbosa du Bocage, see Bocage
Barbour, Thomas, 61, 63, 74, **84-85**, 88, 90, 92, 93, 99, 112, 114, 121, 128
Barrio, Avelino, **133**
Barros, see Donoso-Barros
Bartram, William, 26
Başoğlu, Muhtar, **132**
Baur, Georg Hermann Carl Ludwig, 47, 48, **60-61**, 62, 128
Bechstein, Johann Matthaeus, 14
Bedriaga, Jacques Vladimir von (Jacob Vladimirovich), **58**
Bell, Thomas, 35, **35-36**
Bellairs, Angus d'Albini, 86
Berg, Carlos (Friedrich Wilhelm Carl), **64**
Bibron, Gabriel, 14, 20, 21, 25, 30, 31, 32, **32-33**, 36, 43
Bishop, Sherman Chauncey, 75, 76, **94-95**
Blair, Albert Patrick, 130, 131
Blair, William Franklin, **130-131**
Blanchard, Charles Émile, 59
Blanchard, Frank Nelson, 63, 74, 75, **95-96**, 110, 113

Blanchard, Frieda Cobb, 96
Blatter, Ethelbert, 125
Blomhoff, Jan Cock, 21
Blumenbach, Johann Friedrich, 15, 21, 22, 28
Blyth, Edward, **38-39**, 54
Bobeau, Georges, 127
Bocage, José Vincente Barbosa du, **51-52**
Bocourt, Marie-Firmin, 44, 106
Bodmer, Carl, 22
Boettger, Oskar, **56-57**, 70, 77, 98
Bogdanov, Anatoli Petrovich, 58
Bogert, Charles Mitchill, 91, 98, 117, 119
Boie, Friedrich, 21, 22
Boie, Heinrich, **21-22**, 25, 30, 77
Bojanus, Ludwig Heinrich, **20-21**
Bolkay, Stephan Joseph Ladislaws, 65, 85
Bonaparte, Charles-Lucien-Jules-Laurent, 25, **29-30**
Bonnaterre, Pierre-Joseph, 20
Boring, Alice Middleton, 94, **107-108**, 123
Boulenger, Edward George, 55
Boulenger, George Albert, 32, 46, 48, 54, 55, **55-56**, 57, 61, 62, 65, 65, 66, 67, 70, 79, 86, 93, 106, 106, 122
Bourret, René Léon, 87, **107**
Boveri, Theodor, 108
Brandt, Johann Friedrich (Fedor Fedorovich), 49
Brattstrom, Bayard Holmes, 98, 117
Brazil (Mineiro da Campanha), Vital, 88
Breder, Charles Marcus, Jr., 113
Brehm, Alfred Edmund, 30, 57, 71
Brehm, Christian Ludwig, 30
Breijer-de Rooij, see Rooij
Broadley, Donald G., 122
Brocchi, Paul-Louis-Antoine, 44
Brodkorb, Pierce, 131
Brongniart, Alexandre, 18, 19, 28, 32
Bronn, Heinrich Georg, 39
Brown, Walter Creighton, 114
Brühl, Carl Bernhard, 69
Buffon, Georges-Louis Leclerc, 11, 13, 14, 15, 19, 20
Burden, William Douglas, 93
Burmeister, Carl Hermann Conrad, **45**, 64
Burt, Charles Earle, 74

- C -

Cagle, Fred Ray, 131, 131
Cahn, Alvin Robert, 75
Calinescu, Raul I., 128
Camerano, Lorenzo, 29, 66, 67
Camp, Charles Lewis, **80-81**, 90, 128
Campo, see Martín del Campo
Cantor, Theodore Edward, 39
Čarevskij, see Tsarevsky
Carr, Archibald Fairly, Jr., 76, 84, 85, **114-115**, 116
Carr, Marjorie Harris, 115
Carroll, Robert Lynn, 128
Carus, Julius Victor, 60
Carvalho, Antenor Leitão de, 88, 114
Casas Andreu, Gustavo, 119
Catesby, Mark, 12

Cei, José Miguel, 64
Cepède, see Lacepède
Chang, Tso-kan, 108
Charas, Moyse (Moïse), 8, 12
Chernov, Sergius Alexandrovich, 69, **103**, 104
Children, John George, 34
Chou, Shu-ch'un, 108
Chun, Carl, 98
Clark, Henry James, 40
Clark, William, 22
Clay, William Marion, 96
Cliff, Frank Samuel, 114
Cochran, Doris Mable, 63, **110-111**, 116, 130
Cocteau, Jean-Théodore, 33
Cogger, Harold George, 100, 101
Colbert, Edwin Harris, 117
Collins, James Paul, 132
Conant, Roger, 111, 113
Condamin, Michel, 125
Confucius (K'ung Fu-tzu), 123
Congdon, Justin Daniel, 132
Conklin, Edwin Grant, 108
Cope, Edward Drinker, 41, **46-49**, 56, 60, 61, 61, 62, 90, 93, 95
Copernicus, Nicolaus, 7
Corkill, Norman L., 105
Cosgriff, John, 81
Cowles, Raymond Bridgman, 76, **116-117**
Cristoforis, Giuseppe De, 49
Curran, Charles Howard, 118
Cuvier, Frédéric, 18
Cuvier, Georges (Jean-Léopold-Nicolas-Frédéric-Dagobert), **17-19**, 19, 20, 27, 28, 31, 33, 35, 37, 38, 39, 40, 44, 51, 59

- D -

d'Alton, see Alton
Dante (Dante Alighieri), 8
Darevsky, Ilya Sergeevich, 103, 124
Darwin, Charles Robert, 18, 25, 35, 36, 39, 39, 40, 46, 65
Daubenton, Louis-Jean-Marie, 14, 18, 20
Daudin, Adele, 20
Daudin, François-Marie, 19, **20**, 25
Davis, Delbert Dwight, 92
Degenhardt, William George, 118
DeKay, James Ellsworth, 33
Denburgh, see Van Denburgh
Denisova, Maria Nikolaevna, 124
Deraniyagala, Paulus Edward Pieris, 84, **120-121**
De Rooij, see Rooij
De Witte, see Witte
Dice, Lee Raymond, 130
Dickerson, Mary Cynthia, 81, 90, 91, 92
Ding, see Ting
Ditmars, Raymond Lee, **73-74**, 84, 88, 117
do Amaral, see Amaral
Donoso-Barros, Roberto, 64, **126-127**, 130
Doria, Giacomo, 38
Dowling, Herndon Glenn, Jr., 93
Drosdov, Nikolai Nikolaevich, 124
Duerden, James Edwin, 80
Dugès, Alfredo (Alfred Auguste Delsescautz), **59-60**

Dugès, Antoine Louis Delsescautz, 59
Dugès, Eugenio (Eugène), 59
Duméril, André-Marie-Constant, 14, 18, 19, 20, 21, 25, 30, **31-32**, 32, 33, 43, 59
Duméril, Auguste-Henri-André, 30, 31, 32, **43-44**, 59
Dunham, Arthur Earl, 132
Dunkle, David Hosbrook, 128
Dunn, Emmett Reid, 63, 81, 82, 83, 84, 91, **92-93**
Duvernoy, Georges-Louis, 18, 19

- E -

Eigenmann, Carl H, 113
Emerson, Alfred Edwards, 92
Espada, see Jiménez de la Espada

- F -

Fabre, Jean-Henri, 67
Fedchenko (or Fedtschenko), Aleksei Pavlovich, 69
Fejérváry von Komlós-Keresztes, Géza Gyula (Géza Julius Emerich), 65, **85-86**
Fejérváry-Lángh, Aranka Mária, 85
Ferdinand II, Grand Duke of Tuscany, 8
Ferguson, Gary Wright, 132
Fitzinger, Leopold Joseph Franz Johann, **24-25**, 30, 54
FitzSimons, Desmond Charles, 121
FitzSimons, Frederick William, 121
FitzSimons, Vivian Frederick Maynard, **121-122**
Flint, Vladimir Evgenevich, 124
Flores Villela, Oscar, 119
Fontana, Felice, 8
Forcart, Lothar Hendrich Emil Wilhelm, 76
Ford, George Henry, 36
Forskål, Pehr (Petrus), 11
Forsten, Eltio Elegondus, 21
Fowler, Henry Weed, 93
Fuhn, Ion Eduard, **127-128**
Funkhouser, William Delbert, 75

- G -

Gaige, Helen Beulah Thompson, 74, 75, **81-82**
Galileo (Galileo Galilei), 8
Gans, Carl, 79, 118, 128, 132
Garman, Samuel Walton, **61-62**, 84, 85
Garman, William Harrison, 62
Gay, Claude, 53, 64
Gegenbauer, Carl, 58
Geoffroy Saint-Hilaire, Étienne-François, 14, 18, 39, 51
Gessner, Conrad, **7-8**, 11, 12
Gibbons, James Whitfield, 132
Gilbert, Charles Henry, 72
Girard, Charles Frédéric, 41, **42-43**, 48, 62
Glauert, Ludwig, **101**
Gloyd, Howard Kay, 96, **112-113**
Gmelin, Johann Friedrich, 9, 11
Goeldi, Emílio August (Emil August), **64-65**, 88
Goin, Coleman Jett, 111, 115, **115-116**
Goin, Olive Lynda Bown, 116
Gomes, João Florencio de Salles, 88
Gordon, Kenneth Llewellyn, 76
Gorman, Joe (Joseph B., Jr.), 81

Grandidier, Guillaume, 59
Grandison, Alice Georgie Cruickshank, 106
Gravenhorst, Johann Ludwig Christian Carl, 13, **28**
Gray, Edward Whitaker, 17
Gray, John Edward, 19, **34-35**, 35, 46, 47, 55, 106
Green, Jacob, 33
Green, John, 55
Gregory, William King, 81, 90, 128
Griffith, Edward, 19
Grinnell, Joseph, 80, 97
Grobman, Arnold Brams, 95
Guérin-Méneville, Félix-Édouard, 19
Guibé, Jean, 107
Gundlach, Juan (Johannes Christopher), **53**
Günther, Albert Carl Ludwig Gotthilf, 35, **45-46**, 47, 55, 55, 106
Gyllenborg, Count Carl, of Sweden, 12

- H -

Haacke, Wulf Dietrich, 122
Haas, Georg, 70, **101-102**
Haeckel, Ernst Heinrich Philipp August, 58, 64
Hairston, Nelson George, 94
Hallowell, Edward, 33
Hamilton, William John, Jr., 76
Hardwicke, Thomas, 35
Harlan, Richard, 26, **27**, 33
Harper, Francis, 75
Hartweg, Norman Edouard, 74, 130
Hasselquist, Fredrik, 11
Hasselt, Johan Coenraad van, 21, 77
Hay, Oliver Perry, 61, 81
Heckel, Johann Jakob, 30
Hellmich, Walter, 78, 132
Hemprich, Friedrich Wilhelm, 25
Henshaw, Samuel, 84
Héron-Royer, Louis-François, 67
Hesse, Richard, 92
Hewitt, John, **80**
Hibbard, Claude William, 83
Highton, Richard, 118
Hilaire, see Geoffroy Saint-Hilaire
Hoge, Alphonse Richard, **127**
Hoge, Sylvia Alma R. W. de Lemos Romano, 127
Holbrook, John Edwards, frontispiece, 4, 27, **33-34**, 72
Holbrook, Silas Pinckney, 34
Holm, Åke, 12
Hoover, Herbert Clark, President of the United States, 63
Hora, Sunder Lal, 79
Houttuyn, Martinus, 11
Hu, Shu-qing, 123
Huang, N.-z., 123
Hubbs, Carl Leavitt, 82
Humboldt, Friedrich Wilhelm Heinrich Alexander von, 23, 37, 37, 39, 63

- I -

Inger, Robert Frederick, 92
Ionides, Constantine John Philip, 105
Iordansky, Nikolai Nikolaevich, 124
Ischenko, Vladimir Georgyevich, 124

- J -

Jacquin, Joseph Franz von, 24
James, Edwin, 26
Jan, Giorgio, 47, **49-50**, 50
Jiménez de la Espada, Marcos, **52-53**, 114
Jonston, John, 7, 12
Jordan, David Starr, 79, 113

- K -

Kalm, Pehr (Peter), 11
Kammerer, Paul, 91
Kampen, Pieter Nicolaas van, 77
Kauffeld, Carl Frederick, 91, **117-118**
Kaup, Johann Jakob, 25, 30
Keast, James Allen, 100
Kellaway, Charles Halliley, 101
Kelly, Howard Atwood, 97
Kennicott, Robert, 41, 42
Kerr, Robert, 14
Keulemans, John Gerrard, 30
Kezer, Leonard James, 76
Kielmeyer, Karl Friedrich, 18
Kinghorn, James Roy, **100-101**
Kingsley, John Sterling, 95
Kiritzescu (or Kirițescu), Constantin, 128
Kirtland, Jared Potter, 33
Klauber, Laurence Monroe, **97-98**, 113, 116
Klein, Jakob Theodor, **9-10**
Klemmer, Konrad, 98
Klingelhöffer, Wilhelm Karl August, **102-103**
Kochva, Elazar, 102
Komai, Taku, 109
Konopický, Eduard, 54
Kopstein, Felix, **99-100**
Kosswig, Curt, 132
Kramer, Gustav, 132
Kraus, Rodolpho (Rudolph), 70
Krefft, Gerard (Johann Gerhard Louis), **51**
Kuhl, Heinrich, 21, 30, 77

- L -

Lacepède, Bernard-Germain-Étienne de la Ville-sur-Illon, **14**, 16, 18, 20, 31, 59
Lal Hora, see Hora
Lamarck, Jean-Baptiste-Pierre-Antoine de Monet de, 18, 19
Lamotte, Maxime Georges, 106, 107
Lataste, Fernand, 55, 67
Latreille, Pierre-André, 16, 18, **19**, 20, 28, 32
Laurent, Raymond Ferdinand, 122
Laurenti, Josephus Nicolaus, **12-13**, 14
Lear, Edward, 35
Leidy, Joseph, 47, 48
Leitaõ de Carvalho, see Carvalho
Leonardo da Vinci, 8
Lessona, Michele, 66
LeSueur, Charles Alexandre, 22, 26
Leuckart, Carl Georg Friedrich Rudolf, 60
Leviton, Alan Edward, 114
Lewis, Meriwether, 22

Leydig, Franz, 10, 104
Lichtenstein, Martin Hinrich Carl, 28, 37, 63
Link, Heinrich Friedrich, 63
Linnaeus, Carl (Carl von Linné), 9, 9, 10, 10, **11-12**, 12, 15, 15, 68, 68
Linné, see Linnaeus
Liu, Cheng-chao, 76, 108, **123-124**
Livezey, Robert Lee, 76
Löfling, Petrus, 11
Logier, Eugene Bernard Shelley, **109-110**
Long, Stephen Harriman, 26, 27
Lönnberg, Axel Johan Einar, 12, **67-68**, 68
Loveridge, Arthur, 80, 84, 85, **111-112**
Lowe, Charles Herbert, Jr., 98, 117
Lütken, Christian Frederik, 44
Lutz, Adolpho, **87**, 111, 119, 120
Lutz, Bertha Maria Julia, 87, 111, **119-120**
Lutz, Gualter Adolpho, 87, 120

- M -

MacClelland, John, 39
Macklot, Heinrich Christian, 21
Macleay, William John, 51
Maclure, William, 26
Main, Albert Russell, 101
Makeev, V. M., 124
Maki, Moichirō, **109**
March, Douglas D. H., 89
María, see Nicéforo Maria
Marsh, Othniel Charles, 47, 48, 60, 61
Martín del Campo y Sánchez, Rafael, **119**
Martius, Carl Friedrich Philipp von, 23, 24, 39
Maslin, Thomas Paul, Jr., 114
Maximilian (Ferdinand Maximilian Joseph), Archduke of Austria and Emperor of Mexico, 44
Maximilian, see Wied-Neuwied
McCann, Yule Mervyn Charles, **125**
McCoy, Frederick, 51
McCulloch, Allan Riverstone, 100
Medem, Federico (Friedrich Johann), **132-133**
Méhelÿ, Lajos (Ludwig von), **65**, 85
Merian, Maria Sibylla, 10
Merrem, Blasius, **15-16**
Merriam, Clinton Hart, 62
Mertens, Robert Friedrich Wilhelm, 78, **98-99**
Merton, Hugo, 76
Methuen, Paul Ayshford, 80
Meyen, Franz Julius Ferdinand, 29
Michelangelo (Michelangelo Buonarroti), 8
Mikan, Johann Christian, 23
Milne-Edwards, Henri, 37, 59
Miranda Ribeiro, Alipio de, **87-88**
Miranda Ribeiro, Paulo de, 88
Mitchell, Francis John, **126**
Mocquard, François, 44, 106, 107
Moore, George Azro, 131
Morgan, Thomas Hunt, 107
Mosauer, Walter, 70, 117
Müller, Fritz (Friedrich), **57-58**, 76
Müller, Johannes Peter, 32, 37, 38, 46, 48, 58
Müller, Lorenz, **77-78**, 99
Müller, Philipp Ludwig Statius, 11, 14

Müller, Salomon, 21, 30
Myers, George Sprague, **113-114**

- N -

Namiye, Motokichi, 79
Napoleon I (Napoleon Bonaparte), Emperor of the French, 14, 18, 22, 23, 29, 31, 49, 51, 54
Napoleon III (Louis Napoleon), Emperor of the French, 44
Narayan Rao, see Rao
Nash, Charles William, 109
Nason, Elias, 4
Natterer, Johann, 23, 25, 54
Necker, Walter Ludwig, 113
Neiva, Arthur, 88
Netting, Morris Graham, 74
Nevo, Eviatar, 102
Nicander of Colophon, 13
Nicéforo María, Brother (Antoine Rouaire Siauzade), **89-90**
Nicholls, George Edward, 90
Nichols, John Treadwell, 113
Nightingale, Florence, 36
Nikolsky, Alexander Mikhailovich, 49, 58, **68-69**, 103
Noble, Gladwyn Kingsley, 81, 84, **90-91**, 91, 94, 113, 117, 128
Nodder, Frederick Polydore, 17
Norris, Kenneth Stafford, 117

- O -

Ochoterena, Isaac, 119
Okada, Yaichirō, **108-109**
Oken (formerly, Ockenfuss), Ludwig Lorenz, 24, 25, 36, 39
Oliver, James Arthur, 74
Olson, Everett Claire, 128
Oppel, Nikolaus Michael, 31
Orejas-Miranda, Braulio R., 126, 130
Ortenburger, Arthur Irving, 74
Osbeck, Pehr, 11
Ōshima, Masamitsu, **79-80**
Owen, Richard, 27
Özeti, Neclâ, 132

- P -

Pallas, Peter Simon, 15
Park, Orlando, 92
Park, Thomas, 92
Parker, George Howard, 121
Parker, Hampton Wildman, **106**
Parsons, Thomas Sturges, 128
Pearl, Raymond, 108
Pedro II (Dom Pedro), Emperor of Brazil, 64
Peña, see Uribe Peña
Peracca, Mario Giacinto, **66**, 67
Perkins, Clarence Basil, 97
Peter the Great (Peter I), Emperor of Russia, 9, 49
Peters, James Arthur, 110, 111, 126, **129-130**
Peters, Wilhelm Carl Hartwig, **37-38**, 53
Philippi, Rodolfo Amando (Rudolph Amandus), **63-64**
Pidgeon, Edward, 19
Pieris, see Deraniyagala
Pistoia, Louis Joseph, 118

Pitman, Charles Robert Senhouse, **105**
Poey (y Aloy), Felipe, 53
Pope, Clifford Hillhouse, 91, 92, **94**, 108, 117
Pope, Sarah Haydock Davis, 94
Powell, John Wesley, 61
Prashad, Baini, 79
Price, Llewellyn Ivor, 129
Procter, Joan Beauchamp, 106
Przewalski (or Przhevalsky), Nikolai Mikhailovich, 49, 58

- R -

Raddi, Guiseppe, 23
Rafinesque(-Schmaltz), Constantine Samuel, **25-26**, 27
Ramsden, Charles Theodore, 85
Rao, C. R. Narayan, 79
Rapp, Wilhelm Ludwig von, 45
Ray (or Wray), John, 8, 11
Redi, Francesco, **8**, 12
Reeve, Wayne Lee, 83
Reinhardt, Johannes Christopher Hagemann, 44, 45
Reinhardt, Johannes Theodor, **44-45**
Reinwardt, Caspar Georg Carl, 21, 30
Rendahl, Carl Hialmar, 68
Ribeiro, see Miranda Ribeiro
Richard (or Richards), John H., 33
Robinson, Douglas Clark, 93
Roesel von Rosenhof, August Johann, **10-11**
Roesel von Rosenhof, Wilhelm, 10
Rogers, James Speed, 114, 115
Rollinat, Pierre André Marie Raymond, **66-67**
Romano Hoge, see Hoge, Alma Romano
Romer, Alfred Sherwood, 81, 90, **128-129**
Rondelet, Guillaume, 7
Rondon, Colonel Cândido Mariano da Silva, 88
Rooij (later, Breijer-de Rooij), Nelly de (Petronella Johanna), **76-77**
Rooy, see Rooij
Rosenhof, see Roesel von Rosenhof
Rouaire Siauzade, Antoine, see Nicéforo María
Roux, Jean, **76**
Rüppell, Eduard Wilhelm Peter Simon, 57, 99
Russell, Alexander, 16
Russell, Claud, 16
Russell, Findlay Ewing, 98
Russell, Patrick, **16-17**
Rustamov, Anver Kejusevich, 124
Ruthe, Johann Friedrich, 28
Ruthven, Alexander Grant, 63, **74-75**, 81, 82, 96

- S -

Sagra, Ramón de la, 33
Saint-Hilaire, see Geoffroy Saint-Hilaire
Sarasin, Fritz (Carl Friedrich), 58, 76
Sarasin, Paul Benedikt, 58
Sato, Ikio, **122-123**
Savage, Jay Mathers, 93, 114
Say, Thomas, 22, 26, **26-27**, 27
Schenkel, Ehrenfried, 57
Scherpner, Christoph, 102
Scheuchzer, Johann Jakob, 18
Schinz, Heinrich Rudolf, 19, 36, 39

Schlegel, Hermann, 21, **30-31**, 36, 47
Schmidt, Karl Patterson, 76, 82, 90, 91, **91-92**, 94, 113
Schneider, Johann Gottlob Theaenus, **13**, 28
Schoepff, Johann David, **14-15**
Schreber, Johann Christian Daniel, 10, 15
Schreibers, Carl Franz Anton von, 24, 30
Scortecci, Giuseppe, **104-105**
Seba, Albertus, **9**, 10, 11, 12, 49
Seibert, Henri Cleret, 93
Sera, J., 33
Seuss, Eduard, 54
Shaw, Charles Edward, 98
Shaw, George, **17**
Shcherbak, see Szczerbak
Shreve, Benjamin, 112
Siebenrock, Friedrich, 54, **69-70**
Siebold, Karl Theodor Ernst von, 47, 60
Siebold, Philipp Franz von, 30
Slevin, Joseph Richard, 73
Smit, Joseph, 30
Smit, Pierre Jacques, 55, 122
Smith, Andrew, **36**
Smith, Hobart Muir, 76, 83, 103, 113
Smith, Malcolm Arthur, 83, **86-87**, 107
Smith, Philip Wayne, 129
Sonnini (de Manoncourt), Charles-Nicolas-Sigisbert, 19, 20
Sordelli, Ferdinando, 50, **50**
Sowerby, James deCarle, 35
Speakman, John, 26
Spix, Johann Baptist von, **23**, 24, 25, 39, 77
Stansbury, Howard, 41
Stebbins, Robert Cyril, 111, 117
Steindachner, Franz, **54**, 69, 70
Steinheil, Fritz, 78
Stejneger, Leonhard Hess, 61, **62-63**, 79, 85, 92, 92, 96, 99, 110, 113, 130
Sternfeld, Richard, 98
Steyn, Willem J., 121
Stickel, William Henson, 96
Stoliczka, Ferdinand, 54
Storer, David Humphreys, 33
Storr, Glen Milton, 101
Strauch, Alexander Alexandrovich, **49**, 58, 68
Stuart, Laurence Cooper, 74
Stull (later, Davis), Olive Griffith, 74
Summerville, William Harry, 118
Szczerbak, Nikolai Nikolaevich, 124

- T -

Takakuwa, Yosiyuki, 109
Tanner, Wilmer Webster, 83
Taylor, Edward Harrison, **82-84**
Temminck, Coenraad Jacob, 21, 30
Terentjev, Paul Victorovich, 103, **103-104**, 116
Thompson, Crystal, 75, 82
Thompson, Helen, see Gaige, Helen
Thompson, Joseph Cheesman, 73
Thunberg, Carl Peter, 11
Tian, Wan-shu, 123
Tiedemann, Friedrich, 21, 39
Tihen, Joseph Anton, 95
Tilley, Stephen George, 132

Ting, Han-po, 108
Tinkle, Donald Ward, **131-132**
Tirant, Gilbert, 107
Tokunaga, Shigeyasu, 109
Toner, George Clive, 110
Topsell, Edward, 7
Traill, Thomas Stewart, 30
Trapido, Harold, 76
Troost, Gerard, 26, 27
Troschel, Franz Hermann, 46
Tsarevsky, Sergius F., 103
Tschudi, Friedrich von, 36
Tschudi, Johann Jakob von, 19, **36-37**

- U -

Uribe Peña, Zeferino, 119

- V -

Vaillant, Léon-Louis, 44, **58-59**, 106
Valenciennes, Achille, 18, 31, 33, 44
Vancea, Stefan, 128
Vandelli, Domenico, 51
Van Denburgh, John, **72-73**, 97
Van Hasselt, see Hasselt
Van Kampen, see Kampen
Vanzolini, Paulo Emílio, 12, 68
Versluys, Jan, 101
Vesalius, Andreas, 7
Victoria (Alexandrina Victoria), Queen of the United Kingdom and Empress of India, 36
Villela, see Flores Villela
Villiers, André, **124-125**
Vitt, Laurie Joseph, 132
Vogel, Zdeněk, **118-119**
Voigt, Friedrich Siegmund, 19
Von Rosenhof, see Roesel von Rosenhof
Vosmaer, Arnout, 9

- W -

Wagler, Johann Georg, 23, **23-24**, 25, 39, 77
Waite, Edgar Ravenswood, **71**
Walcott, John, 11
Walker, Charles Frederic, 74, 82, 93
Wall, Frank, **71-72**, 79, 121
Wallace, Alfred Russell, 35
Wang, Hao-ting, 94
Watson, David Meredith Seares, 129

Webb, Robert Graven, 83
Weller, Worth Hamilton, 93
Wellington, First Duke of (Arthur Wellesley), 36
Werler, John Ernest, 118
Wermuth, Heinz, 99
Werner, Franz Joseph Maria, 65, **70-71**, 101
Werner, Yehudah Leopold, 102
Wettstein(-Westersheimb), Otto von, 69, 70, 99, 101
Whipple, see Wilder, Inez
White, Gilbert, 35, 38
White, John, 17
Wied-Neuwied, Alexander Philipp Maximilian zu, **22-23**, 23, 47
Wiegmann, Arend Friedrich, 28
Wiegmann, Arend Friedrich August, **28-29**, 63
Wiegmann, Carl Arend Friedrich, 28
Wilbur, Henry Miles, 132
Wilder, Burt Green, 75
Wilder, Harris Hawthorne, 67, 93
Wilder, Inez Luanne Whipple, 93
Wilkes, Charles, 41, 43
Williams, Ernest Edward, 80, 112
Williston, Samuel Wendell, 128
Winterl, Jakob Joseph, 12
Witte, Gaston-François de, **122**
Wolf, Joseph, 30
Wolterstorff, Willy, 67, 77, 127
Wray, see Ray
Wright, Albert Hazen, **75-76**, 91, 95, 116, 123
Wright, Anna Allen, **75-76**

- Y -

Yang, Fu-hua, 123
Yarrow, Henry Crècy, 62
Yung, Emile, 76

- Z -

Zappalorti, Robert Thomas, 118
Zarevskij, see Tsarevsky
Zetek, James, 84
Zhao, Er-mi, 123
Zhu, C.-g., 123
Zichy, Eugen (Jenö), 65
Zittel, Karl Alfred von, 60
Zucconi, David, 118
Zug, George Robert, 116, 130
Zweifel, Richard George, 117

INDEX OF AUTHORS IN TAXONOMIC HERPETOLOGY

by John S. Applegarth

As a contribution to the history of herpetology, this index is intended to include the complete names, dates, countries, and orders for all authors who have participated in the naming of one or more taxa, at the level of genus or below, within the living families of amphibians and reptiles. Also included are authors who have had a taxon (within the living families of amphibians and reptiles) named in their honor *and* who have authored at least one contribution to herpetology.

Names.—Last (family) names and initials of first (given) names are in boldface print. Names that authors have seldom used are included for the benefit of historians and librarians but are given without boldface. Accent marks are included but ignored in the alphabetizing of names. When the mother's family name is given after the father's family name, the two surnames are alphabetized as a unit. Family names consisting of more then one word are also aphabetized as a unit. An "initial" letter without a period after it is not an abbreviation; for example, in Carl H Eigenmann the H does not stand for any name. Equivalent names are in parentheses and are preceded by an equal sign (=). Diminutives and nicknames that have appeared in publications are placed within quotation marks and treated as equivalent names. Non-academic titles (which are not capitalized in order to distinguish them from names) and personal relationships to other authors in this list are in parentheses after an author's first names.

Many of the living authors prefer to be known by certain parts of their complete name, and some avoid or dislike other parts. Please respect the feelings of these people and address them using the form they prefer. If one does not know the form an author prefers, one might use only initials for their first names and only of those names that in this index have the first letter in boldface print.

Dates.—Solid dots separate categories of information (who, when, where, and what). The first dot separates an author's name (and related information) from their date of birth (and, if applicable, death). Birth and death dates are written symmetrically to facilitate reading only the years. If there is only one date or year, that is the date of birth and the person is presumed to be living. A year that has not been found is represented by a question mark (?). If neither birth nor death year are known and the author published over 80 years ago, the known year(s) of their publications are given in parentheses with dashes and question marks on both sides.

Countries.—After the second dot is a short name for the country or geographic region where the author was born or resided. Known multiple places of residence are separated by a slash mark (/) and are given in chronological order.

Orders.—Following the third dot in each author's entry are the names of the herpetological orders in which taxa were proposed by, for, or by means of that author. *If* the first letter of an order is capitalized, this indicates that it represents living taxa within that order. If the first letter of an order is *not* capitalized and there is an asterisk (*) in front of that order, this indicates that it represents only fossil or extinct taxa. If an order is both capitalized and preceded by an asterisk (*), then it represents both living and fossil taxa.

If one or more taxa in an order were named *for* an author (to honor that person) *and* that author did not personally name any taxa in that order, then parentheses are placed around that order. If the proposed taxa were published by someone else within the work of an author *and* that author did not otherwise participate in naming any taxa in that order, then the order representing those taxa is in square brackets. Therefore if an order is within both parentheses and square brackets, it means that one or more taxa within that order were named for that author and were published in works by that author (not necessarily the same taxa) *and* that the author never named any taxa in that order.

Note.—It is the personal opinion of the compiler that the methods and recent writings of Richard W. Wells and C. Ross Wellington are inconsistent with acceptable practices of taxonomy, and that such writings should be rejected by the International Commission on Zoological Nomenclature. Therefore Mr. Wells and Mr. Wellington are not included in this compilation. For further opinions on this matter see *Herpetological Review* 16: 4-7 and 69, and *Australian Entomological Society News Bulletin* 21: 66-69.

Acknowledgments.—I am indebted to the many people who have helped in many ways. Kraig Adler generously provided author information and editorial guidance. I thank everyone who answered my data requests, especially the following people who took time to gather biographical information about others: T. Alvarez-S., M. Atatür, S. C. Ayala, R. Barbault, J. Bons, B. Brattstrom, R. Conant, A. J. Coventry, R. I. Crombie, Y. Develey, S. K. Dutta, R. Etheridge, R. Formas, D. R. Frost, E. S. Gaffney, J. M. Gallardo, H. W. Greene, K. Haker, H. S. Harris, H. Heatwole, T. Hikida, J. J. Kirk, E. Kramer, M. A. Labra, R. Lawson, C. J. McCoy, I. Mercadal, K. Miyazaki, M. Penna-V., J. C. Rage, P. M. Ruiz-C., H. Saint Girons, T. Seto, N. N. Shcherbak, J. E. Simmons, A. F. Stimson, B. W. Thomas, F. Tiedemann, H. G. Tunner, M. J. Tyler, Y. Werner, E. E. Williams, and especially E.-m. Zhao (who gathered data for over 60 Chinese authors!). For local assistance I am very grateful to James Kezer, Susan Crawford, and the staff of the University of Oregon libraries (especially R. H. Felsing, P. D. Morrison, and B. K. Wycoff). My special thanks to the Oregon Herpetological Society for a grant-in-aid toward postage expenses.

Corrections.—My apologies to anyone who was overlooked, incorrectly or incompletely cited, or otherwise offended. I anticipate producing a revised edition, so I would greatly appreciate any additions or corrections. Please send all pertinent information to: Box 532, Lorane Route, Cottage Grove, Oregon 97424, USA. Copies of your taxonomic publications would be helpful. Thank you.

TAXONOMIC AUTHORS

- A -

Aagaard, Carl Johan Ove Mønster • 1882–? • (?Denmark) / Siam • (Serpentes)
Abalos, Jorge Washington • ? • Argentina • Serpentes
Abbott, Charles Conrad • 4 June 1843–1919 July 27 • USA • Sauria
Abbott, William Louis • 23 February 1860–1936 April 2 • USA • (Anura, Sauria)
Abe, Yoshio • 3 January 1891–1960 April 22 • Japan • Caudata
Acharji, M. N. • ? • India • Serpentes
Adams, Andrew Leith • 1827–1882 August • Scotland / England / Ireland • *testudines
Adams, Leverett Allen • 23 September 1877–? • USA • *caudata
Adams, M. • ? • Australia • Anura
Adler, Kraig Kerr • 6 December 1940 • USA • Anura, Caudata, (Sauria), Testudines
Adriani, Giuseppe • ? • Italy • Sauria
Adrover, Rafael • ? • Spain • Anura
Aellen, Villy • 4 December 1926 • Switzerland • Caudata
Agassiz, Jean Louis Rodolphe • 28 May 1807–1873 December 14 • Switzerland / USA • Anura, Caudata, *Crocodilia, Sauria, Serpentes, Testudines
Ahl, Ernst • 1898–World War II • Germany • Anura, Gymnophiona, Sauria, Serpentes, *Testudines
Akhmedov, M. I. • ? • USSR • Sauria
Alberch, Pere • 2 November 1954 • Spain / USA • (Anura)
Albino, Adriana M. • ? • Argentina • *serpentes
Alcala, Angel Chua • 1 March 1929 • Philippines • Anura, Sauria, (Serpentes)
Alcock, Alfred William (lieutenant colonel) • 23 June 1859–1933 March 24 • England • Gymnophiona, Serpentes
Alcover, Josep Antoni • ? • Spain • *anura
Aldrovandi, Ulisse • 11 September 1522–1605 November 10 • Italy • (Sauria)
Alekperov, A. M. • ? • USSR • *testudines
Alekseyev (=Alexejew), Aleksei Karpovich • ? • Russia • *sauria
Aleman G., César • ? • Venezuela • (Anura), Serpentes
Alexander, Alexander Allan • 19 July 1928 • USA • Sauria
Alfaro Gonzáles, Anastasio • 16 February 1865–1951 January 20 • Costa Rica • (Caudata, Sauria)
Allen, Ensil Ross • 2 January 1908–1981 May 17 • USA • (Anura), Caudata, Serpentes
Allen, Glover Morrill • 8 February 1879–1942 February 14 • USA • (Sauria), Serpentes
Allen, Joel Asaph • 19 July 1838–1921 August 29 • USA • (Serpentes)
Allen, Morrow J. • 24 November 1909 • USA • Sauria
Allison, Allen • ? • USA / Papua New Guinea • Anura, Sauria
Alluaud, Charles • 1861–? • France • (Anura, Sauria, Serpentes)
Altig, Ronald Gail • 8 June 1941 • USA • Anura, Serpentes

Altobello, Giuseppe • ? • Italy • Caudata
Álvarez, Antenor • 1864–? • Argentina • (Serpentes)
Alvarez del Toro, Miguel • 23 August 1917 • Mexico • (Caudata), Sauria
Álvarez Solorzano, José Ticul • 26 February 1935 • Mexico • Caudata, Sauria
Amaral, Afrânio Pompílio Bastos do, 1 December 1894–1982 November 29 • Brazil • Sauria, Serpentes
Ambrosetti, Juan Bautista • 22 August 1865–1917 May 28 • Argentina • *crocodilia, *sauria
Ameghino, Florentino (=Fiorino) • 18 September 1854–1911 August 6 • Italy / Argentina • *anura, *sauria
Amiet, Jean-Louis • ? • Cameroons • Anura, (Sauria)
Ammon, Ludwig Johann Georg Friedrich von • 14 December 1850–1922 July 26 • Germany • *testudines
Ananjeva, Natalia B. • ? • USSR • Sauria, [Serpentes]
Anderson, James Donald, Jr. • 16 August 1930–1976 November 20 • USA • Caudata, Sauria, Serpentes
Anderson, Jeromie A. • ? • (?USA) / Pakistan • Sauria
Anderson, John • 4 October 1833–1900 August 15 • Scotland / British India / England • Anura, Caudata, Sauria, Serpentes, Testudines
Anderson, Robert A. • ? • USA • Sauria
Anderson, Steven Clement • 7 September 1936 • USA • Sauria
Andersson, Lars Gabriel • 22 February 1868–1951 February 13 • Sweden • Anura, Caudata, Sauria, Serpentes
Andrade, Gilda V. • ? • Brazil • Anura
Andrén, Claes • ? • Sweden • Anura, Sauria, Serpentes
Andrews, Charles William • 1866–1924 May 25 • England • *crocodilia, *serpentes, *testudines
Andrews, Edward Wyllys, IV • 11 December 1916–1971 July 3 • USA • Serpentes
Andrews, Ethan Allen • 10 September 1859–? • USA • (Anura)
Andrzejowski, Antoni • 1785–1868 • (?Poland) • Caudata, Sauria, Serpentes
Angel, Fernand • 2 February 1881–1950 July 13 • France • Anura, (Gymnophiona), Sauria, Serpentes
Annandale, Thomas Nelson • 15 June 1876–1924 April 10 • Scotland / British India • Anura, Gymnophiona, Sauria, Serpentes, Testudines
Anstis, Marion • ? • Australia • Anura
Aoki, Riosuke • 8 September 1954 • Japan • *crocodilia
Aplin, Ken • ? • Australia • Anura
Arambourg, Louis Joseph Camille • 3 February 1885–1969 November 19 • France • (*crocodilia, *sauria), *testudines
Archey, Gilbert Edward (sir) • 9 August 1890–1974 • England / New Zealand • (Anura)
Ardila Robayo, María Cristina • 15 February 1947 • Colombia • Anura
Aristotle • BC 384–322 BC • Greece • (Serpentes)
Armstrong, Barry L. • ? • USA • Serpentes
Armstrong-Ziegler, Judy Gail • ? • USA • *serpentes
Arndt, Rudolf Gerhard Ernst • 28 October 1941 • Germany / USA • Serpentes
Arnold, Douglas L. • ? • USA • Sauria

Arnold, Edwin Nicholas • 16 October 1940 • England • (Anura), Sauria, Serpentes, (Testudines)
Arnoult, Jacques • 25 February 1914 • France • (Anura, Sauria)
Arthaber, Gustav Adolph (edler) von • 21 October 1864–1943 April 29 • Austria • *testudines
Ashe, James • ? • Kenya • Serpentes
Astre, Gaston Prosper • 16 April 1896 • France • *crocodilia
Ataev, Chary Ataevich • ? • USSR • Sauria
Atatür, Mehmet Kutsay • 15 February 1947 • Turkey • Anura, Caudata
Audouin, Jean-Victor • 27 April 1797–1841 November 9 • France • Anura, Sauria
Audubon, John James • 26 April 1785–1851 January 27 • Haiti / France / USA / Scotland / USA • (Sauria)
Auffenberg, Walter • 6 February 1928 • USA • *anura, *caudata, Sauria, *Serpentes, *testudines
Augé, Marc • 21 April 1950 • France • *sauria
Avery, David Franklin • 30 January 1939 • USA • Sauria
Ávila-Pires, Teresa Cristina Sauer de • ? • Brazil • Sauria, Serpentes
Axtell, Ralph William • 20 April 1928 • USA • Sauria
Ayala, Stephen Charles • 13 June 1942 • USA / Ecuador / Colombia / USA • Sauria
Ayarzagüena, Jose • ? • Venezuela • Anura
Aymard, Auguste • 1808–1889 June 26 • France • *anura

- B -

Babbitt, Lewis Hall • 1902–1972 September 28 • USA • Serpentes
Babcock, Harold Lester • 30 May 1886–1953 January 21 • USA • Testudines
Bachmayer, Friedrich • 10 September 1913 • Austria • *serpentes, *testudines
Bacon, James Patterson, Jr. • 22 August 1939–1986 April 4 • USA • Sauria
Bacqué, Alfred • ?–? • Paraguay • Serpentes
Badham: see Caughley, J. A. B.
Badmaeva, V. I. • ? • USSR • Sauria
Báez, E. C. • ? • Argentina • Serpentes
Bagnara, Joseph Thomas • 26 July 1929 • USA • Anura
Baikie, William Balfour • 27 August 1825–1864 December 12 • England / Nigeria • Crocodilia
Bailey, A. • ? • (?Australia) • [Sauria]
Bailey, John Wendell (lieutenant colonel) • 9 January 1895 • USA • Sauria
Bailey, Joseph Randle • 17 September 1913 • USA • (Anura), Serpentes
Bailey, Vernon Orlando • 21 June 1864–1942 April 20 • USA • (Sauria)
Baird, Spencer Fullerton • 3 February 1823–1887 August 19 • USA • Anura, Caudata, Sauria, Serpentes, Testudines
Bakde, Ramesh • ? • India • Anura
Baker, James Kenneth • ? • USA • Caudata
Bakradze, Michael Aleksandrovich • 1939 • USSR • Sauria
Balletto, Emilio • ? • Italy • Anura
Balli, Antonio • ? • Italy • Serpentes

Ballinger, Royce Eugene • 21 February 1942 • USA • Sauria
Ballion, Ernst E. von • 1816–1901 • Russia • Caudata
Baloutch, M. • ? • Iran • Sauria
Bangs, Outram • 12 January 1863–1932 September 22 • USA • (Testudines)
Banta, Benjamin Harrison • 2 January 1927 • USA • Sauria, Serpentes
Baran, İbrahim • 15 January 1940 • Turkey • Anura, Caudata, Sauria, Serpentes
Barbault, Robert • 24 January 1943 • France • Sauria
Barbour, Roger William • 5 April 1919 • USA • Caudata, (Testudines)
Barbour, Thomas • 19 August 1884–1946 January 8 • USA • Anura, (Caudata), Gymnophiona, *Sauria, Serpentes, Testudines
Barbosa Rodrigues: see Rodrigues, J. B.
Barboza du Bocage: see Bocage, J. V. B. du
Barnes, Ben • 1903 • Germany • *serpentes
Barnes, Daniel Henry • 1785–1828 • USA • Caudata
Barrio, Avelino • 10 August 1920–1979 June 30 • Spain / Argentina • Anura, (Serpentes)
Barron, P. A. R. • ? • (?England) / Siam • (Serpentes)
Barry, John Chase • 21 January 1946 • USA • [Sauria]
Bartenjev, A. • ? • USSR • Sauria
Bartholomew, George Adelbert, Jr. • 1 June 1919 • USA • (Serpentes)
Bartlett, Edward • (*circa* 1836)–1908 April • England • Sauria, Testudines
Barton, Benjamin Smith • 10 February 1766–1815 December 19 • USA • Caudata
Bartram, William • 9 April 1739–1823 July 22 • USA • (Anura, Testudines)
Baskin, Jonathan Noel • 25 July 1939 • USA • Sauria
Bas Lopez, Santiago • ? • Spain • Serpentes
Başoğlu, Muhtar • 6 April 1913–1981 February 21 • Turkey • Caudata, (Sauria), Serpentes
Bataller Calatayud, José Ramón • 10 August 1890 • Spain • *crocodilia, *testudines
Bate, Dorothea Minola Alice • ?–? • England • *testudines
Battersby, James Clarence • 11 April 1901 • England • (Anura, Gymnophiona), Sauria, Serpentes
Bauer, Aaron Matthew • 27 June 1961 • USA • Sauria
Baumann, Franz • 25 December 1885–1961 May 2 • Switzerland • Anura, Sauria, Serpentes
Baur, George (=Georg Hermann Carl Ludwig) • 4 January 1859–1898 June 25 • Germany / USA • *sauria, *testudines
Bavay, Arthur René Jean Baptiste • ?–? • (?France) • Sauria, Serpentes
Baylor, Edward Randall • 21 January 1914 • USA • Anura
Beargie, Kathleen • ? • USA • Sauria
Beauvois: see Palisot (baron de Beauvois), A.-M.-F.-J.
Beçak, Willy • 1932 • Brazil • Anura
Beccari, Nello • 11 January 1883–1957 March 20 • Italy • (Anura, Sauria)
Bechstein, Johann Matthäus • 11 July 1757–1822 February 23 • Germany • Caudata, Serpentes
Beck, Rollo Howard • 26 August 1870–1950 November 22 • USA • (Sauria, Testudines)
Beddard, Frank Evers • 19 June 1858–1925 July 14 • England • Anura

Beddome, Richard Henry (colonel) • 11 May 1830–1911 February 23 • British India / England • Anura, Gymnophiona, Sauria, Serpentes
Bedot, Maurice • 1859–? • Switzerland • (Serpentes)
Bedriaga, Jacques de (=Yakov Vladimirovich Bedryaga) • 1854–1906 • Russia • Anura, Caudata, Sauria, Serpentes, Testudines
Beebe, Charles William • 29 July 1877–1962 June 4 • USA • (Anura, Sauria)
Behrndt, A. C. • ? • Australia • Sauria
Beireis, Gottfried Christoph • 28 February 1730–1809 September 18 • Germany • Anura
Bélanger, Charles-Paulus • 29 May 1805–1881 November 18 • France / Martinique • [Anura, Serpentes, Testudines]
Belding, Lyman • 12 June 1829–1917 November 22 • USA • (Anura, Sauria)
Bell, L. Neil • ? • USA • (Caudata), Serpentes
Bell, Thomas • 11 October 1792–1880 March 13 • England • Anura, Caudata, Sauria, Serpentes, *Testudines
Belluomini, Hélio Emerson • ? • Brazil • Serpentes
Bendiscioli, Giuseppe • ?–? • Italy • Serpentes
Beneden, Pierre Joseph van • 19 December 1809–1894 January 8 • Belgium • *testudines
Bennett, Edward Turner • 6 January 1797–1836 August 21 • England • Crocodilia, (Sauria), Testudines
Bennett, George F. • 31 January 1804–1893 September 29 • Australia • [*sauria]
Bequaert, Joseph Charles Corneille • 24 May 1886–1982 • Belgium / USA • (Anura)
Berg, Carlos (=Frederick Wilhelm Carl) • "21 March" (=2 April) 1843–1902 January 19 • Russia / Argentina • Anura, Serpentes
Berg, Dietrich E. • 1932 • Germany • *crocodilia
Berg, Johannes • ?–? • Germany • (Sauria)
Berger, Thomas Joseph • ? • USA • Anura
Bergounioux, Edmond André Frédéric-Marie (père) • 15 October 1900 • France • *testudines
Berman, David S. • 10 January 1940 • USA • *sauria
Berry, Charles Thompson • 6 November 1906 • USA • *testudines
Berry, James Frederick • 22 December 1947 • USA • Testudines
Berry, P. Y. • ? • Malaya • Anura
Berthelot, Sabin • 4 April 1794–1880 November 10 • France • [Sauria]
Berthold, Arnold Adolph • 26 February 1803–1861 February 3 • Germany • Anura, Caudata, Gymnophiona, Sauria, Serpentes, Testudines
Besharse, Joseph Culp • 21 January 1944 • USA • Caudata
Beshkov, Vladimir A. • ? • Bulgaria • (Caudata)
Bethencourt Ferreira: *see* Ferreira, J. B.
Betta, Francesco Edoardo (nobile) de • 1822–1896 • Italy • Anura, Sauria, Serpentes, (Testudines)
Beurden: *see* van Beurden, E. K.
Beutler, Axel • ? • Germany • Sauria
Beyer, George Eugene • 9 September 1861–1926 June 2 • Germany / USA • (Caudata)
Bezy, Robert Lee • 26 September 1941 • USA • Sauria
Bhaduri, Jeetendra Lal • ?–? • India • Anura

Bianconi, Giovanni Giuseppe • 31 July 1809–1878 October 18 • Italy • Anura, Sauria, Serpentes
Bibron, Gabriel • 1806–1848 March 27 • France • Anura, Caudata, Gymnophiona, Sauria, Serpentes, *Testudines
Biedermann-Imhoof, W. G. Adolf • ?–(1862-1876)–? • Switzerland • *testudines
Bielz, Eduard Albert • 4 February 1827–1898 May 27 • Hungary • [Sauria]
Bigelow, Jacob • 27 February 1786–1879 January 10 • USA • Serpentes
Bird, Roland Thaxter • 29 December 1899–1978 January 24 • USA • *crocodilia
Bischoff, Wolfgang • ? • Germany • Sauria
Bishop, Sherman Chauncey • 18 November 1887–1951 May 28 • USA • Caudata
Biswas, S. • ? • India • Sauria, Serpentes
Black, Dennis • ? • Papua New Guinea • Sauria
Blainville, Henri-Marie Ducrotay de • 12 September 1777–1850 May 1 • France • *crocodilia, Sauria, Serpentes, Testudines
Blair, Albert Patrick (*brother of* W. F. Blair) • 3 November 1913 • USA • Anura
Blair, William Franklin (=Frank, *brother of* A. P. Blair) • 25 June 1912–1984 February 9 • USA • (Anura, Serpentes)
Blake, Anthony John Dyson • ? • Australia • Anura
Blanc, Charles P. • ? • Madagascar / France • (Anura), Sauria
Blanchard, Charles Émile • 7 March 1819–1900 February 11 • France • Caudata, Sauria
Blanchard, Frank Nelson • 19 December 1888–1937 September 21 • USA • (Anura, Sauria), Serpentes
Bland-Sutton, John (sir) • 21 April 1855–1936 December 20 • England • (Anura)
Blaney, Richard M. • 10 March 1942 • USA • Serpentes
Blanford, William Thomas • 7 October 1832–1905 June 23 • England / British India / England • Anura, Sauria, Serpentes
Blasius, Johann Heinrich • 7 October 1809–1870 May 26 • Germany • Anura
Blatchley, Willis Stanley • 6 October 1859–1940 May 28 • USA • Caudata
Bleeker, Pieter • 1819–1878 • Holland • Anura, Gymnophiona, Sauria, Serpentes
Blomberg, Rolf David • 11 November 1912 • Sweden • (Anura)
Blomefield: *see* Jenyns, L.
Blommers-Schlösser, Rose M. A. • 30 January 1944 • Holland • Anura
Bloxam, Q. • ? • England • Sauria
Blumenbach, Johann Friedrich • 11 May 1752–1840 January 22 • Germany • Anura, Caudata, Serpentes
Blyth, Edward • 23 December 1810–1873 December 27 • England / British India / England • Anura, Sauria, Serpentes, Testudines
Bocage, José Vicente Barboza du • 2 May 1823–1907 November 3 • Portugal • Anura, Caudata, Gymnophiona, Sauria, Serpentes
Bocourt, Marie-Firmin • 19 April 1819–1904 February 3 • France • (Anura), Crocodilia, Sauria, Serpentes, Testudines

Bocquentin Villanueva, Juan (=Jean) • ? • Venezuela • *crocodilia
Boda, Anton • ? • Hungary • *testudines
Boddaert, Pieter • 1730–1796 • Holland • Anura, (Sauria, Serpentes), Testudines
Bodenheimer, Friedrich Simon • 1897 • (?Germany) • Caudata
Boehme: see Böhme, W.
Boettger (=Böttger), Caesar Rudolf (*nephew of* O. Boettger) • 20 May 1888–? • Germany • Sauria
Boettger (=Böttger), Oskar (*uncle of* C. R. Boettger) • 31 March 1844–1910 September 25 • Germany • Anura, Caudata, Gymnophiona, *Sauria, Serpentes, Testudines
Bogachev, Vladimir V. • 1881–? • USSR • *testudines
Bogert, Charles Mitchill • 4 June 1908 • USA • Anura, Caudata, Sauria, (*sauria), Serpentes, Testudines
Bohlin, Anders Birger • 26 March 1898 • Sweden • *testudines
Bohls, Johann Friedrich Wilhelm • 1863–? • Germany • (Sauria), Testudines
Böhme (=Boehme), Wolfgang • 21 November 1944 • Germany • Anura, *Sauria, Serpentes
Bohn, Péter • ? • Hungary • *testudines
Boie, Friedrich (*brother of* H. Boie) • 4 June 1789–1870 March 3 • Germany • Gymnophiona, Sauria, Serpentes
Boie, Heinrich (*brother of* F. Boie) • 4 May 1794–1827 September 4 • Germany / Holland / Java • Caudata, Serpentes, (Testudines)
Bois-Reymond, Claude du • 21 December 1855–1925 • Germany / China • (Anura)
Böker, Hans • 14 November 1886–1939 April 23 • Mexico / Germany • (Sauria)
Bokermann, Werner Carl August • ? • Brazil • Anura, (Gymnophiona)
Bolkay, Stephan Joseph Ladislaws • 29 March 1887–1930 August 17 • Hungary / Yugoslavia • *anura, Caudata, *Sauria, *Serpentes
Bologna, Marco • ? • Italy • Caudata
Bölsche, Wilhelm • 1843–1893 • Germany • Rhynchocephalia
Bonaparte, Carlo Luciano (=Charles Lucien Jules Laurent, prince) • 24 May 1803–1857 July 29 • France / Italy / USA / Italy / France • Anura, Caudata, Sauria, Serpentes
Bonnaterre, Pierre-Joseph (l'abbé) • 24 November 1751–1804 September 21 • France • Anura, Caudata, Crocodilia, Sauria, Serpentes, Testudines
Bonpland, Aimé-Jacques Alexandre "Goujaud" • 28 April 1773–1858 March 11 • France / Argentina / Brazil • [Serpentes]
Bons, Jacques • 1933 • Morocco / France • Anura, Sauria, Serpentes
Boring, Alice Middleton • 22 February 1883–1955 September 18 • USA / China / USA • Anura, (Caudata)
Borkin, Leo Ya. • 31 August 1949 • USSR • [Serpentes]
Börner, Achim-Rüdiger • 1955 • Germany • Sauria
Borre, Alfred Preud'homme de • 1833–1905 • Belgium • Crocodilia, Sauria
Borroughs: see Burrows, R. J.
Borsuk-Białynicka, Magdalena • ? • Poland • *anura, *sauria

Bory de Saint-Vincent, Jean-Baptiste-Geneviève-Marcellin (baron de) • 6 July 1778–1846 December 22 • France • Anura, Caudata, Crocodilia, Sauria, Serpentes, Testudines
Bosc d'Antic, Louis-Augustin-Guillaume • 29 January 1759–1828 July 10 • France • (Anura), Sauria, (Testudines)
Boscá y Casanoves, Eduardo • ?–? • Spain • Anura, Caudata, Sauria, Serpentes
Bossu, Jean-Bernard • 29 September 1720–1792 May 4 • France • [Serpentes]
Bostic, Dennis L. • ?–? • USA • Sauria, (Serpentes)
Botta, Paolo Emilio (=Paul Émile) • 1802–1870 • Italy / France • (Serpentes)
Böttger: see Boettger
Boubée, Nérée • 12 May 1806–1863 • France • Anura
Bouet, Georges-Théodore-Louis • 1869–? • France • (Sauria, Serpentes)
Boulenger, Edward George (*son of* G. A. Boulenger) • 8 May 1888–1946 April 30 • England • Anura, Caudata, Sauria
Boulenger, George Albert (*father of* E. G. Boulenger) • 19 October 1858–1937 November 23 • Belgium / England / Belgium • Anura, Caudata, Crocodilia, Gymnophiona, Sauria, Serpentes, *Testudines
Bour, Roger Henri • 9 July 1947 • France • Testudines
Bourgat, Robert • ? • Madagascar / Togo • (Anura)
Bourret, René Léon • 28 January 1884–1957 July 28 • France / French Indochina / France • Anura, Caudata, Sauria, Serpentes, Testudines
Boutan, Louis Marie Auguste • 1859–? • France • Sauria
Bowdich, Thomas Edward • 20 June 1791–1824 January 10 • England • [Sauria, Serpentes]
Boycott, Richard C. • ? • Swaziland • Anura
Braconnier, Séraphin • ?–? • France • (Anura), Sauria, (Serpentes)
Brady, Maurice Kirby • ? • USA • Serpentes
Bragg, Arthur Norris • 18 December 1897–1968 August 27 • USA • Anura
Brain, Charles Kimberlin • ? • South Africa • Sauria, Serpentes
Braithwaite, Richard W. • ? • Australia • Sauria
Bräm, Heinrich • ? • Switzerland • *testudines
Bramble, Dennis Marley • ? • USA • Testudines
Brame, Arden Howell, Jr. II • 19 March 1934 • USA • *Caudata
Braña Vigil, Florentino • ? • Spain • Serpentes
Brandon, Ronald Arthur • 3 December 1933 • USA • Caudata, Sauria
Brandt, Bartholomew Brandner • 30 July 1898 • USA • Anura
Brandt, Johann Friedrich • 25 May 1802–1879 July 3 • Germany / Russia • Serpentes, Testudines
Brattstrom, Bayard Holmes • 3 July 1929 • USA • *anura, *sauria, *serpentes, *testudines
Brauer, August Bernhard • 3 April 1863–1917 September 10 • Germany • (Sauria)
Braun, Cristina Assuncão Sirangelo (*wife of* P. C. Braun) • ? • Brazil • Anura
Braun, Max (=Maximilian Gustav Christian Carl) • 30 September 1850–1930 • Germany • Sauria

Braun, Pedro Canisio (*husband of* C. A. S. Braun) • ? • Brazil • Anura
Bravard, Pierre-Joseph-Auguste • 18 June 1803–1861 March 20 • France / Argentina • *crocodilia, *testudines
Bravo, Telesforo • ? • Canary Islands • *sauria
Brazenor, Charles Walter • 2 April 1897–1979 April • Australia • Serpentes
Brazil, Vital • 28 April 1865–1950 May 8 • Brazil • (Serpentes)
Breder, Charles Marcus, Jr. • 25 June 1897–1983 October 28 • USA • (Anura)
Brelih, Savo • ? • Yugoslavia • Sauria
Bréthes, Jean (=Juan) • 24 February 1871–1928 July 2 • France / Argentina • Serpentes
Breuil, Michel • ? • France • Caudata
Briceño Rossi, Antonio Leocadio • ? • Venezuela • Serpentes
Brickell, John, • (?1749)–(?1809 or 1810) • USA • Serpentes
Brimley, Clement Samuel (*brother of* H. H. Brimley) • 18 December 1863–1946 July 23 • England / USA • (Anura), Caudata
Brimley, Herbert Hutchinson (*brother of* C. S. Brimley) • 7 March 1861–1946 April 4 • England / USA • (Caudata)
Brinkerink, Johann P. • ? • Holland • *testudines
Brizzi, Rossana • ? • Italy • Sauria
Broadley, Donald G. • ? • Rhodesia • Anura, Sauria, Serpentes, Testudines
Brocchi, Paul-Louis-Antoine • 1839–1898 • France • Anura, Caudata, (Sauria), Serpentes
Brock, Vernon Eugene • 24 June 1912 • USA • Serpentes
Brode, John Martin • 31 May 1933 • USA • Caudata
Brodie, Edmund Darrell, Jr. • 29 June 1941 • USA • Caudata, Serpentes
Broili, Ferdinand • 11 April 1874–1946 April 30 • Germany • (Caudata)
Broin, France de • 5 October 1938 • France • *testudines
Bromme, Traugott • 1802–1866 September 4 • Germany • [Anura, Serpentes]
Brongersma, Leo Daniël • 17 May 1907 • Holland • (Anura), *Sauria, Serpentes
Brongniart, Alexandre • 5 February 1770–1847 October 7 • France • (*crocodilia, Sauria, Testudines)
Bronn, Heinrich Georg • 3 March 1800–1862 July 5 • Germany • *testudines
Brooker, Michael G. • ? • Australia • (Sauria)
Brookes, Joshua • 24 November 1761–1833 January 10 • England • Caudata, (Sauria)
Broom, Robert • 30 November 1866–1951 April 6 • England / South Africa • *rhynchocephalia, Sauria, (Serpentes)
Brown, Arthur Erwin • 14 August 1850–1910 October 29 • USA • Anura, Serpentes
Brown, Barnum • 12 February 1873–1963 February 5 • USA • (*crocodilia)
Brown, Bryce Cardigan • 7 May 1920 • USA • Sauria, Serpentes
Brown, Jill R. (*wife of* L. E. Brown) • 20 February 1943 • USA • Anura
Brown, Lauren Evans (*husband of* J. R. Brown) • 4 September 1939 • USA • Anura
Brown, Walter Creighton • 18 August 1913 • USA • Anura, Caudata, Sauria, Serpentes
Brüggemann, Friedrich • ?–? • Germany • (Sauria)
Bruguières, Jean-Guillaume • 1750–1799 October 1 • France • Serpentes
Brunner, Georg • ? • Germany • *caudata, *sauria, *serpentes
Bruno, Silvio • ? • Italy • Caudata
Brushko, Z. K. • ? • USSR • Sauria
Brygoo, Edouard Raoul • 22 April 1920 • France / Madagascar / France • (Anura), Sauria, (Testudines)
Buchholz, Karl F. • 1 February 1911–1967 July 1 • Germany • Sauria, Serpentes
Buchholz, Reinhold Wilhelm • 2 October 1837–1876 April • Germany • Anura, Sauria, Serpentes
Buck, Emil August • 20 April 1840–1899 December 17 • Germany • Sauria
Buckeridge, John St James Stewart • 6 December 1949 • New Zealand • *testudines
Budak, Abidin • 9 November 1943 • Turkey • Sauria
Buden, Donald William • 1943 • USA • Serpentes
Budgett, John Samuel • 16 June 1872–1904 January 19 • England • Anura
Buffetaut, Eric Charles Nicolas • 19 November 1950 • France • *crocodilia
Buffon, Georges-Louis Leclerc (=le Clerc, comte de) • 7 September 1707–1788 April 16 • France • [Serpentes]
Buller, Walter Lawry (sir) • 9 October 1838–1906 July 19 • New Zealand • Rhynchocephalia, Sauria
Bullini, Luciano • ? • Italy • Anura, Caudata
Bullock, D. J. • ? • Scotland • Sauria
Bulmer, Ralph Neville Hermon • 3 April 1928–1988 July 18 • England / New Guinea / New Zealand • (Anura)
Bumzahem (*changed to* Campbell), Carlos B. • ? • USA • Anura, Caudata, Sauria
Burbidge, Andrew A. • ? • Australia • (Sauria)
Burchell, William John • 23 July 1781–1863 March 23 • England / South Africa / Brazil / England • (Sauria), Serpentes
Burden, William Douglas • 24 September 1898–1978 November 14 • USA • (Sauria)
Buresch, Iwan • 1885–? • Bulgaria • Caudata, (Sauria), Serpentes
Bürger, Heinrich Otto Wilhelm • 4 May 1865–1945 January 18 • Germany / Chile / Germany • (Anura, [Sauria])
Burger, William Leslie • 1925 • USA / Japan / USA • Anura, Caudata, Sauria, Serpentes
Burley, F. William • ? • USA • Sauria
Burmeister, Karl Hermann Konrad (=Carlos Germán Conrado) • 15 January 1807–1892 May 2 • Germany / Argentina • Anura, *crocodilia, Sauria
Burns, Douglas Murray • 7 February 1935–1976 November • USA / Thailand • Caudata
Burrowes, Patricia A. • 8 April 1961 • Colombia / USA / Puerto Rico • Anura
Burrows (*not* Borroughs), Robert Jonathan • 29 January 1958 • USA • Anura
Burt, Charles Earle (*husband of* M. D. Burt) • 12 August 1904–1963 July 13 • USA • Sauria
Burt, May Danheim (*wife of* C. E. Burt) • 19 January 1904 • USA • Sauria

Burton, Edward • ?–(1836)–? • England • Sauria
Burton, Thomas C. • ? • Australia • Anura
Busack, Stephen Dana • 3 October 1944 • USA • Anura, Sauria
Bush, Francis Marion • 5 September 1933 • USA • Caudata
Busse, Klaus • ? • Germany • Anura
Bustard, H. Robert • ? • Australia • Sauria
Bustos O., Eduardo • ? • Chile • [Anura]
Butler, Amos William • 1 October 1860–1937 August 5 • USA • (Serpentes)
Butler, Arthur Lennox • ?–? • Malaya • (Anura, Sauria), Serpentes
Butler, William Henry (=Harry) • 1930 • Australia • (Sauria)

- C -

Cadle, John Everett • ? • USA • Anura
Cagle, Fred Ray • 9 October 1915–1968 August 8 • USA • Testudines
Cai, Chunmo • 8 February 1941 • China • Caudata
Calabresi, Enrica • ? • Italy • Anura, Sauria
Caldwell, David Keller • 6 August 1928 • USA • Testudines
Caldwell, Janalee Paige • 20 July 1942 • USA • Anura
Călinescu, R. J. (?=Raul I.) • ? • Romania • Anura, Caudata
Call, Richard Ellsworth • 13 May 1856–1917 • USA • *crocodilia, Serpentes
Callison, George • 1 June 1940 • USA • *rhynchocephalia
Camerano, Lorenzo • 9 April 1856–1917 November 22 • Italy • Anura, Sauria, Serpentes
Camp, Charles Lewis • 12 March 1893–1975 August 14 • USA • *Anura, Caudata, Sauria
Campbell, Jonathan Atwood • 13 May 1947 • USA • Anura, Gymnophiona, Sauria, Serpentes
Campbell: see also Bumzahem, C. B.
Cann, John • ? • Australia • (Serpentes), Testudines
Cannatella, David Charles • 25 December 1954 • USA • Anura
Cantor, Theodore Edward (esquire) • 6 January 1809–1860 • Denmark / Germany / British India • Anura, Sauria, Serpentes, Testudines
Capellini, Giovanni • 23 August 1833–1922 May 28 • Italy • *crocodilia, *testudines
Capolongo, Domenico • ? • Italy • Sauria, Serpentes
Capula, Massimo • ? • Italy • Anura
Cara, Alberto • 1847–? • Italy • Sauria
Caramaschi, Ulisses • ? • Brazil • Anura
Cardoso, Adão José • 13 January 1951 • Brazil • Anura
Čarevskij: see Zarevsky, S.
Carfi, S. • ? • Italy • Sauria
Carlleyle, A. C. L. • ?–(1869)–? • (?England) / British India • Sauria, Serpentes
Carpenter, Charles Congden • 2 June 1921 • USA • (Sauria)
Carr, Archie (=Archibald Fairly, Jr.) • 16 June 1909–1987 May 21 • USA • (Anura), Caudata, (*Sauria), Testudines, (*testudines)
Carrillo de Espinoza, Nelly • 2 November 1932 • Peru • Serpentes

Carrington da Costa: see Costa, J. C. S. da
Carruthers, Vincent Craig • 16 June 1942 • South Africa • Anura
Carter, Henry John • 1813–1895 • British India • (Sauria), *testudines
Carvalho, Antenor Leitão de • ? • Brazil • Anura, Crocodilia, (Sauria), Serpentes
Casamiquela, Rodolfo M. • ? • Argentina • *anura
Castelnau, Francis L. de Laporte ("comté de," a pseudonym) • 1812–1880 February 4 • England / North America / South America / France / Australia • (Sauria, Serpentes)
Castro, L. P. • ? • Argentina • Sauria
Castro H., Fernando • ? • Colombia • Sauria
Castroviejo, Javier • ? • Spain • Sauria
Catesby, Mark • 3 April 1683–1749 December 23 • England / Carolina / England • (Anura, Serpentes), Testudines
Cattaneo, Augusto • ? • Italy • Serpentes
Cattoi, Noemi V. • ? • Argentina • *testudines
Caughley, Judith A. Badham • ? • Australia • Sauria
Caup: see Kaup, J. J.
Cautley, Proby Thomas (sir) • 3 January 1802–1871 January 25 • England / British India / England • *crocodilia, *testudines
Cei, José Miguel • 23 March 1918 • Italy / Argentina / Portugal • Anura, Sauria
Cépède: see Lacépède, B. G. É. de
Cerdas, Luis • ? • Costa Rica • Serpentes
Černov: see Chernov, S. A.
Cetti, Francesco • 9 August 1726–1778 November 20 • Italy • (Sauria, Serpentes)
Chabanaud, Paul • 1876–1959 February 27 • France • Anura, Sauria, Serpentes
Chanda, S. K. • ? • India • Anura
Chang, Mang-ven (formerly Ling-yu) • 5 July 1903 • China • Anura, Caudata, Serpentes
Chang, Tso-kan • ? • China • Anura
Chang, Tsong-han • ? • China • Testudines
Chang: see also Zhang
Chani, José María • ? • Argentina • *sauria
Channing, Alan • ? • South Africa • Anura
Chantell, Charles Joseph • 19 May 1931 • USA • *anura, *caudata
Chao: see Zhao
Chapin, James Paul • 9 July 1889–1964 April 5 • USA • (Sauria)
Chapman, Andrew • ? • Australia • (Sauria)
Charas, Moïse • 2 April 1619–1698 January 21 • France • (Serpentes)
Cheke, Anthony S. • ? • England • Sauria
Chelazzi, Guido • ? • Italy • (Serpentes)
Chen, Fuguan • 3 August 1929 • China • Serpentes
Chen, Gongxin • 1936 • China • *testudines
Chen, Huojie • ? • China • Anura
Chen, L.-s. • ? • China • Anura
Chen, S.-h. • ? • Taiwan • Sauria
Cheng, Hsien-yu • 1953 • Taiwan • Sauria
Cheng: see also Zheng
Cherchi, Maria Adelaide • 27 June 1927–1985 July 18 • Italy • Anura, Sauria
Cherlin, V. A. • ? • USSR • Serpentes

Chermock, Ralph Lucien • 25 August 1918–1977 • USA • (Caudata)
Chernov, Sergius Aleksandrovich • 28 July 1903–1964 January 2 • USSR • Anura, *caudata, Sauria, Serpentes, (*testudines)
Chiszar, David Alfred • 21 October 1944 • USA • (Sauria), [Serpentes]
Chkhikvadze (=Čkhikvadze, =Tchikvadze), Viacheslav Mikhailovich • 1940 • USSR • *caudata, *sauria, *Testudines
Chopra, Shiv Raj Kumar • 1931 • India • *crocodilia
Chosatzky: *see* Khosatzky, L. I.
Chow, Ming-chen (=Zhou, Mingzhen) • 9 November 1918 • China • *testudines
Chrapliwy, Peter Stanley • 20 October 1923 • USA • Sauria, Serpentes
Christoph, Hugo Theodor • 1831–1894 • (?Russia) • Serpentes
Christov, L. • ? • (?USSR) • *testudines
Chumakov: *see* Tschumakov, I. S.
Ciobanu, Mihai • ? • Romania • *testudines
Cirer, M.-A. • ? • (?Spain) • Sauria
Cisternas, Rafael • ?–? • Spain • (Anura)
Clark, Herbert Charles • 11 November 1877–1960 November 8 • USA • (Serpentes)
Clark, Hubert Lyman • 9 January 1870–1947 July 31 • USA • (Serpentes)
Clark, John • 26 July 1909 • USA • *testudines
Clark, William Bullock • 15 December 1860–1917 July 27 • USA • *crocodilia, *testudines
Clarke, Barry Thomas • 6 June 1951 • England • Anura
Clausen, Robert Theodore • 26 December 1911 • USA • Caudata
Clay, Brian T. • ? • Australia • (Sauria)
Clay, William Marion • 3 October 1906–1983 April 1 • USA • Serpentes
Clergue-Gazeau, Monique Andrée • 20 November 1931 • France • Caudata
Cliff, Frank Samuel • 3 April 1928 • USA • Serpentes
Clift, William • 14 February 1775–1849 June 20 • England • *crocodilia, (Sauria)
Cloquet, Hippolyte • 10 March 1787–1843 • France • Caudata, Sauria
Cochran, Doris Mable • 18 May 1898–1968 May 22 • USA • Anura, (Caudata), Sauria, Serpentes
Cocroft, Reginald Bifield, III • 5 January 1960 • USA • Anura
Cocteau, Jean-Théodore • 15 March 1798–1838 May 13 • France • Sauria, Serpentes
Cocude-Michel, Marguerite • ? • France • *rhynchocephalia
Cogger, Harold George • 1935 • Australia • Sauria, (Serpentes)
Colbert, Edwin Harris • 28 September 1905 • USA • *crocodilia
Cole, Leon Jacob • 1 June 1877–1948 February 17 • USA • Sauria, Serpentes
Colenso, William (reverend) • 17 November 1811–1899 February 10 • England / New Zealand • Rhynchocephalia, Sauria
Collett, Robert • 2 December 1842–1913 January 27 • Norway • (Sauria)

Collin de Plancy, Victor Émile Marie Joseph • 1853–? • France • (Anura)
Collins, Robert E Lee • 22 October 1898–1952 April 22 • USA • *testudines
Colombelli, Brunella • 25 April 1952 • Switzerland • Anura
Combaz, J. • ? • (?Belgium) • Anura
Comeau, Noël-M. • ?–1976 • Canada • Caudata
Conant, Isabelle de Peyster Hunt (*wife of* C. Kauffeld, ?–? / R. Conant, 1947–1976) • 17 May 1901–1976 November 3 • USA • (Serpentes)
Conant, Roger (*husband of* I. H. Conant, 1947–1976) • 6 May 1909 • USA • (Caudata), Serpentes
Condon, H. T. • ? • Australia • Anura, Sauria
Constable, John Davidson • ? • (?USA) • Sauria, Serpentes
Cook, Harold James • 31 July 1887–1962 September 29 • USA • *caudata, (*sauria)
Cooper, Byrum W. • ? • USA • Anura
Cooper, Edwin Lowell • 23 December 1936 • USA • (Gymnophiona)
Cooper, James Graham • 19 June 1830–1902 July 19 • USA • Caudata, (Sauria), Serpentes, Testudines
Cope, Edward Drinker • 28 July 1840–1897 April 12 • USA • *Anura, Caudata, *Crocodilia, Gymnophiona, *Sauria, *Serpentes, *Testudines
Copland, Stephen J. • ? • Australia • Anura, Sauria
Corben, Chris J. • ? • Australia • Anura
Cordeiro, C. L. • ? • (?Brazil) • Serpentes
Cornalia, Emilio • 25 August 1824–1882 June 8 • Italy • Anura, Testudines
Corredor E., Vladimir • ? • Colombia • Sauria
Corsi, M. • ? • Italy • Sauria
Cosgriff, John William, Jr. • 10 November 1931–1985 April 28 • USA • (*serpentes)
Costa, João Carrington Simões da • 1891 • Portugal • *testudines
Costa, Oronzo Gabriele • 26 August 1787–1867 November 7 • Italy • Caudata, *rhynchocephalia, Sauria
Cott, Hugh Bamford • ?–1987 April 18 (age 86) • England / British East Africa • (Anura), Sauria
Coues, Elliott • 9 September 1842–1899 December 25 • USA • Serpentes
Courtice, Gillian P. • ? • Australia • Anura
Coutinho: *see* Silva Coutinho, J. M. da
Couturier, Gustavo A. • ? • Argentina • Serpentes
Covacevich, Jeanette • ? • Australia • Sauria
Coventry, Albert John • 24 April 1936 • Australia • Sauria
Cowles, Raymond Bridgman • 1 December 1896–1975 December 7 • South Africa / USA • (Sauria, Serpentes)
Cragin, Frank (=Francis) Whittemore • 4 September 1858–1937 • USA • *testudines
Cranbrook, Gathorne Gathorne-Hardy (5th earl of) • 20 June 1933 • England / Malaya / England • Sauria
Creaser, Edwin Philip • 6 February 1907 • USA • (Testudines)
Crenshaw, John Walden, Jr. • 17 May 1923 • USA • Caudata
Crespo, Eduardo G. • ? • Portugal • Anura, (Caudata)
Creveld, ? • ?–(1809)–? • Germany • Sauria

Crombie, Ronald Ian • 28 January 1949 • USA • Anura, Sauria
Crouzel, Fernand Jacques • 1913 • France • *testudines
Crump, Martha Lynn • 23 August 1946 • USA • Anura
Cruz, Carlo Alberto Gonçalves da • ? • Brazil • Anura
Csiki, Ernö (=Ernestus) • 22 October 1875–1954 July 10 • Hungary • (Anura), Serpentes
Cuesta Terrón, Carlos • ?–? • Mexico • Serpentes
Cullom, Shelby J. • ? • USA • Sauria
Cunha, Osvaldo Rodrigues da • ? • Brazil • Sauria, Serpentes
Curry-Lindahl, Kai • 10 May 1917 • Sweden • (Serpentes)
Cuvier, Georges Jean-Léopold-Nicolas-Frédéric Dagobert (baron) • 23 August 1769–1832 May 13 • Germany / France • Anura, *Caudata, Crocodilia, Gymnophiona, Sauria, Serpentes, Testudines, (*testudines)
Cyrén, Carl August Otto • 9 April 1878–1946 September 23 • Sweden • (Caudata), Sauria, (Serpentes)
Czarewskii: *see* Zarevsky, S. F.
Czechura, Gregory Vincent • 11 January 1953 • Australia • Sauria
Czernov: *see* Chernov, S. A.

- D -

da: *see* Cruz *and* Cunha
Dalman, Johan Wilhelm (=Vilhelm) • 4 November 1787–1828 July 12 • Sweden • Serpentes
Daly, John William • 8 June 1933 • USA • Anura
Dam, G. P. F. van • ? • South Africa • Sauria
Dammerman, Karel Willem • 1885–1951 • Dutch East Indies (Java) • (Anura)
Danforth, Stuart Taylor • 23 September 1900–1938 November 25 • USA / Puerto Rico • (Sauria, Serpentes)
Daniel, (hermano) • ? • Colombia • (Anura)
Daniel, J. C. • 9 July 1927 • India • Anura, (Sauria)
Danielyan, Felix Danielanovich • 1938 • USSR • Sauria
Danilevski, B. • ? • ? • (Sauria)
Darevsky, Ilya Sergeevich • 18 December 1925 • USSR • (Anura), *caudata, *Sauria, Serpentes, (*testudines)
Darling, Donald M. • ? • USA • Caudata
Darlington, Philip Jackson, Jr. • 14 November 1904 • USA • (Anura, Sauria, Serpentes)
Darwin, Charles Robert • 12 February 1809–1882 April 19 • England • ([Anura], Crocodilia, Sauria, Serpentes, Testudines)
Dasmann, Marlene M. • ? • USA • Sauria
Dattatri, S. • ? • India • Serpentes
Daubenton, Louis Jean Marie • 29 May 1716–1800 January 1 • France • Caudata
Daudin, François-Marie • 24 March 1774–1804 • France • Anura, Caudata, Crocodilia, Gymnophiona, Sauria, Serpentes, Testudines
David, Armand (l'abbé) • 7 September 1826–1900 November 10 • France / China / France • Anura, Caudata, (Serpentes), [Testudines]
Davidge, Christine • ? • Australia • (Sauria)
Davies, Margaret • 8 November 1944 • Australia • Anura
Davis, Delbert Dwight • 30 December 1908–1965 February 6 • USA • Anura
Davis, John • 25 January 1761–1847 January 14 • USA • Serpentes
Davis, Nathan Smith, Jr. • 5 September 1858–1920 December 21 • USA • Anura
Davis, William B. • 14 March 1902 • USA • Anura, Sauria
Davis: *see also* Stull, O. G.
de: *see* Betta, Borre, Filippi, Grijs, Jong, Lacépède, Plancy, Rooij, Schauensee, Villiers, Witte
Decary, Raymond • 1891 • (?France) / Madagascar • (Anura, Sauria)
Deckert, Kurt • 1907 • Germany • Anura
Deckert, Richard F. • 5 December 1878–? • USA • (Serpentes)
Degen, Edward • ?–? • (?England / Uganda) • (Serpentes)
Degenhardt, William George • 16 April 1926 • USA • Sauria
Degerbøl, Magnus • 1895 • Denmark • Serpentes
Dehne, A. • ?–(1856)–? • Germany • Sauria, Serpentes
De Kay, James Ellsworth • 12 October 1792–1851 November 21 • Portugal / USA • Caudata, *crocodilia, (Serpentes)
Delean, Steven • ? • Australia • (Sauria)
Delfortrie, Eugène-Edmond • 28 March 1816–1885 September 3 • France • *testudines
Dell, John • ? • Australia • (Sauria)
Del Prato: *see* Prato, A. del
Dely, Olivér György • ? • Hungary • Caudata
Demathieu, Georges Raymond • 7 February 1920 • France • *crocodilia
Demidov, Anatoly Nikolaevich (prince) • 1812–1870 • Italy / Russia / France • [Sauria, Serpentes]
Dendy, Arthur • 1865–1925 March 24 • England / Australia / New Zealand / South Africa / England • (Sauria)
Deng, Qixiang • 24 October 1932 • China • Caudata
Dennis, David Michael • 30 November 1939 • USA • Anura, Caudata
Dennys, Nicholas Belfield • ?–? • Singapore • (Anura, Serpentes)
Dent, James Norman • 10 May 1916 • USA • Anura
Depéret, Charles-Jean-Julien • 25 June 1854–1929 May 17 • France • *anura, *sauria, *serpentes, *testudines
Deppe, Wilhelm • ?–(1830)–? • Germany • Sauria
de Queiroz, Kevin • ? • USA • Sauria
Deraniyagala, Paulus Edward Pieris • 8 May 1900–1973 December 1 • Ceylon • *Crocodilia, Sauria, Serpentes, *Testudines
Deriugin (=Derjugin), Konstantin Mikhailovich • 1878–1938 • Russia • (Caudata)
Derleyn, P. P. • ? • (?Belgium / Burundi) • Serpentes
de Sá, Rafael Omar • 4 March 1961 • Uruguay / USA • Anura
Desjardins, Julien-François • 27 July 1799–1840 August 18 • France • Sauria
Des Lauriers, James Robert • 21 May 1939 • USA • *testudines
De Sola, Clarence Ralph • 26 July 1908 • USA • Testudines
Despax, Raymond • ? • France • Anura, Sauria, Serpentes
Despott, Giuseppe • 26 July 1878–1936 September 22 • Malta • (Sauria)

Deuve, J. • ? • France • Serpentes
De Vis, Charles Walter • 9 May 1829–1915 April 30 • Australia • Anura, *crocodilia, *Sauria, Serpentes, *testudines
DeWeese, James Edward • ? • USA • Anura
Diaz, Nelson F. • ? • Chile • Anura, Sauria
Dice, Lee Raymond • 15 July 1887–1977 January 31 • USA • (Sauria)
Dickerson, Mary Cynthia • 7 March 1866–1923 April 8 • USA • Sauria
Ding (=Ting), Hanpo • 3 November 1912 • China • Anura
Dinnik, N. Ya. • ? • Russia • (Serpentes)
Distant, William Lucas • 12 November 1845–1922 February 4 • England • (Sauria, Serpentes)
Ditmars, Raymond Lee • 20 (?or 22) June 1876–1942 May 12 • USA • (Sauria)
Dixon, James Ray • 1 August 1928 • USA • Anura, Caudata, Sauria, Serpentes
Djao: *see* Zhao, E.-m.
Doi, Hironobu • ? • Japan • Sauria
Dole, Jim Walter • 28 May 1935 • USA • Anura
Dollo, Louis-Antoine-Marie-Joseph • 27 December 1857–1931 April 29 • Belgium • *crocodilia, *sauria, Serpentes, *testudines
Domergue, Charles A. • ? • France / Madagascar • (Anura), Sauria, Serpentes, (Testudines)
Donadio, Oscar Enrique • 8 March 1954 • Argentina • *sauria
Dong, Qian • 6 February 1927 • China • [Anura]
Dong, Zhiming • ? • China • *crocodilia, *sauria
Donndorff, Johann August • 23 March 1754–1837 November 22 • Germany • Sauria, Serpentes, Testudines
Donnezan, Albert • ?–? • France • *testudines
Donoso Barros, Roberto • 5 October 1922–1975 August 2 • Chile • Anura, Sauria, Serpentes, (Testudines)
D'Orbigny: *see* Orbigny, A. de *and* C. de
Doria, Giacomo (marchese) • 1840–1913 • Italy • Anura, Sauria, Serpentes
Dorson, Edward E. • ? • USA • Sauria
Douglas, Athol Mardon • ? • Australia • (Sauria)
Douglass, Earl • 28 October 1862–1931 January 13 • USA • *sauria, *serpentes, (*testudines)
Doumergue, François • 11 May 1858–1939 December 23 • France / Algeria • Sauria
Dowling, Herndon Glenn, Jr. • 2 April 1921 • USA • Serpentes
Downs, Floyd Leslie • ? • USA • Serpentes
Drewes, Robert Clifton • 14 February 1942 • USA • Anura, Serpentes
Drewry, George E. • ? • USA • Anura
Dring, Julian C. M. • ? • England • Anura, Sauria
Dryden, Gilbert Littleton • 23 July 1938 • USA • Serpentes
du: *see* Bois-Reymond *and* Toit
Dubinin, V. B. • ? • USSR • Sauria
Dubois, Alain • ? • France • Anura, Caudata
Dubois, Eugene • 1858–1940 • Holland • *crocodilia
Duda, P. L. • ? • India • Sauria
Duellman, William Edward (*husband of* L. Trueb) • 6 September 1930 • USA • Anura, Cauadata, Sauria, Serpentes

Duerden, James Edwin • 1869–1937 September 4 • England / Jamaica / USA / South Africa • (Serpentes), Testudines
Duff-MacKay, Alex • ? • Kenya • Anura
Dugès, Alfred Auguste Delsescautz • 16 April 1826–1910 January 7 • France / Mexico • Anura, Caudata, Sauria, Serpentes, Testudines
Dujardin, Félix • 5 April 1801–1860 April 8 • France • *crocodilia
Duméril, André-Marie-Constant (*father of* A-H-A. Duméril) • 1 January 1774–1860 August 14 • France • Anura, Caudata, Crocodilia, Gymnophiona, Sauria, Serpentes, *Testudines
Duméril, Auguste-Henri-André (*son of* A-M-C. Duméril) • 30 November 1812–1870 November 12 • France • Caudata, Crocodilia, Gymnophiona, Sauria, Serpentes, Testudines
Duncan, Robert • ? • USA • Caudata
Duncker, Georg • 1870–? • Germany • (Serpentes)
Dundee, Harold A. • 23 August 1924 • USA • Caudata
Dunger, Gerald T. • ? • Nigeria • Sauria, Serpentes
Dunkle, David Hosbrook • 9 September 1911 • / USA • *sauria, (Serpentes)
Dunlap, Helen Louise (*wife of* E. R. Pianka, 1965–1980) • 9 April 1940 • USA • (Sauria)
Dunn, Alta Merle Taylor (*wife of* E. R. Dunn, 1930–1956) • ? • USA • (Anura), Sauria
Dunn, Emmett Reid (*husband of* A. M. Dunn, 1930–1956) • 21 November 1894–1956 February 13 • USA • Anura, Caudata, (*caudata), Gymnophiona, Sauria, Serpentes, (*serpentes, Testudines)
Du Pasquier, Louis • 17 March 1941 • France / Switzerland • Anura
Duperrey, Louis Isidore • 21 October 1786–1865 September 10 • France • [Anura, Sauria, Serpentes]
Durant, Pedro • ? • Venezuela • Anura
Dürigen, Bruno • 1853–1930 • Germany • Anura, Caudata, Sauria, Serpentes, Testudines
Dury, Ralph • 1899 • USA • (Caudata)
Dutta, Sushil Kumar • 23 July 1952 • India • Anura
Duval, Julian J. • ? • USA • Sauria
Duvernoy, Georges-Louis • 7 August 1777–1855 March 1 • France • Serpentes
Dvigubsky, Jean (=Ivan Alekseevich Dwigubskij) • 30 December 1771–1839 January 11 • Russia • Serpentes
Dybowski, Benedykt Ivanovich • 30 April 1833–1930 January 31 • Russia / Poland / Siberia / Poland • (Anura), Caudata

- E -

Eason, Perri Kaye • 26 September 1958 • USA • Sauria
Eberhardt, Hans-Joachim • 6 July 1920–1980 • Germany • Caudata
Eberle, W. Gary • 31 March 1944 • USA • Serpentes
Echelle, Alice F. • ? • USA • Sauria
Echelle, Anthony Allan • 9 September 1940 • USA • Sauria
Echternacht, Arthur Charles • 3 September 1939 • USA • Sauria
Edeling, A. C. J. • ?–? • East Indies • Anura, Sauria

Edgren, Richard Arthur, Jr. • 28 May 1925 • USA • Serpentes
Edwards, George • 3 April 1693–1773 July 23 • England • (Sauria, Serpentes)
Edwards, Stephen Richard • 12 January 1942 • USA / Switzerland • Anura
Edwards: *see also* Milne-Edwards, H.
Efimov (=Yefimov), Michael Borisovich • 1947 • USSR • *crocodilia
Ehmann, Harry F. W. • ? • Australia • Sauria
Ehrenberg, Christian Gottfried • 19 April 1795–1876 June 27 • Germany • (Anura, Sauria)
Eibl-Eibesfeldt, Irenäus • 15 June 1929 • Austria / Germany • Sauria, (Serpentes)
Eichwald, Carl Eduard Ivanovich von • 4 July 1795–1876 November 10 • Latvia / Russia • Anura, Sauria, Serpentes
Eigenmann, Carl H • 9 March 1863–1927 April 24 • Germany / USA • Caudata, (Sauria)
Eimer, Gustav Heinrich Theodor • 22 February 1843–1898 May 29 • Switzerland / Germany • Sauria
Eiselt, Josef • 3 May 1912 • Austria • (Anura), Caudata, (Gymnophiona), Sauria, Serpentes
Eisen, Gustav (=Gustavus Augustus) • 2 August 1847–1940 October 29 • Sweden / USA • (Serpentes)
Eisentraut, Martin • 21 October 1902 • Germany • Sauria
Elbel, Robert Edwin • 8 July 1925 • USA / Thailand • Anura, Sauria, Serpentes
Elias, Paul • ? • USA / England • Caudata
Elliot, Daniel Giraud • 7 March 1835–1915 December 22 • USA • [Serpentes]
Elliott, Walter (sir) • 16 January 1803–1887 March 1 • Scotland / British India / Scotland • (Sauria), Serpentes
Elpatjevskij, V. S. • ? • Russia • Serpentes
Emelianov, Aleksander Adrianovich • 27 June 1878–1946 January 7 • USSR • (Anura), Serpentes
Emin Pasha: *see* Schnitzer, E.
Emlen, John Thompson, Jr. • 28 December 1908 • USA • Anura, Sauria
Emmons, Ebenezer • 16 May 1799–1863 October 1 • USA • *crocodilia
Endo, Riuji • 1891 • Japan • *rhynchocephalia
Engels, William Louis • 28 October 1905 • USA • Serpentes
Entz, Géza • 1842–1919 • Hungary • (Anura)
Eremchenko, Valery Konstantinovich • ? • USSR • Sauria
Erickson, Bruce Ronald • 1930 • USA • *crocodilia, *testudines
Ernst, Carl Henry • 28 September 1938 • USA • Testudines
Ersch, Johann Samuel • 23 June 1766–1828 January 16 • Germany • [Serpentes]
Eschscholtz, Johann Friedrich von • 1 November 1793–1831 May 7 • Russia • Caudata, Sauria, Testudines
Eshelman, Ralph Ellsworth • 1947 • USA • *sauria
Espada: *see* Jiménez de la Espada, M.
Espinoza: *see* Carrillo de Espinoza, N.
Essex, Robert • ? • South Africa • Sauria
Estes, Richard Dean • 9 May 1932 • USA • *anura, *caudata, *gymnophiona, *sauria
Estrada, Alberto R. • ? • Cuba • Anura

Etheridge, Richard Emmett • 16 September 1929 • USA • (Anura), *Sauria, (Serpentes)
Evans, George Henry (colonel) • 28 November 1863–1948 June 1 • British India • Serpentes
Evans, Howard Edward • 22 September 1922 • USA • Anura
Evermann, Barton Warren • 24 October 1853–1932 September 27 • USA • (Sauria)
Eversmann, Eduard Friedrich Aleksandrovich • "12 January" (=24 January) 1794–1860 April 14 • Germany / Russia • Sauria
Eydoux, Joseph Fortuné Théodore • 1802–1841 • France • Sauria, Serpentes

- F -

Fabiani, Ramiro • 15 May 1879–? • Italy • *crocodilia
Fahr, Aenny • ? • Germany • (Sauria)
Falanruw, Marjorie V. C. • ? • USA • Sauria
Falconer, Hugh • 29 February 1808–1865 July 31 • Scotland • *crocodilia, *sauria, *testudines
Falla, Robert Alexander (sir) • 21 July 1901 • New Zealand • (Sauria)
Fan, Tsang-how • ? • China • Sauria, Serpentes, Testudines
Fang, Junjiu • 25 August 1934 • China • [Serpentes]
Fang, Pingwen • ? • China • Caudata, Testudines
Fang, Rengsheng • 25 July 1932 • China • Anura
Fatio, Paul Victor • 1838–1906 • Switzerland • Anura, Caudata, Sauria, Serpentes
Fauvel, Albert-Auguste • 7 November 1851–1909 November 3 • France • Crocodilia
Fayrer, Joseph (sir) • 6 December 1824–1907 May 21 • England • Serpentes
Feener, D. J. • ? • USA • Anura
Fehlmann, Herman Adair • 31 January 1917 • USA • Sauria
Fei, Liang • 3 July 1936 • China • Anura, Caudata
Fejérváry von Komlós-Keresztes, Géza Jyula (=Géza Gyula Imre, báró) • 25 June 1894–1932 June 2 • Hungary • (Anura), *anura, Caudata, Sauria, (*testudines)
Ferguson, Denzel Edward • 28 August 1929 • USA • Caudata
Ferguson, Gary Wright • 15 April 1941 • USA • Sauria
Ferguson, William • 1820–1887 July 31 • England / Ceylon • Anura, Sauria, Serpentes
Ferrari Perez, Fernando • ?–1927 • Mexico • (Sauria), [Serpentes]
Ferreira, Julio Bethencourt • ?–? • Portugal • Anura, Sauria, Testudines
Fetcho, Joseph Robert • ? • USA • Serpentes
Filhol, Henri • 11 May 1843–1902 April 23 • France • (*caudata), *crocodilia, *sauria, *serpentes
Filippi, Filippo de • 20 April 1814–1867 February 9 • Italy • Sauria, (Serpentes)
Finn, Frank • 1868–1932 October 1 • England • Serpentes
Firschein, Irwin Lester • ? • USA • Anura, (Caudata)
Fischberg, Michail • 2 June 1918–1988 June 26 • Switzerland • Anura
Fischer, Johann Gustav • 1 March 1819–(?1889) • Germany • (Caudata), Gymnophiona, Sauria, Serpentes

Fischer, Johann Gotthelf (ritter von Waldheim) • 13 October 1771–1853 October 18 • Germany / Russia • Caudata, Sauria, Serpentes
Fisher, Albert Kendrick • 21 March 1856–1948 June 12 • USA • (Anura)
Fitch, Henry Sheldon • 25 December 1909 • USA • Sauria, (Serpentes)
Fitzinger, Leopold Joseph Franz Johann • 13 April 1802–1884 September 20 • Austria • Anura, Caudata, Gymnophiona, *rhynchocephalia, Sauria, Serpentes, *Testudines
FitzSimons, Frederick William (*father of* V. F. M. FitzSimons) • 6 August 1870–1951 March 25 • Ireland / South Africa • (Sauria)
FitzSimons, Vivian Frederick Maynard (*son of* F. W. FitzSimons) • 7 February 1901–1975 August 1 • South Africa • Anura, Sauria, Serpentes, Testudines
Fleischmann, Friedrich Ludwig • ?–(1831)–? • Germany • Serpentes
Fleming, John • 10 January 1785–1857 November 18 • England • Caudata, Sauria, Serpentes, Testudines
Fletcher, Joseph James • 1850–1926 • Australia • Anura
Flores, Glenn • ? • USA • Anura
Flower, Stanley Smyth (major) • 1 August 1871–1946 February 3 • England / Siam / England • (Anura, Sauria, [Serpentes])
Flury, Alvin Godfrey • 1 November 1920 • USA • Serpentes
Folkerts, George William • 26 November 1938 • USA • Testudines
Forcart, Lothar Hendrich Emil Wilhelm • 10 December 1902 • Switzerland • Anura, (Gymnophiona, Serpentes)
Ford, Julian Ralph • ? • Australia • Sauria
Ford, Linda Sherill • ? • USA • Anura, Serpentes
Formas, J. Ramón • 25 December 1939 • Chile • Anura
Forskål, Per (=Petrus) • 11 January (?1732–1736)–1763 July 11 • Sweden / Denmark / Arabia • Sauria, Serpentes, Testudines
Forster, Johann Reinhold • 22 October 1729–1798 December 9 • Germany / England • Serpentes
Fouquette, Martin John, Jr. • 14 June 1930 • USA • Anura, Serpentes
Fourtau, René • ? • France • *crocodilia
Fowler, Henry Weed • 23 March 1878–1965 June 2 • USA • (Anura), Caudata, Sauria, Testudines
Fowler, James Abbott • 15 July 1916 • USA • Caudata
Fowler, Samuel Page • 22 April 1800–1888 • USA • (Anura)
Fox, Rufus Wade, Jr. • 2 June 1920–1964 September 20 • USA • Serpentes
Fox, Stanley Forrest • 21 March 1946 • USA • Sauria
Fraas, Oscar Friedrich von • 17 January 1821–1897 November 22 • Germany • *serpentes
Frank: *see* van Frank, R.
Franz, Richard • ? • USA • Anura
Franzen, Michael • ? • Germany • Caudata
Fraser, Douglas Fyfe • 17 June 1941 • USA • Serpentes
Fraser, N. C. • ? • England • *rhynchocephalia
Freiberg, Marcos A. • 23 April 1911 • Argentina / USA • (Anura), Crocodilia, Sauria, Serpentes, *Testudines
Fréminville, Christophe-Paulin de La Poix de • 23 January 1787–1848 January 12 • France • (Serpentes)

Freycinet, Louis-Claude de Saulces de • 7 August 1779–1842 August 18 • France • (Anura, [Sauria])
Freyer, Heinrich • 2 July 1802–1867 August 21 • Slovenia (*now in* Yugoslavia) • (Caudata), Serpentes
Freytag, Günther E. • 14 November 1918 • Germany • Caudata
Friant, Madeleine • ? • France • *anura
Fries, Carl, Jr. • 30 September 1910 • USA / Mexico • *sauria
Fries, Isabelle Scott • 5 December 1947 • USA • Serpentes
Fritts, Thomas Harold • 23 April 1945 • USA • Anura, Sauria, Serpentes
Fritze, Adolf • ?–(1894)–? • Germany • Serpentes
Frogner, Karl John • 7 June 1941 • USA • Anura
Frost, Charles • ?–? • Australia • Sauria, Serpentes
Frost, Darrel Richmond • 12 June 1951 • USA • Anura
Frost, John Stephen • 22 May 1945 • USA • Anura
Fry, Dene Barrett • (?1894)–1917 • Australia • Anura, Sauria, Serpentes, Testudines
Fuchs, E. • ? • Germany • *testudines
Fuchs, Karlheinz • ? • Germany • Crocodilia
Fuentes: *see* Jiménez Fuentes, E.
Fuhn, Ion Eduard • "23 January" (=5 February) 1916–1987 August 31 • Romania • Caudata, Sauria
Fuhrmann, Otto • 1871–1945 • Switzerland • (Anura), Gymnophiona, (Serpentes)
Fülleborn, Friedrich Georg Hans • 1866–? • Germany • (Anura)
Fuller, Phillip J. • ? • Australia • (Sauria)
Funkhouser, Anne Black (*wife of* J. W. Funkhouser) • 10 October 1929 • USA • Anura
Funkhouser, John William (*husband of* A. Funkhouser) • 28 August 1926–1974 December 3 • USA • Anura

- G -

Gabunia, Leo Kalistratovich • ? • USSR • [*crocodilia], *testudines
Gachet, Antoine-Hippolyte • 15 November 1798–1842 November 22 • France • Serpentes
Gadow, Hans Friedrich • 8 March 1855–1928 May 16 • Germany / England • Anura, (Caudata), Sauria, *testudines
Gaffney, Eugene Spencer • 12 August 1942 • USA • (*testudines)
Gaiduchenko, L. L. • ? • USSR • *testudines
Gaige, Helen Beulah Thompson (=mrs. Frederick McMahon Gaige) • 24 November 1890–1976 October 24 • USA • Anura, Caudata, Sauria, (Serpentes, Testudines)
Gaimard, Joseph-Paul • 31 January 1793–1858 December 10 • France • Sauria, (Serpentes)
Galbreath, Edwin Carter • 18 March 1913 • USA • (*serpentes), *testudines
Gàlgano, Mario • ? • Italy • (Anura)
Gali, Frank • ? • USA • Sauria
Gallagher, Daniel Stephen, Jr. • ? • USA • Sauria
Gallardo, José María Alfonso Félix • 1 August 1925 • Argentina • Anura, Sauria
Galleguillos G., Ricardo • ? • Chile • Anura
Galvan, Mark Allan • 31 May 1965 • USA • Sauria

Gamero, Maria Lourdes Diaz de • ? • Venezuela • *testudines
Gans, Carl • 7 September 1923 • Germany / USA • Sauria, Serpentes
Gao, Kequin • ? • China • *anura
García, Evaristo • 1845–? • Colombia • Serpentes
Garden, Alexander • January 1730–1791 April 15 • USA • Caudata
Garman, Samuel Walton (*brother of* W. H. Garman) • 5 June 1843–1927 September 30 • USA • Anura, Sauria, Serpentes, Testudines
Garman, William Harrison (*brother of* S. W. Garman) • 27 December 1856–1944 August 7 • USA • Caudata
Garnier, John Hutchinson • 1823–1898 February 1 • France / Canada / India • Caudata
Garrido, Orlando H. • ? • Cuba • Sauria, Serpentes
Garrido R., A. • ? • Venezuela • Sauria
Gartside, D. F. • ? • Australia • Anura
Gasc, Jean-Pierre • ? • France • Gymnophiona, Serpentes
Gasser, François • ? • France • Caudata
Gauthier, Jacques Armand • 7 June 1948 • USA • *sauria
Gay, Claude • 18 March 1800–1873 November 29 • France • Sauria
Gayda, Henrik St. • ? • (?Germany) • Anura, Caudata, (Serpentes)
Gehlbach, Frederick Renner • 5 July 1935 • USA • Caudata
Geisenheyner, Ludwig • 1841–1926 • Germany • Sauria
Gené, Carlo Giuseppe • 1800–1847 • Italy • Anura, Caudata, Sauria
Geoffroy Saint-Hilaire, Étienne-François • 15 April 1772–1844 June 19 • France • *Crocodilia, Sauria, Serpentes, Testudines
Georgi, Johann Gottlieb • 31 December 1729–1802 October 27 • Germany / Russia • [Serpentes]
Gervais, François-Louis-Paul • 26 September 1816–1879 February 10 • France • Caudata, *crocodilia, [*rhynchocephalia], *Sauria, Serpentes, *Testudines
Gesner, Conrad (=Cunrat Gessner) • 26 March 1516–1565 December 13 • Switzerland • (Caudata)
Gibbes, Lewis Reeve • 4 August 1810–1894 November 21 • USA • Caudata
Gibbes, Robert Wilson • 8 July 1809–1866 October 15 • USA • Caudata
Gibbons, John R. H. • ?–1986 November 15 • Australia / Fiji • Sauria
Giebel, Christoph Gottfried Andreas • 13 September 1820–1881 November 14 • Germany • *caudata, *crocodilia, Sauria, (Testudines)
Giglioli, Enrico (=Henry Hillyer) • 13 June 1845–1909 December 16 • England / Italy • (Caudata), Sauria
Gillam, Mike W. • ? • Australia • Sauria
Gillespie, John • ? • USA • Serpentes
Gilliams, Jacob • 1784–1868 February 4 • USA • Caudata, Sauria
Gilmore, Charles Whitney • 11 March 1874–1945 December 27 • USA • (*anura), *caudata, *crocodilia, *sauria, *serpentes, *rhynchocephalia, *testudines
Ginés, (hermano) • ? • Venezuela • (Anura)
Ginsburg, Léonard • 1927 • France • *crocodilia, (*testudines)
Girard, Charles Frédéric • 9 March 1822–1895 January 29 • France / USA / France • Anura, Caudata, Sauria, Serpentes, Testudines
Gislén, Torsten Richard Emanuel • 21 June 1893–1954 • Sweden • [Caudata]
Gistel, Johannes von Nepomuk Franz Xaver • 11 August 1809–1873 • Germany • Anura, Caudata, Sauria, Serpentes
Glass, Bryan Pettigrew • 21 August 1919 • USA • Anura, Testudines
Glauert, Ludwig • 5 May 1879–1963 February 1 • England / Australia • (Anura), Sauria, Serpentes, Testudines
Gleadow, Frank • ?–? • British India • (Sauria)
Gloor, Hans Jacob • 8 February 1917 • Switzerland • Anura
Gloyd, Howard Kay • 12 February 1902–1978 August 7 • USA • Serpentes
Glückselig, August Maria • ?–(1851)–? • Czechoslovakia • Sauria
Gmelin, Carl Christian • 18 March 1762–1837 July 26 • Germany • Serpentes
Gmelin, Johann Friedrich • 8 August 1748–1804 November 1 • Germany • Crocodilia, Sauria, Serpentes, Testudines
Gmelin, Samuel Gottlieb • 4 July 1744–1774 July 27 • Germany / Russia • Testudines
Godman, Frederick DuCane • 15 January 1834–1919 February 19 • England • ([Anura], Sauria, Serpentes)
Goeldi: *see* Göldi, E. A.
Goff, Carlos Clyde • 2 October 1905–1939 January 13 • USA • Serpentes
Goin, Coleman Jett (*husband of* O. B. Goin) • 25 February 1911–1986 May 12 • USA • Anura, (*anura), *Caudata, Serpentes
Goin, Olive Lynda Bown (*wife of* C. J. Goin) • 2 December 1912 • USA • Anura
Goldfuss, Georg August • 18 April 1782–1848 October 2 • Germany • *caudata, *rhynchocephalia, Sauria, Serpentes
Göldi, Emil August (=Emilio Augusto Goeldi) • 28 August 1859–1917 July 5 • Switzerland / Brazil / Switzerland • (Anura), Testudines
Golliez, Henri • ?–(1889)–? • Switzerland • *testudines
Golubev, Mikhail Leonidovich • 1947 • USSR • Sauria
Gomes, João Florencio de Salles • 1886–1919 May 29 • Brazil • Serpentes
Gonçalves, J. Moura • ? • Brazil • Serpentes
Gonzáles Sponga, M. A. • ? • Venezuela • Serpentes
Good, David Andrew • 6 August 1956 • USA • Sauria
Goode, George Brown • 13 February 1851–1896 September 6 • USA • (Sauria)
Gordon, Kenneth Llewellyn • 7 June 1899–1983 October 29 • USA • (Caudata)
Gorham, Stanley W. • ? • Canada • Anura
Gorman, Joe (=Joseph B, Jr.) • 25 April 1918 • USA • Anura, Caudata
Gorzula, Stefan J. • ? • Venezuela • Anura
Gosse, Philip Henry • 6 April 1810–1888 August 23 • England • Anura, Sauria
Gough, Lewis Henry • ?–? • South Africa • (Anura), Serpentes
Gouvêa, E. • ? • Brazil • Anura
Gow, Graeme Francis • ? • Australia • Sauria, Serpentes

Graham, Eugene D., Jr. • ? • USA • Sauria
Grandidier, Alfred • 20 December 1836–1921 September 13 • France / Madagascar / France • Anura, *Crocodilia, Sauria, Serpentes, Testudines, (*testudines)
Grandison, Alice Georgie Cruickshank • 25 March 1927 • England • Anura, (Gymnophiona), Sauria
Granger, Walter Willis • 7 November 1872–1941 September 6 • USA • (*crocodilia, *sauria), *testudines
Grant, Chapman (major) • 22 March 1887–1983 January 5 • USA • Sauria, Serpentes, (Testudines)
Gravenhorst, Johann Ludwig Christian Carl • 14 November 1777–1857 January 14 • Germany • Anura, Caudata, Sauria
Graves, L. • ?–? • (?Belgium) • Crocodilia
Gray, Francis Calley • 19 September 1790–1856 December 29 • USA • Serpentes
Gray, John Edward • 12 February 1800–1875 March 7 • England • Anura, Caudata, *Crocodilia, Rhynchocephalia, Sauria, Serpentes, *Testudines
Gray, Peter • 23 May 1953 • USA • Anura
Greeff, Richard • 14 March 1829–1892 August 30 • Germany • (Sauria)
Green, Jacob • 26 July 1790–1841 February 1 • USA • Caudata, Sauria
Green, Norman Bayard • 15 November 1905–1988 May 23 • USA • Caudata
Green, Robert H. (="Bob") • ? • Australia • (Sauria)
Greene, Harry Walter • 26 September 1945 • USA • *sauria
Greer, Allen E., Jr. • ? • USA / Australia • Sauria, Serpentes
Gressitt, Judson Linsley • 16 June 1914–1982 April 26 • USA • Anura, Sauria, Serpentes
Grey, George (sir) • 14 April 1812–1898 September 19 • Australia • (Sauria), [Serpentes]
Griffin, Lawrence Edmonds • 10 September 1874–1949 September 12 • USA • Sauria, Serpentes
Griffith, Edward • 1790–1858 January 8 • England • [Caudata, Sauria, Testudines]
Griffiths, Ivor • ? • England / Tanganyika • Anura
Grigg, Gordon Clifford • 15 August 1942 • Australia • Anura
Grijs (=Grys), Pedro de • ?–1940 February 3 • Germany • Anura, Sauria, (Serpentes)
Grismer, Larry Lee • 19 November 1955 • USA • Sauria
Gritis, Paul Alexander • 22 December 1952 • USA • Anura
Grobman, Arnold Brams • 28 April 1918 • USA • Caudata, Serpentes
Groessens-van Dyck, Marie-Claire • ? • (?Belgium) • *testudines
Grohmann, Francesco Saverio • ?–(1832)–? • Sicily • Sauria
Gronow (=Gronovius), Lorenz Theodor • 1730–1777 • Holland • (Sauria, Serpentes)
Grubant, V. N. • ? • USSR • Serpentes
Gruber, Hans Jürgen • ? • Germany • Sauria
Gruber, Johann Gottfried • 29 November 1774–1851 August 7 • Germany • [Serpentes]
Gruber, Ulrich F. • 7 May 1932 • Germany • Sauria, Serpentes, Testudines
Grys: *see* Grijs, P. de
Gudynas, Eduardo • ? • Uruguay • Sauria

Gueldenstaedt: *see* Güldenstädt, J. A. von
Guérin-Méneville, Felix-Édouard • 12 October 1799–1874 January 26 • France • Gymnophiona, Sauria, Serpentes, Testudines
Guibé, Jean • 18 February 1910 • France • Anura, Sauria, Serpentes
Guichenot, Alphonse (?or Adolphe) • ?–? • France • Sauria, Serpentes
Guillette, Louis Joseph, Jr. • ? • USA • Sauria
Güldenstädt, Johann Anton von • 26 April 1745–1781 March 23 • Latvia / Russia • Serpentes
Gulia, Giovanni • ? • Italy • *Sauria
Gundlach, Juan (=Johannes Christopher) • 17 July 1810–1896 March 15 • Germany / Cuba • (Anura), Sauria, Serpentes
Günther, Albert Carl Ludwig Gotthilf • 3 October 1830–1914 February 1 • Germany / England • Anura, Caudata, Gymnophiona, (Rhynchocephalia), *Sauria, Serpentes, *Testudines
Günther, Rainer • 11 September 1941 • Germany • Anura
Gürich, Georg Julius Ernst • 25 September 1859–1938 August 16 • Germany • *crocodilia
Gyldenstolpe, Nils Carl Gustav Fersen (greve) • 30 September 1886–? • Sweden • (Sauria), Serpentes

- H -

Haacke, Wulf Dietrich • 15 December 1936 • South Africa • Sauria, Serpentes
Haas, Georg • 19 January 1905–1981 September 13 • Austria / Germany / Israel • Sauria, Serpentes, *testudines
Haase, Erich • ?–? • Germany / Siam • (Sauria)
Haast, William E. • 1910 • USA • (Serpentes)
Haddad, Célio F. B. • ? • Brazil • Anura
Hahn, Donald Edgar • 2 September 1939 • USA • Sauria, Serpentes
Haile, Neville Seymour • ? • Malaysia • Anura
Hairston, Nelson George • 16 October 1917 • USA • Caudata
Haldeman, Samuel Stehman • 12 August 1812–1880 September 10 • USA • (Caudata, Serpentes)
Hall, Charles William • 8 February 1922 • USA • Serpentes
Hall, David H. • ? • USA • Anura
Hall, William Purington, III • ? • USA • Sauria
Hallowell, Edward • 14 September 1808–1860 February 20 • USA • Anura, Caudata, Sauria, Serpentes
Haly, Amyrald • ?–(1888)–? • Ceylon • Serpentes
Han, Defen • 1933 • China • *testudines
Hanitsch, Karl Richard • 22 December 1860–1940 August 11 • Germany / England • (Anura, Serpentes)
Hanlon, Mark (=Timothy Marcus Stephen) • ? • Australia • (Sauria)
Hardwicke, Thomas (major-general) • 1756–1835 March 3 • British India • Sauria, Serpentes
Hardy, Graham Stuart • 25 April 1947 • New Zealand • Sauria
Hardy, Jerry David, Jr. • 1929 • USA • Anura, Sauria
Hardy, Laurence McNeil • 24 February 1939 • USA • Serpentes

Harlan, Richard • 19 September 1796–1843 September 30 • USA • Anura, Caudata, (*crocodilia), Sauria, Serpentes, Testudines
Harold, Gregory • ? • Australia • (Sauria)
Harper, Francis • 17 November 1886–? • USA • Anura, Sauria
Harris, Dennis Martin • 1950 • USA • Sauria
Harris, Herbert Stanley, Jr. • 30 June 1942 • USA • Serpentes
Harrison, Julian Ravenel, III • 23 August 1934 • USA • Anura
Harrison, Launcelot • 1880–1928 • Australia • Anura
Hartert, Ernst Johann Otto • 29 October 1859–1933 November (?11 or 14) • Germany / England / Germany • Sauria
Hartmann, Max (=Maximilian) • 7 July 1876–1962 October 11 • Germany • Sauria
Hartmeyer, Robert Hermann • 19 May 1874–1923 October 13 • Germany • [Anura, Sauria, Serpentes]
Hartweg, Norman Edouard • 20 August 1904–1964 February 16 • USA • Anura, (Caudata), Sauria, Serpentes, Testudines
Harvey, Chris • ? • Australia • Sauria
Hasselquist, Frederik • 3 January 1722–1752 February 9 • Sweden • (Sauria)
Hasselt, Alexander Willem Michiel van • 1814–1902 • Holland • Serpentes
Hasselt, Johan Coenraad van • 1797–1823 • Holland / East Indies • Anura, Gymnophiona, Sauria, Serpentes
Hassler, William Grey • 1 September 1906–? • USA • Anura, Sauria
Hatt, Robert Torrens • 17 July 1902 • USA • [*sauria]
Haughton, Sidney Henry • 7 May 1888 • England / South Africa • *anura, *testudines
Häupl, Michael • ? • Austria • Sauria
Hay, Oliver Perry • 22 May 1846–1930 November 2 • USA • Caudata, *Testudines
Hay, William Perry • 8 December 1871–1947 January 26 • USA • Testudines
Hayden, Ferdinand Vandiveer • 7 September 1829–1887 December 22 • USA • (Serpentes)
Haynes, David Polk • 13 January 1942 • USA • Testudines
Heang, Kiew Bong • ? • Malaysia • Anura
Heatwole, Audry Ann Yoder (*wife of* H. Heatwole) • 22 September 1932 • USA / Puerto Rico / Australia • Anura
Heatwole, Harold Franklin (*husband of* A. Heatwole) • 2 December 1934 • USA / Puerto Rico / Australia • Anura, Sauria
Hebrard, James Jack • 1944 • USA / Kenya • Sauria
Hechenbleikner, Herbert • ? • USA • Sauria
Hecht, Günther • 1902 • Germany • Sauria, Serpentes
Hecht, Max Knobler • 15 February 1925 • USA • *anura, (Sauria), *sauria, *serpentes
Heck, Johann Georg • ?–1857 • (?Germany) • [Caudata]
Hedges, Stephen Blair • ? • USA • Anura, Sauria
Heilprin, Angelo • 31 March 1853–1907 July 17 • Hungary / USA • (Anura), Serpentes
Heinze, Albert A. • ? • USA • Caudata
Heller, Edmund • 21 May 1875–1939 July 18 • USA • Sauria, (Serpentes)
Heller, Florian • 1905 • Germany • *serpentes

Hellmich, Walter • 1906 • Germany • Anura, Caudata, (Gymnophiona), Sauria, (Serpentes)
Hemmer, Helmut • 7 December 1940 • Germany • Anura
Hemprich, Friedrich Wilhelm • 24 June 1796–1825 June 30 • Poland / Germany • Sauria, Serpentes
Henderson, John Robertson • 21 May 1863–1925 October 26 • England • Testudines
Henderson, Robert W. • 1945 • USA • Sauria, Serpentes
Hendricks, Fred Samuel • 4 October 1946 • USA • Sauria, Serpentes
Hendrickson, John Roscoe • 26 August 1921 • USA • Anura
Henle, Klaus • 5 January 1955 • Germany • Anura
Hennig, Emil Hans Willi • 20 April 1913 • Germany • Sauria, Serpentes
Hensel, Reinhold Friedrich • 1826–1881 • Germany • Anura
Henshaw, Henry Wetherbee • 3 March 1850–1930 August 1 • USA • Anura, (Sauria, Serpentes)
Hensley, Marvin Max • 5 April 1922 • USA • Serpentes
Heppich, Susanna • 3 December 1953 • Austria • Anura
Hermann, Johannes (=Jean) • 31 December 1738–1800 October 8 • France • Sauria, Serpentes, Testudines
Hernández, Francisco • 1517–1587 January 28 • Spain • (Sauria)
Hernández Camacho, Jorge I. • ? • Colombia • Anura
Hernández Pacheco, Eduardo • 1872–? • Spain • *testudines
Hernández Sampelayo: *see* Sampelayo, P. H.
Héron-Royer, Louis François • ?–1891 December • France • Anura
Herre, Albert William Christian Theodore • 16 September 1868–1962 January 16 • USA • (Anura, Sauria)
Herre, Karl Wolf • 3 May 1909 • Germany • *Caudata
Herrera Lopez, Alfonso Luis • 1868–1942 • Mexico • (Serpentes, Testudines)
Hershkovitz, Philip • 12 October 1909 • USA • Gymnophiona
Hertz, Paul Eric • 8 July 1951 • USA • Sauria
Hesse, Paul • 3 February 1857–1938 February 26 • Germany • (Sauria)
Heude, Pierre-Marie • 25 June 1836–1902 January 2 • France / China • Testudines
Heuglin, Martin Theodor (hofrath) von • 20 March 1824–1876 November 5 • Germany • (Anura)
Hewitt, John • 23 December 1880–1961 August 4 • England / Sarawak / South Africa • Anura, [Crocodilia], Sauria, Testudines
Heyden, Carl Heinrich Georg von • 20 January 1793–1866 January 7 • Germany • Anura, Sauria
Heyer, William Ronald • 29 August 1941 • USA • Anura, Sauria
Heymons, Richard Friedrich Wilhelm Carl • 29 May 1867–1943 December 1 • Germany • (Anura)
Hibbard, Claude William • 21 March 1905–1973 October 9 • USA • (*anura, *caudata), *sauria, (*serpentes, [*testudines])
Hidalgo, Hugo • 1951 • USA • Sauria
Higginbottom, John • 1788–1876 • England • Caudata
Higgins, S. B. • ?–(1873)–? • USA • Serpentes
Highton, Richard • 24 December 1927 • USA • Caudata, Serpentes

Hikida, Tsutomu • 9 March 1951 • Japan • Sauria
Hilgendorf, Franz Martin • 5 December 1839–1904 July 5 • Germany / Japan / Germany • *sauria, Serpentes
Hillenius, Dirk (="Dick") • 1927–1987 May 4 • Holland • (Anura), *Sauria
Hillis, David Mark • 21 December 1958 • USA • Anura
Hilton, John William • 9 September 1904 • USA • (Serpentes, Testudines)
Hilton, William Atwood • 27 June 1878–? • USA • Caudata
Hinckley, Mary H. • 6 April 1845–1944 • USA • Anura
Hirsch, Karl Franz • 20 March 1921 • USA • *crocodilia
Hoard, Robert S. • ? • USA • Serpentes
Hodrová, Marcela (=mrs. Marvanová) • ? • Czechoslovakia • *caudata
Hoernes, Rudolf • 7 October 1850–1912 August 20 • Austria • *testudines
Hoeven, Jan van der • 9 February 1801–1868 March 10 • Holland • *Caudata, Sauria,
Hoevers, Leo G. • ? • Holland • Serpentes
Hoffman, Abraham Carel • 1903 • South Africa • Anura, Serpentes
Hoffman, Richard Lawrence • 25 September 1927 • USA • Caudata, Sauria
Hoffmann, Joseph • ? • Luxemburg • Anura
Hoffstetter, Robert Julien • 11 July 1908 • France • *caudata, *sauria, *serpentes
Hofmann, Adolf • 1853–? • Austria • *crocodilia, (*sauria)
Hoge, Alphonse Richard (*husband of* S. A. Romano Hoge) • 5 September 1912–1982 December 25 • Brazil / Belgium / Brazil • Serpentes, (Testudines)
Hogg, John • 1800–1869 September 16 • England • Caudata
Holbrook, John Edwards • 30 December 1794–1871 September 8 • USA • Anura, Caudata, Sauria, Serpentes, Testudines
Holl, Friedrich • ?–(1820-1834)–? • Germany • *caudata
Holland, Richard L. • ? • USA • Serpentes
Holland, William Jacob • 16 August 1848–1932 Dec. 13 • Jamaica / USA • *crocodilia, (Serpentes, *testudines)
Holman, J Alan • 24 September 1931 • USA • (*anura), *Caudata, *Sauria, *serpentes, *testudines
Holmer, Lars Magnus • 1764–1800 • Sweden • Sauria
Holsinger, John Robert • 6 April 1934 • USA • Caudata
Hombron, Jacques Bernard • 15 January 1798–1852 September 16 • France • Sauria, Serpentes
Honnorat-Bastide, Édouard F. • ?–? • France • (Anura)
Hoofien, Jacob Haim • 23 October 1913 • Israel • Sauria
Hoogmoed, Marinus Steven • 19 March 1942 • Holland • Anura, Gymnophiona, Sauria, Serpentes
Hooijer, Dirk (="Dick") Albert • ? • Holland • *testudines
Hooley, Reginald Walter • 5 September 1865–1923 May 5 • England • *testudines
Horn, Hans Georg • ? • Germany • (Sauria)
Hornaday, William Temple • 1 December 1854–1937 March 6 • USA • Crocodilia
Horner, P. G. • ? • Australia • Sauria
Hornstedt, Claës Fredrik • 1758–1809 May • Sweden • Serpentes
Horowitz, Samuel Boris • 26 Aug. 1927 • USA • Sauria
Horst, Rutger • ?–(1883)–? • Holland • Anura
Horton, David Robert • 4 April 1945 • Australia • Sauria

Horváth, Géza • 1847–1937 • Hungary • (Sauria)
Hosmer, William • ? • Australia • Anura, Sauria
Hotz, Hansjürg • ? • Switzerland / USA • Anura
Hou, Lianhai • ? • China • *sauria
Houston, Terry Francis • ? • Australia • Sauria
Houttuyn, Martin (=Martinus) • 1720–1798 • Holland • Caudata, Sauria
Howell, Kim M. • ? • USA / Tanzania • Anura
Howell, Thomas Raymond • 17 June 1924 • USA • Serpentes
Hoy, Philo Romayne (?*or* Romeyn) • 3 November 1816–1892 December 9 • USA • (Testudines)
Hoyt, Dale L. • ? • USA • Anura
Hsu, Chunhua • ? • China • *testudines
Hsu, Hsi-fan • 9 March 1906 • China / USA • Anura, Testudines
Hu, Buqing • 4 April 1915–1979 September 30 • China • Serpentes
Hu, Qixiong • 25 April 1945 • China • Anura, Caudata
Hu, Shuqin (=Shu-chin, *wife of* Liu, C.-c.) • 13 February 1914 • China • Anura, Caudata, Sauria, Serpentes
Huang, Kangcai • 23 Dec. 1935 • China • Anura, Sauria
Huang, Meihua • 25 November 1930 • China • Serpentes
Huang, Qingyun • 18 June 1938 • China • Sauria, Serpentes
Huang, Yongzhao • 24 June 1939 • China • Anura, Serpentes
Huang, Zhengyi • 3 November 1933 • China • Anura, Caudata, Serpentes
Huang, Zhujian • 1932 • China • [Anura], Serpentes
Hubrecht, Ambrosius Arnold Willem • 1853–1915 • Holland • Sauria, Serpentes, Testudines
Hubricht, Leslie • 11 January 1908 • USA • (Caudata)
Huene, Friedrich von • 22 March 1875–1969 April 4 • Germany • *crocodilia, *rhynchocephalia
Huey, Raymond Brunson • 14 Sept. 1944 • USA • Sauria
Hughes, Richard Chester • 21 July 1900 • USA • Caudata
Huheey, James Edward • 2 Aug. 1935 • USA • Serpentes
Hulke, John Whitaker • 6 November 1830–1895 February 19 • England • *crocodilia
Hulse, Arthur Charles • 20 December 1945 • USA • Sauria
Hulselmans, Jan L. J. • ? • Belgium • Anura
Humbert, Aloïs • 1829–1887 • Switzerland • *testudines
Humboldt, Friedrich Wilhelm Heinrich Alexander von • 14 September 1769–1859 May 6 • Germany • (Anura, Caudata), Serpentes
Hummel, Karl A. • 14 October 1889–1945 April 7 • Germany • *testudines
Hummelinck, Pieter Wagenaar • 1907 • (?Holland) / West Indies • Sauria
Humphrey, Frances L. • ? • USA • Anura, Serpentes
Humphries, Robert B. • ? • Australia • (Sauria)
Hunsaker, Don, II • 6 April 1930 • USA • (Sauria)
Hunt, Lawrence E. • ? • USA • Sauria
Hurter, Julius • 2 July 1842–1917 November • Switzerland / USA • (Anura)
Hutterer, Rainer • ? • Germany • *sauria
Hutton, A. F. • ? • (?England) / India • (Serpentes)
Hutton, Frederick Wollaston • 16 November 1836–1905 October 29 • England / New Zealand • Sauria
Huxley, Thomas Henry • 4 May 1825–1895 June 29 • England • *crocodilia

- I -

Ihering (=Jhering), Hermann von (*father of* R. v. Ihering) • 9 October 1850–1930 February 24 • Germany / Brazil / Germany • (Anura, Sauria, Serpentes)
Ihering, Rodolpho von (*son of* H. v. Ihering) • 1883–1939 September 15 • Brazil • Gymnophiona, Serpentes
Incháustegui Miranda, Sixto J. • ? • Dominican Republic • Sauria, (Serpentes)
Ineich, Ivan • ? • French Polynesia • Sauria
Inger, Robert Frederick • 10 September 1920 • USA • Anura, Caudata, Sauria, Serpentes
Ingoldby, C. M. • ?–? • British India • Sauria, Serpentes
Ingram, Glen J. • ? • Australia • Anura, Sauria
Ingram, William, III • 1945 • USA • Sauria
Ionides, Constantine John Philip • 1900–1968 September 22 • England / Tanganyika • (Sauria, Serpentes)
Ishikawa, Chiyomatsu • 8 January 1860–1935 January 17 • Japan • (Anura), Caudata
Iskandar, Djoko T. • ? • Indonesia • Anura
Isle (=l'Isle) du Dreneuf, Arthur de la • ?–? • France • *anura, Caudata
Ito, Riu • ? • Japan • Anura
Iturra Constant, Patricia • 22 Sept. 1947 • Chile • Anura
Iverson, John Burton, III • 4 October 1949 • USA • Testudines
Iwan: *see* Lepekhin, I. I.
Izecksohn, Eugenio • ? • Brazil • Anura

- J -

Jackson, Dale Robert • 1949 • USA • *testudines
Jackson, Herbert William • 5 January 1911 • USA • (Caudata)
Jackson, James Frederick • 11 Nov. 1943 • USA • Sauria
Jacobs, Jeremy Fischer • 3 May 1955 • USA • Caudata
Jacobsen, N. H. G. • ? • South Africa • Sauria, Serpentes
Jacquin, Joseph Franz (freiherr) von • 7 February 1766–1839 December 9 • Austria • Sauria
Jacquinot, Honoré • 1814–? • France • Sauria, Serpentes
Jaekel, Otto Max Johannes • 21 February 1863–1929 March 6 • Poland / Germany • *rhynchocephalia
Jain, Sohan Lall • 1931 • India • *testudines
James, Edwin • 27 August 1797–1861 October 28 • USA • [Anura, Caudata, Sauria, Serpentes]
Jameson, David Lee • 3 June 1927 • USA • Anura
Jan, Giorgio (=Georg) • 21 December 1791–1866 May 7 • Austria / Italy • Sauria, Serpentes
Janensch, Werner • 10 November 1878–1969 October 20 • Germany • *rhynchocephalia
Jarocki, Feliks Paweł na Jaroczynie • 1790–1865 • Poland • Caudata, Serpentes, Testudines
Jeitteles, Ludwig Heinrich • 1830–1883 • Austria • Sauria
Jensen, Adolf Severin • 1866–? • Denmark • Serpentes
Jenyns (=Blomefield *after 1871*), Leonard • 1800–1893 • England • Caudata
Jepsen, Glenn Lowell • 4 March 1903 • USA • *sauria
Jerdon, Thomas Claverhill • 1811–1872 June 12 • England / India / England • Anura, Sauria, Serpentes, Testudines
Jeude: *see* Lidth de Jeude, T. W. van
Jhering: *see* Ihering

Jiang, Yaoming • 12 January 1932 • China • Anura, Sauria, Serpentes
Jim, Jorge • ? • Brazil • Anura
Jiménez, Alfonso • ? • (?Costa Rica) • Serpentes
Jiménez de la Espada, Marcos • 5 March 1831–1898 October 3 • Spain • Anura, Caudata
Jiménez Fuentes, Emiliano • ? • Spain • *testudines
Joger, Ulrich • ? • Germany • Sauria, Serpentes
Johnson, Murray Leathers • 16 October 1914 • USA • Serpentes
Johnson, R. D. O. • ? • USA / Colombia / USA • (Anura)
Johnston, C. Stuart • ?–1939 • USA • *testudines
Johnston, Gregory Ronald • 28 September 1962 • Australia • Anura
Johnstone, Ronald Eric • 1949 • Australia • (Sauria)
Joleaud, Léonce • 1880–1938 • France • *crocodilia
Jolly, Gordon Gray (lieutenant-general, sir) • 1886–1962 March 1 • England / British India • (Serpentes)
Jones, Clyde Joe • 3 March 1935 • USA • Caudata
Jones, J. Paul • ? • USA • Sauria
Jones, Kirkland Lee • 1 Oct. 1941 • USA • Anura, Sauria
Jones, Richard Evan • 13 May 1940 • USA • (Sauria)
Jong, Jan Kornelis de • ?–? • Holland • Sauria, Serpentes
Jonnès: *see* Moreau de Jonnès, A.
Jopson, Harry Gorgas Michener • 23 June 1911 • USA • Caudata
Jordan, David Starr • 19 January 1851–1931 September 19 • USA • ([Anura], Caudata)
Jörg, Erwin • 10 January 1917–1977 February 11 • Germany • *sauria
Jourdan, ? • ?–? • France • *rhynchocephalia, Serpentes
Joynson, H. W. • ? • (?England) / Siam • (Serpentes)
Juliá Zertuche, Jordi • ?–1986 • Mexico • Serpentes

- K -

Kail, Johann Alois • 21 April 1856–? • Austria • *croc.
Kälin, Josef Alois • 1903 • Switzerland • *crocodilia
Kallert, Eduard • 1888–? • Germany • (Caudata)
Kalm, Pehr (=Peter) • 6 March 1716–1779 November 16 • Finland • (Anura)
Kamei, Tadao • 1925 • Japan • *crocodilia
Kamita, Tsuneichi • ? • Japan • Sauria
Kammerer, Paul • 17 August 1880–1926 September 23 • Austria • ([Sauria])
Kampen, Pieter Nicolaas van • 1878–1937 July 3 • Holland • Anura, (Sauria)
Kanberg, Hans • ? • Germany • (Anura)
Karaman, Stanko L. • 1889 • Yugoslavia • Anura, Caudata, Sauria
Karg, Josef Maximilian • ?–(1786-1805)–? • Germany • (*serpentes)
Karges, John Paul • 22 April 1955 • USA • Sauria
Karl, Hans-Volker • ? • Germany • *testudines
Karlin, Alvan A. • 3 May 1950 • USA • Caudata
Kashchenko (=Kaščenko), Nikolai Feofanovich • ?–(1899)–? • Russia • Anura, Sauria
Kattan, Gustavo H. • 26 July 1953 • Colombia *and* USA • Anura
Kattinger, Emil • 1902 • Germany • Sauria

Kaudern, Walter Alexander • 24 March 1881–1942 • Sweden • Sauria, Serpentes

Kauffeld, Carl Frederick (*husband of* I. H. Kauffeld, ?–?) • 17 April 1911–1974 July 10 • USA • Serpentes

Kauffeld, I. H.: *see* Conant, I. H.

Kaup, Johann Jakob • 20 April 1803–1873 July 4 • Germany • (Gymnophiona), Sauria

Kauri, Hans • ? • Sweden • Caudata

Kawano, Usaburo • ? • Japan • Anura

Kay, Fenton Ray • 10 October 1942 • USA • Sauria

Keferstein, Wilhelm Moritz • 7 June 1833–1870 • Germany • Anura, Caudata

Keith, Ronalda • ? • USA • Anura

Kelaart, Edward Frederick (major) • 21 November 1819–1860 August 31 • Ceylon / England / Gibraltar / Ceylon • Anura, Sauria, Serpentes

Kellogg, Arthur Remington • 5 October 1892–1969 May 8 • USA • Anura

Kennedy, Joseph Patrick • 9 March 1932 • USA • (Anura)

Kennicott, Robert • 13 November 1835–1866 May 13 • USA • Serpentes

Kent, William Saville • ?–1908 • England • (Serpentes)

Kerster, Harold William • 1932 • USA • Sauria

Kessler, Karl Fedorovich • 1815–1881 • Russia • Caudata, Sauria

Ketchersid, Chesley Arthur • ? • USA • Sauria

Key, George Franklin • 11 April 1942 • USA • (Sauria)

Keyserling, Alexander F. M. L. Andreevich (graf) • 15 August 1815–1891 May 8 • Latvia / Russia / Estonia • Anura, (Caudata)

Khan, Muhammad Sharif • 11 August 1939 • Pakistan • Anura, Sauria, Serpentes

Kharin, V. E. • ? • USSR • Serpentes

Khosatzky, Leo (=Lev Isaakovich Chozatskij) • 1913 • USSR • *anura, *testudines

Kiester, A. Ross • ? • USA • Sauria

Kiew: *see* Heang, K. B.

Kimakowicz, Moritz von • ?–(1888)–? • Romania • Sauria

King, Dennis R. • ? • Australia • (Sauria)

King, F. Wayne • ? • USA • Sauria

King, Fred Willis • 24 May 1908 • USA • Caudata

King, Max • ? • Australia • Sauria

King, Philip Parker (rear-admiral) • 13 December 1791–1856 February 26 • Norfolk Island / England / Australia • ([Sauria], Serpentes)

Kinghorn, James Roy • 12 October 1891–1983 March 4 • Australia • (Anura), Sauria, Serpentes

Kinkelin, Georg Friedrich • 15 July 1836–1913 August 13 • Germany • *serpentes

Kiriţescu (=Kiritzesco), Constantin • 1876–? • Romania • Caudata

Kirk, John (sir) • 19 December 1832–1922 January 15 • Scotland / East Africa / England • (Gymnophiona, Sauria)

Kirtisinghe, P. • ? • Ceylon • Anura

Kirtland, Jared Potter • 10 November 1793–1877 December 10 • USA • Serpentes

Kishida, Kyukichi • 25 August 1888–1968 October 4 • Japan • Sauria

Kisteumacher, Geraldo • ? • Brazil • Anura

Klappenbach, Miguel Angel • 4 May 1920 • Uruguay • Anura, Sauria

Klauber, Laurence Monroe • 21 December 1883–1968 May 8 • USA • (Caudata), Sauria, Serpentes

Klausewitz, Wolfgang • 20 July 1922 • Germany • Sauria

Klaver, Charles J. J. • 12 December 1946 • Holland • Sauria

Klembara, J. • ? • Czechoslovakia • *sauria

Klemmer, Konrad • ? • Germany • (Anura), Sauria, Serpentes

Klewen, Reiner Franz • 1956 • Germany • Caudata

Klingel, Gilbert Clarence • 4 October 1908–1983 May 16 • USA • Sauria

Klinikowski, Ronald F. • ? • USA • (Anura), Sauria

Kloss, Cecil Boden • 28 March 1877–1949 August 19 • England / British Malaya / England • (Sauria, Serpentes)

Kluge, Arnold Girard • 27 July 1935 • USA • *anura, Sauria

Knaap-van Meeuwen, Maria Sophia • ? • Holland • Serpentes

Kneeland, Samuel • 1 August 1821–1888 September 27 • USA • Caudata

Kneller, Mathias • ? • Germany • Anura

Knight, James L. • ? • USA • Serpentes

Knight, R. Alec • ? • USA • Sauria

Knoepffler, Louis-Philippe • 11 March 1926–1984 November 15 • France • Serpentes

Knox, Frederick John • ?–1873 (age 82) • Scotland / New Zealand • Sauria

Kobel, Hans Rudolf • 18 Aug. 1936 • Switzerland • Anura

Koch, Carl • 1 April 1894–1970 June 14 • Germany • (Sauria)

Koch, Carl (=Karl Jakob Wilhelm) • 1 June 1827–1882 April 18 • Germany • Anura

Kochva, Elazar • 16 June 1926 • USSR (Bessarabia) / Israel • Sauria

Kofron, Christopher P. • ? • USA • Serpentes

Kolombatovič, Gjuro (=Georg) • ?–? • (?Croatia) • (Anura), Caudata, Sauria

Kopstein, Felix • 4 June 1893–1939 April 14 • Austria / Dutch East Indies / Holland • Sauria, Serpentes

Kormos, Theodor (=Tivadar) • 1881–? • Hungary • *sauria, *testudines

Koshy, Mammen • ? • India • Anura, Sauria

Koslowsky (=Kozlovski), Julio Germán • 15 September 1866–1923 September 23 • Russia / Argentina • Anura, Sauria, Serpentes

Kostin, Anatoly A. • ? • (?USSR) / Manchuria • Anura, (Caudata)

Kotsakis, Tassos • 1945 • Italy • *sauria

Kou, Zhitong • 2 October 1935 • China • Anura, Caudata, Serpentes

Kraglievich, Jorge Lucas • ? • Argentina • *sauria

Kramer, Eugen • 9 January 1921 • Switzerland • Serpentes

Krassawzeff, Boris (=Borys Arkadjevich Krassavtzev) • 3 December 1909–1943 June • USSR • Serpentes

Krassovsky, D. B. • ? • USSR • Sauria

Kraus, Edward Frederick • 12 June 1959 • USA • Caudata

Krausse, Anton Hermann • 1878–1929 • (?Germany) / Sardinia • Sauria

Krebs, Salome Litwin • ? • USA • Caudata

Krefft, Johann Ludwig Louis Gerard • 17 February 1830–1881 February 19 • Germany / Australia • Anura, Crocodilia, (Sauria), Serpentes, Testudines
Krefft, Paul • ?–? • Germany • Caudata
Kreyenberg, Martin • ? • Germany • (Serpentes)
Krieg, Hans • 18 June 1888–1970 October 5 • Germany • (Anura, Sauria)
Kripalani, M. B. • ? • India • Anura
Kroll, James Clarence • 5 November 1946 • USA • Sauria
Krinitsky, Ivan Andreevich (=Jean Krynicki) • "2" (=13) July 1797–1838 "September 30" (=October 12) • Russia • Anura, Sauria
Kubykin, R. A. • ? • USSR • Sauria
Kues, Barry Stephen • 15 March 1946 • USA • *crocodilia
Kuhl, Heinrich • 17 Sept. 1797–1821 Sept. 14 • Germany / East Indies (Java) • Anura, Sauria, Serpentes, Testud.
Kuhn, Oskar • 1908 • Germany • *anura, *crocodilia, *sauria, *serpentes
Kükenthal, Willy Georg • ?–? • Germany • (Sauria, Serpentes)
Kuliga, Paul • 1878–? • (?Germany) • (Sauria)
Kupriyanova, Larissa A. • ? • USSR • Sauria
Kuramoto, Mitsuru • 24 October 1934 • Japan • Anura
Kushakevich (=Kušakevič), Sergius • ?–? • Russia • (Sauria)
Kuss, Siegfried Ernst • ? • Germany • *testudines
Küster (=Kuester), Heinrich Carl • 1807–1876 • Germany • Anura
Kuznetsov (=Kusnezov), Valentin Vasilevich • 1923 • USSR • *testudines
Kyriakopoulou-Sklavounou, Pasqualina • ? • Greece • Anura

- L -

Labra Lillo, Maria Antonieta • 9 December 1963 • Chile • Sauria
Lacépède, Bernard Germain Étienne de la Ville sur Illon (comte de) • 26 December 1756–1825 October 6 • France • (Caudata), Sauria, Serpentes, (Testudines)
Lacerda, João Baptista de • 1846–? • Brazil • Serpentes
Lafrentz, Karl August E. • ?–1945 April 26 • Germany • (Anura), Caudata
Laidlaw, Frank Fortescue • ?–? • England • Serpentes
Lamar, William Wylly, III • 18 June 1950 • USA • Sauria
La Marca, Enrique • ? • Venezuela • Anura
Lamb, J. • ? • Australia • Anura
Lambe, Lawrence Morris • 27 August 1863–1919 • Canada • *crocodilia, *serpentes
Lamborot, Madeleine • ? • Chile • Sauria
Lamotte, Maxime Georges • 26 June 1920 • France • Anura, (Gymnophiona)
Lampe, Eduard • ?–(1902)–? • Germany • [Serpentes]
Lancici V., Abdem Ramón • ? • Venezuela • Sauria, Serpentes
Landbeck, Christian Ludwig • ?–(1861)–? • (?Germany) • [Sauria]
Lando, R. V. • ? • USA • Serpentes
Landy, Macreay John • 1937 • USA • Serpentes
Lane, Henry Higgins • 17 February 1878–1965 • USA • (Sauria)

Lang, Herbert Otto Henry • 24 March 1879–? • Germany / USA • (Sauria)
Langebartel, David A. • 1929 • USA • Anura, *Sauria, Serpentes
Langhammer, James Kenneth • 9 March 1936 • USA • Serpentes
Langone, José A. • ? • Uruguay • Anura
Langston, Wann, Jr. • 10 July 1921 • USA • *crocodilia
Lannom, Joseph R., Jr. • ? • USA • (Serpentes)
Lantz, Louis Amédée • 20 March 1886–1953 February 3 • France / Russia / England • Caudata, Sauria
Lanza, Benedetto • ? • Italy • Anura, Caudata, Sauria, Serpentes
Lara Góngora, Guillermo • ? • Mexico • Sauria
Largen, Malcolm J. • ? • (?England) / Ethiopia / England • Anura
La Rivers, Ira John • 1 May 1915–1977 October 11 • USA • *anura
Larsell, Olof • 13 March 1886–1964 • Sweden / USA • (Caudata)
Larsen, Kenneth R. • ? • USA • Sauria
Lartet, Édouard Armand Isidore Hippolyte • 15 April 1801–1871 January 28 • France • *caudata, *sauria, *serpentes, *testudines
Lataste, Fernand • 5 March 1847–1934 January 6 • France / Chile / France • Anura, Caudata, Sauria, Serpentes, Testudines
Latifi, Mahmood • ? • Iran • (Serpentes)
Latreille, Pierre-André • 29 November 1762–1833 February 6 • France • Anura, Caudata, Sauria, Serpentes, Testudines
Laube, Gustav Carl • 9 January 1839–1923 April 12 • Bohemia / Czechoslovakia • *caudata, *testudines
Laufe, Leonard Edward • 22 July 1924 • USA • Sauria, Serpentes
Laurent, Raymond Ferdinand • 16 May 1917 • Belgium / Argentina • Anura, Sauria, Serpentes, Testudines
Laurenti, Josephus Nicolaus (=Joseph Nicolas) • 4 December 1735–1805 February 17 • Austria • Anura, Caudata, Crocodilia, Sauria, Serpentes
Lavocat, René • ? • France • *crocodilia
Lawrie, Bruce C. • ? • Australia • Sauria
Lazell, James Draper, Jr. • 5 September 1939 • USA • *Sauria, Serpentes
Lazzarini, Alfredo • ?–? • Italy • Caudata
Leach, William Elford • 1790–1836 August 25 • England • Crocodilia, Sauria, Serpentes
Leakey, Jonathan H. E. • ? • Kenya • (Serpentes)
Le Conte, John Eatton (major, *father of* J. L. Le Conte) • 22 February 1784–1860 November 21 • USA • Anura, Caudata, Testudines
Le Conte, John Lawrence (*son of* J. E. Le Conte) • 13 May 1825–1883 November 15 • USA • (Serpentes), Testudines
Lee, Anthony Kingston • 1933 • Australia • Anura
Lee, Daniel K. • ? • USA • Sauria
Lee, Julian Carlee • 7 September 1943 • USA • (Anura)
Leenhardt, Henri • ? • France • *sauria
Legler, John Marshall • 9 September 1930 • USA • (Anura), Serpentes, Testudines
Lehmann Valencia, F. Carlos • ?–1974 August 15 • Colombia • (Anura, Sauria, Serpentes)

Lehrs, Philipp • 9 August 1881–1956 April 20 • Germany • Sauria
Lei, Yizhen • 1944 • China • *testudines
Leidy, Joseph • 9 September 1823–1891 April 29 • USA • *crocodilia, *sauria, *testudines
Leighton, Gerald Rowley • 12 December 1868–1953 September 8 • England / New Zealand / Scotland / England • (Serpentes)
Leith-Adams: *see* Adams, A. L.
Lema, Thales de • ? • Brazil • Serpentes
Lenz, Harald Othmar • 27 February 1798–1870 January 13 • Germany • Serpentes
Lenz, Heinrich Wilhelm Christian • 1846–1913 • Germany • (Sauria)
Lepekhin, Ivan Ivanovich (=Iwan Lepechin) • 10 September 1740–1802 April 6 • Russia • Serpentes
Lescure, Jean • 1932 • France • Anura, Gymnophiona
Lesson, René-Primevère • 20 March 1794–1849 April 28 • France • Anura, Croc., Sauria, Serpentes, Testudines
Lessona, Michele • 1823–1894 • Italy • Anura
Lesueur, Charles-Alexandre • 1 January 1778–1846 December 12 • France / USA / France • (Anura), Sauria, Serpentes, Testudines
Leuckart, Friedrich Sigismund • 26 August 1794–1843 September 25 • Germany • *Caudata, Sauria
Leviton, Alan Edward • 11 January 1930 • USA • Anura, Caudata, Sauria, Serpentes
Lewis, Charles Bernard, Jr. • 9 April 1913 • USA / Jamaica • (Anura, Sauria)
Lewis, Thomas Howard • 28 July 1919 • USA • Sauria
Leybold, Friedrich (=Federico) • 1827–1879 • Germany / Chile • Serpentes
Li, Dejun • 23 November 1936 • China • [Anura], Sauria, Serpentes
Li, Fanglin • ? • China • Anura
Li, Jinling • ? • China • *sauria
Li, Shengquan • 20 March 1955 • China • Anura, Sauria
Li, Siming • 20 January 1946 • China • Anura, Sauria
Li (*not* Liu), **Youheng** • ? • China • *anura, *testudines
Li, Zhixun • 24 November 1929 • China • Testudines
Liang, Yun-sheng • ? • Taiwan • Anura, Sauria
Lichtenstein, Martin Hinrich Carl • 10 January 1780–1857 Sept. 3 • Germany • Anura, Sauria, Serpentes
Lidth de Jeude, Theodorus Willem van • 1 February 1853–1937 May 29 • Holland • (Anura), Sauria, Serpentes, Testudines
Lieb, Carl Sears • 27 May 1949 • USA • Sauria, Serpentes
Liebus, Adalbert • 1876–? • Czechoslovakia • *testudines
Liem, David S. • ? • Australia • Anura
Liénard, Élizé (père) • ?–(1842)–? • Mauritius • *sauria
Lilljeborg, Wilhelm (=Vilhelm) • 6 October 1816–1908 • Sweden • Crocodilia, (Serpentes)
Lima Verde, J. S. • ? • Brazil • Serpentes
Limeses, Celia E. • 1918–1974 January 3 • Argentina • Anura
Lin, Jun-yi • ? • Taiwan • Sauria
Lindaker, Johanne Thaddeus • ?–(1791)–? • (?Austria) • Serpentes
Lindholm, Wilhelm A. • ?–? • Russia • Sauria, Serpentes, Testudines, (*testudines)
Liner, Ernest Anthony • 11 Feb. 1925 • USA • Serpentes

Link, Heinrich Friedrich • 2 February 1767–1851 January 1 • Germany • Sauria, Serpentes
Linnaeus: *see* Linné, C. von
Linné, Carl von (=Carolus Linnaeus) • 23 May 1707–1778 January 10 • Sweden • Anura, Caudata, Crocodilia, Gymnophiona, Sauria, Serpentes, Testudines
Lioy, Paolo • 1836–1911 • Italy • *crocodilia
L'Isle: *see* Isle, A. de la
List, James Carl • 6 July 1926 • USA • Anura
Littlejohn, Murray J. • ? • Australia • Anura
Liu, Baohe • ? • China • Anura
Liu, Ch'eng-chao (*husband of* Hu, S.-c.) • 12 September 1900–1976 April 9 • China • Anura, Caudata
Liu, Hsien-t'ing • 1921 • China • *testudines
Liu, Jisheng • 11 June 1937 • China • [Anura]
Liu, Youheng: *see* Li, Y.-h.
Liu, Yuezhen • 4 October 1933 • China • Sauria
Lockington, William Neale • (?1842)–1902 • USA • Serpentes
Löding, Henry Peter • 11 September 1869–1942 February 26 • Denmark / USA • (Caudata, Serpentes)
Loennberg: *see* Lönnberg, A. J. E.
Loftus-Hills, Jasper J. • ?–1974 June 11 (age 28) • Australia / USA • Anura
Long, Stephen Harriman (colonel) • 30 Dec. 1784–1864 Sept. 4 • USA • [Anura, Caudata, Sauria, Serpentes]
Longley, Glenn, Jr. • 2 June 1942 • USA • Caudata
Longman, Heber A. • ? • Australia • *crocodilia, Sauria, Serpentes, Testudines
Lönnberg (=Loennberg), Axel Johan Einar • 24 December 1865–1942 November 21 • Sweden • Anura, Sauria, Serpentes
Loomis, Frederic Brewster • 22 November 1873–1937 July 28 • USA • *crocodilia, *sauria, *testudines
Loomis, Richard Biggar • 18 June 1925 • USA • (Anura), Sauria, Serpentes
López Seoane: *see* Seoane, V. L.
Lortet, Louis-Charles • 1836–1909 • France • Testudines
Loumont-Vigny, Catherine • 6 December 1942 • Switzerland • Anura
Loveridge, Arthur • 28 May 1891–1980 February 16 • Wales / East Africa / USA / Saint Helena • Anura, Gymnophiona, Sauria, Serpentes, Testudines
Low, Timothy • 1956 • Australia • Sauria
Lowe, Charles Herbert, Jr. • 16 April 1920 • USA • Anura, Caudata, Sauria, (*sauria), Serpentes
Lu, Qingwu • ? • China • *testudines
Lucas, Arthur Henry Shakespeare • 7 May 1853–1936 June 10 • England / Australia • Sauria, Serpentes
Lucas, Spencer George • 25 April 1955 • USA • *croc.
Lüderwaldt: *see* Luederwaldt, H.
Ludwig, Rudolph August Birminhold Sebastian • 24 October 1812–1880 Dec. 11 • Germany • *crocodilia
Lue, Kuang-yang • ? • Taiwan • Anura, Sauria
Luederwaldt, Hermann • 1865–1934 • Brazil • Testudines
Lugeon, Maurice • 10 July 1870–1953 October 23 • France / Switzerland • *testudines
Lukina, Galina Pantelejmonovna • 1930 • USSR • Sauria
Lull, Richard Swann • 6 November 1867–1957 April 22 • USA • *crocodilia
Lunau, Hinnerk • ? • Germany • *caudata
Lungu, A. N. • ? • USSR • [*sauria], *testudines

Lütken, Christian Frederik • 4 Oct. 1827–1901 February 6 • Denmark • Anura, Gymnophiona, Sauria, Serpentes
Lutz, Adolpho (*father of* B. Lutz) • 18 December 1855–1940 October 6 • Brazil • Anura
Lutz, Bertha Maria Julia(*daughter of* A. Lutz) • 2 August 1894–1976 September 16 • Brazil • Anura, (Sauria)
Lutz, Daniela • 8 August 1959 • Austria • Sauria
Lydekker, Richard • 25 July 1849–1915 April 16 • England • *crocodilia, *testudines
Lynch, James Francis • 19 Nov. 1942 • USA • Caudata
Lynch, John Douglas • 30 July 1942 • USA • *Anura, Sauria
Lynn, William Gardner • 26 December 1905 • USA • Anura, ([Sauria]), *testudines
Lyon, Marcus Ward, Jr. • 5 February 1875–1942 May 19 • USA • (Serpentes)

- M -

Ma, Desan • ? • China • Anura
Ma, Jifan • 23 October 1940 • China • Serpentes
Mabee, William Bruce • 1 January 1897 • USA • (Caudata)
Macarovici, Neculai • ? • Romania • *testudines
Maccagno, Angiola Maria • ? • Italy • *crocodilia
MacClelland (=M'Clelland), **John** • 1805–1875 • British India • (Serpentes)
Macdonald, James Reid • 22 Aug. 1918 • USA • *sauria
Macdonald, M. A. • ? • Scotland • Sauria
MacDougall, Thomas Baillie • 9 December 1896–1973 January 17 • Scotland / USA / Mexico • (Anura, Caudata, Sauria, Serpentes)
Macey, J. Robert • 6 November 1963 • USA • Anura
Machado, Ottilio • ? • Brazil • (Sauria), Serpentes
Mackey, James Penrose • 28 Feb. 1930 • USA • Anura
Macklot, Heinrich Christian • 20 October 1799–1832 May 12 • Germany / Holland / Dutch East Indies • (Anura), Sauria, (Serpentes)
Macleay, William John (sir, *cousin of* W. S. Macleay) • 13 June 1820–1891 December 7 • Scotland / Australia • Anura, Sauria, Serpentes
Macleay, William Sharp (*cousin of* W. J. Macleay) • 21 July 1792–1865 Jan. 26 • England / Australia • Sauria
Macola, G. S. • ? • Argentina • Sauria
Magalhães, Octávio de • ? • Brazil • Serpentes
Maglio, Vincent Joseph • 2 Oct. 1942 • USA • Serpentes
Mahony, Michael J. • ? • Australia • Anura
Mahendra, Beni Charan • ? • India • Serpentes
Main, Albert Russell • ? • Australia • Anura, (Sauria)
Major, Charles Immanuel Forsyth • 1848–1923 March 25 • England • (Anura)
Maki, Moichiro • 20 February 1886–1959 April 19 • Japan / Taiwan / Japan • Caudata, Sauria, Serpentes
Maldonado Koerdell, Manuel • 25 January 1908–1973 • Mexico • Caudata
Malkin, Borys • ? • Switzerland • Anura
Malkmus, Rudolf • ? • (?Germany) / Portugal / Germany • Caudata
Malnate, Edmond V. • 16 July 1916 • USA • Serpentes
Mansukhani, M. R. • ? • India • Anura
Mao, Jierong • 25 October 1921 • China • Sauria
Marca: *see* La Marca, E.

March, Douglas D. H. • ?–1939 April 3 (age 52) • USA • (Serpentes)
Marchand, Lewis J. • ? • USA • Testudines
Markezich, Allan Louis • 15 May 1949 • USA • Serpentes
Marlow, Ronald William • 9 Feb. 1949 • USA • Caudata
Marsh, Othniel Charles • 29 October 1831–1899 March 18 • USA • *crocodilia, *sauria, *serpentes
Marshall, Joe Truesdell, Jr. • 15 February 1918 • USA • Sauria
Martens, Eduard Carl von • 18 April 1831–1904 August 14 • Germany • Anura, Caudata, Sauria, Serpentes
Martin, Angus Anderson • 20 April 1940 • South Africa / Australia • Anura
Martin, Handel Tongue • 1862–1931 January 15 • USA • ([*caudata])
Martin, Paul Schultz • 22 August 1928 • USA • Sauria
Martin, Robert Lawrence • 18 Nov. 1933 • USA • Sauria
Martin, William Charles Linnaeus • 1798–1864 February 15 • England • Sauria, Serpentes
Martín del Campo y Sánchez, Rafael • 3 January 1910–1987 December 25 • Mexico • Sauria
Martínez, Víctor • ? • Panama • Anura
Martín Frias, Eliezer • 1 July 1939 • Mexico • Caudata
Martín Hidalgo, Aurelio • ? • Canary Islands • Sauria
Martins, Márcio • ? • Brazil • Anura
Martius, Carl Friedrich Philipp von • 17 April 1794–1868 December 13 • Germany • Serpentes
Martof, Bernard Stephen • 21 August 1920–1978 November 28 • USA • Caudata
Maruska, Edward Joseph • 19 Feb. 1934 • USA • Caudata
Marx, Hymen • 27 June 1925 • USA • Sauria, Serpentes
Marx, Kevin Wall • 2 October 1923 • USA • Anura
Maslin, Thomas Paul, Jr. • 27 October 1909–1984 February 26 • USA • Anura, Caudata, Sauria, Serpentes
Mason, George E. • ?–(1888)–? • (?British India) • Serpent.
Massalongo, Abramo Bartolomeo • 1824–1860 • Italy • Anura, Caudata, Sauria, Serpentes
Mather, William Williams • 24 May 1804–1859 February 25 • USA • [Serpentes]
Matheron, Pierre Philippe Émile • 18 October 1807–1900 December 31 • France • *crocodilia
Mathers, Sandra • 18 October 1957 • USA • Sauria
Matschie, Georg Friedrich Paul • 11 August 1861–1926 March 7 • Germany • Anura, Sauria, Serpentes
Matsui, Masafumi • ? • Japan • Anura, Caudata
Matsui, Takaji • 24 March 1925 • Japan • Anura, Sauria
Matsumoto, Eiji • ? • Japan • *crocodilia
Matsumoto, Hikoshichiro • ? • Japan • *testudines
Matthes, Benno • 15 September 1825–1911 April 30 • Silesia / Germany / USA • Caudata
Matthew, William Diller • 19 February 1871–1930 September 24 • USA • (*caudata, *sauria), *testudines
Mauduyt, L. • ?–(1844)–? • France • Serpentes
Mayer, August Franz Joseph Carl • 1787–1865 • Germany • Anura
Mayer, Werner • 27 May 1943 • Austria • Sauria
Mayorga, Horacio • ? • Puerto Rico • Anura
McCann, Yule Mervyn Charles • 4 December 1899–1980 November 29 • India / New Zealand • Sauria
McCarthy, C. J. • ? • England • Anura

McCauley, Robert Henry, Jr. • 13 June 1913 • USA • Sauria
McConkey, Edwin Henry • 1930 • USA • Sauria
McCord, William Patrick • 3 December 1950 • USA • Testudines
McCoy, Clarence John, Jr. • 25 July 1935 • USA • Caudata, Sauria
McCoy (=M'Coy), Frederick (sir) • 1823–1899 May 15 • Ireland / Australia • Anura, Sauria, Serpentes
McCoy, Michael • ? • Australia • Sauria
McCrady, Edward, Jr. • 19 Sept. 1906 • USA • Caudata
McCranie, James Randall • 10 October 1944 • USA • Anura, Sauria, (Serpentes)
McCrystal, Hugh K. • ? • USA • Sauria
McCulloch, Allan Riverstone • 20 June 1885–1925 September 1 • Australia • Anura, Testudines
McDiarmid, Roy Wallace • 18 February 1940 • USA • Anura, Serpentes
McDonald, Keith R. • ? • Australia • Anura, Sauria
McDowell, Samuel Booker, Jr. • 13 September 1928 • USA • Serpentes
McGregor, Richard Crittenden • 24 Feb. 1871–1936 Dec. 30 • Australia / USA / Philippine Islands • (Sauria)
McGrew, Paul Orman • 27 August 1909 • USA • (*crocodilia), [*sauria, *serpentes]
McKenzie, Norman Leslie • ? • Australia • (Sauria)
McKinney, Charles Oran • 10 Aug. 1942 • USA • Sauria
McKown, Ronald Ray • 1940 • USA • Testudines
McLachlan, G. R. • ? • South Africa • Sauria
McLain, Robert Baird • 16 August 1877–1946 November 15 • USA • Serpentes
M'Clelland: *see* MacClelland, J.
McMillan, Robert Peter • 10 August 1921 • Australia • (Sauria)
McMorris, John Robert • 17 Dec. 1946 • USA • Testud.
Meade-Waldo, Edmund Gustavus Bloomfield • 1855–1934 February 24 • England • Anura
Means, Donald Bruce • 9 March 1941 • USA • Caudata
Mearns, Edgar Alexander • 11 September 1856–1916 November 1 • USA • (Anura, Sauria)
Mecham, John Stephen • 29 Feb. 1928 • USA • *Anura
Mechler, B. • ? • Switzerland / ? • Sauria
Medem Medem, Federico (=Friedrich Johann, graf von) • 29 August 1912–1984 May 1 • Latvia / Germany / Switzerland / Colombia • (Anura, Caudata), Crocodilia, (Sauria, Serpentes), Testudines
Medina, Carmen Julia • ? • Venezuela • *crocodilia
Medway (lord): *see* Cranbrook, G. G.-H.
Meek, Seth Eugene • 1 April 1859–1914 June 6 • USA • Anura, Sauria, Serpentes
Meerwarth, Hermann • 9 Oct. 1870–? • Germany • Anura
Méhelÿ, Lajos (=Ludwig von) • 24 August 1862–(?1952 or 1953) • Hungary • Anura, (*anura), Caudata, Sauria, Serpentes
Mehl, Maurice Goldsmith • 25 December 1887–1966 March 30 • USA • *crocodilia
Meier, Harald • 14 Nov. 1922 • Germany • (Anura), Sauria
Meise, Wilhelm • ? • Germany • Serpentes
Meisner, Carl Friedrich August • 1765–1825 • Switzerland • Sauria, Serpentes
Melin, Douglas Edvard • 1895–1946 • Sweden • Anura

Mell, Rudolf • 16 February 1878–? • Germany • (Anura, Sauria), Serpentes
Mellado: *see* Pérez Mellado, V.
Mello Leitão, Candido Firmino de • 1886–1948 • Brazil • Anura, Serpentes
Mena, Cesar E. • ? • USA • Serpentes
Mendelssohn, Heinrich • 31 October 1910 • Germany / Israel • Anura, Sauria
Ménétriés, Edouard • 2 October 1802–1861 April 10 • France • Sauria, Serpentes
Menzies, James I. • ? • England / Sierra Leone / Papua New Guinea / England • Anura
Mercadal de Barrio, Isabel Teresa • 3 October 1954 • Argentina • Anura
Merrem, Blasius • 4 Feb. 1761–1824 Feb. 23 • Germany • Anura, Caudata, Croc., Sauria, Serpentes, Testudines
Merriam, Clinton Hart • 5 December 1855–1942 March 19 • USA • (Sauria)
Mertens, Robert Friedrich Wilhelm • 1 Dec. 1894–1975 August 23 • Russia / Germany • Anura, (*Caudata), Gymnophiona, *Sauria, Serpentes, Testudines
Meszoely, Charles Aladar Maria • 24 April 1933 • Hungary / USA • *sauria
Metaxà, Luigi • 1778–1842 • Italy • Sauria, Serpentes
Metcalf, Zeno Payne • 1 May 1885–1956 January 5 • USA • (Caudata)
Methuen, Paul Ayshford (4th baron) • 29 Sept. 1886–1974 January 7 • England • Anura, Sauria, Serpentes
Metter, Dean Edward • 1 August 1932 • USA • (Testud.)
Meuschen, Friedrich Christian • 1719–1790 • Germany • Serpentes
Meyen, Franz Julius Ferdinand • 28 June 1804–1840 September 2 • Germany • [Anura]
Meyer, Adolf Bernhard • 1840–1911 February 5 • Germany • Anura, Sauria, Serpentes, Testudines
Meyer, Christian Erich Hermann von • 3 September 1801–1869 April 2 • Germany • *caudata, *crocodilia, *rhynchocephalia, *serpentes, *testudines
Meyer, Friedrich Albrecht Anton • 29 June 1768–1795 November 29 • Germany • Crocodilia, Sauria
Meyer, John Raymond • ? • USA • (Sauria), Serpentes
Meyer de Schauensee, Rodolphe • 4 January 1901–1984 April 24 • Italy / USA • (Serpentes)
Meylan, Peter André • 28 May 1953 • USA • *sauria, *serpentes, *Testudines
Michaelsen, Johann Wilhelm • 1860–1937 • Germany • [Anura, Sauria, Serpentes]
Michahelles, Georg Christoph Carl Wilhelm • 5 May 1807–1834 August 15 • Germany / Greece • Anura, Caudata, (Sauria), Serpentes, Testudines
Mikan (=Mickan), Johann Christian • 5 December 1769–1844 December 28 • Czechoslovakia • Gymnophiona, Sauria, (Serpentes), Testudines
Mikšić, Sonja • ? • Yugoslavia • Caudata
Miller, Gerrit Smith, Jr. • 6 December 1869–1956 February 24 • USA • Anura, *sauria
Miller, John Sebastian • (circa 1715)–(circa 1790) • Germany / England • Testudines
Miller, Loye Holmes • 13 October 1874–1970 April 6 • USA • (*testudines)
Miller, Tracy J. • ? • USA • Serpentes

Millet de la Turtaudière, Pierre Aimé • 1784–1874 • France • Anura, Serpentes
Millot, Jacques-J. • 9 July 1897 • France • Anura, (Sauria)
Milne-Edwards, Henri • 23 October 1800–1885 July 29 • Belgium / France • Sauria, [Serpentes]
Milstead, William Wright • 12 June 1928–1974 November 29 • USA • Anura, *testudines
Minton, Sherman Anthony, Jr. • 24 February 1919 • USA • Sauria, Serpentes
Minuth, Walter • ? • Germany • Sauria
Miranda, Marta E. • ? • Argentina • Serpentes
Miranda Ribeiro, Alípio do • 21 February 1874–1939 January 8 • Brazil • Anura, Caudata
Mishra, Vijay Prakash • 1942 • India • *crocodilia
Misuri, Alfredo • 1886–? • Italy • *testudines
Mitchell, A. J. L. • ? • South Africa • Sauria
Mitchell, B. L. • ? • Nyasaland • (Anura, Sauria)
Mitchell, Francis John • 8 August 1929–1970 February 23 • Australia • Sauria, Serpentes
Mitchell, Robert Wetsel • 25 April 1933 • USA • Caudata
Mitchell, Silas Weir • 15 February 1829–1914 January 4 • USA • (Serpentes)
Mitchill, Samuel Latham • 20 August 1764–1831 September 7 • USA • Caudata
Mitsukuri, Kakichi • 1 December 1858–1909 September 17 • Japan • (Sauria)
Mittermeier, Russell A. • 8 Nov. 1949 • USA • Testud.
Mittleman, Myron Budd • 19 December 1918 • USA • Anura, Caudata, Sauria
Mivart, Saint George Jackson • 30 November 1827–1900 April 1 • England • Anura, Caudata, (Sauria)
Miyata, Kenneth Ichiro • 1951–1983 October 14 • USA • Anura, Sauria, (Serpentes)
Miyazaki, Koji • 4 June 1929 • Japan • Caudata
Młynarski, Marian • 29 January 1926 • Poland • [*caudata], *serpentes, *testudines
Mocquard, François • 27 October 1834–1917 March • France • Anura, Sauria, Serpentes
Modigliani, Elio • 1861–? • Italy • Sauria
Moellendorff: see Möllendorff, O. F. von
Mojsisovics, Felix Georg Herman August (edler von Mojsvár) • 1848–1897 August 27 • Austria • [Serpentes], Testudines
Mole, Richard Richardson • ?–? • (?England) / Trinidad • (Sauria)
Moler, Paul Edmunds • 25 March 1945 • USA • Anura
Molina, Juan Ignacio • 20 July 1740–1829 • Chile / Italy • Sauria
Moll, Edward Owen • 30 Nov. 1939 • USA • Serpentes
Moll, Karl Ehrenbert (freiherr) von • 1760–1838 • (?Austria) • [Anura]
Möllendorff, Otto Franz von • 1848–1903 • Germany • (Anura), Caudata, (Sauria, Serpentes)
Molnar, Ralph E. • 1943 • Australia • *crocodilia
Monard, Albert • 2 September 1886–1952 September 27 • Switzerland • Anura, Sauria
Montague, P. D. • ? • (?England) • Serpentes
Montalenti, Giuseppe • 13 Dec. 1904 • Italy • (Anura)
Montanucci, Richard Roman • 13 May 1944 • USA • Sauria
Montrouzier, (père) • ?–(1860)–? • France • Serpentes

Moodie, Roy Lee • 30 July 1880–1934 February 16 • USA • (Anura)
Moody, Richard Thomas Jones • 3 October 1939 • England • *testudines
Moody, Scott Michael • 13 March 1949 • USA • *sauria
Mook, Charles Craig • 7 May 1887–1966 October 10 • USA • *crocodilia
Moore, George Azro • 12 Sept. 1899 • USA • Caudata
Moore, John Alexander • 27 June 1915 • USA • Anura
Moore, John Edward Salvin • 1873–? • England • (Anura)
Moore, John Percy • 17 May 1869–1965 March 1 • USA • Caudata
Mooser Barendun, Oswald • 23 December 1903–1983 July 14 • Switzerland / Mexico / Switzerland • *testudines
Moraes, A. • ? • Portugal • *crocodilia
Morath: see Stemmler-Morath, C.
Moreau de Jonnès, Alexandre • 19 March 1778–1870 March 28 • France • Sauria
Morelet, Pierre Marie Arthur (chevalier) • 1809–1892 • France • (Anura, Caudata, Crocodilia)
Moreno, Luis V. • ? • Cuba • Anura
Moret, Léon Marie Louis • 4 July 1890 • France • *testudines
Mori, Tamezo • 1884–1962 • Japan / Korea and China • Caudata
Morice, Albert • 1848–1877 • France / Indochina • Serpentes
Morton, Samuel George • 26 January 1799–1851 May 15 • USA • *crocodilia
Mosauer, Walter • 5 May 1905–1937 August 10 • Austria / USA / Mexico • (Caudata), Sauria, (Serpentes)
Mosimann, James Emile • 26 Oct. 1930 • USA • Testud.
Mostert, David P. • ? • South Africa • Sauria
Moszkowski, Max • ? • Germany • (Sauria)
Mou, Yung-ping • ? • Taiwan / France • Anura
Mount, Robert Hughes • 25 December 1931 • USA • Sauria, Testudines
Mouton, P. le F. N. • ? • South Africa • Sauria
Müller, Friedrich (=Fritz) • 8 May 1834–1895 March 10 • Switzerland • Anura, Sauria, Serpentes
Müller, Johannes Peter • 14 July 1801–1858 April 28 • Germany • Anura, Serpentes
Müller, Johann Wilhelm (baron) • 4 March 1824–1866 October 24 • Germany • [Serpentes]
Müller, Lorenz • 18 Feb. 1868–1953 Feb. 1 • Germany • Anura, Caudata, *croc., Sauria, Serpentes, Testudines
Müller, Otto Frederik • 11 March 1730–1784 December 26 • Denmark • Anura
Müller, Paul • 11 October 1940 • Germany • Sauria
Müller, Salomon • 7 April 1804–1864 • Holland / Dutch East Indies • Crocodilia, Serpentes, Testudines
Munsterman, Harold Ernest • ? • USA • Anura, Serpentes
Murphy, James Bernard • 2 November 1939 • USA • Serpentes
Murphy, Robert Cushman • 29 April 1887–? • USA • (Sauria)
Murphy, Robert Ward • 22 March 1948 • USA / Canada • Sauria, Serpentes
Murray, Andrew • 19 February 1812–1878 January 10 • England • Crocodilia
Murray, James A. • 1865–1914 • British India • Anura, Sauria, Serpentes

Murray, Keith F. • ? • USA • Caudata
Murray, Leo Tildon • 4 May 1902–1958 March 2 • USA • (Testudines)
Murthy, T. S. N. • ? • India • (Anura), Sauria
Muskhelishvili (=Mushkelischwili), **Teymuras** A. • ? • USSR • Sauria
Musters, C. J. M. • ? • Holland • Sauria
Myers, Charles William • 4 March 1936 • USA • Anura, Caudata, Sauria, Serpentes
Myers, George Sprague • 2 February 1905–1985 November 4 • USA • Anura, Caudata, Sauria

- N -

Náder, Rumio • ? • Argentina • Serpentes
Nagai, Kamehiko • ? • Japan • Serpentes
Nakamura, Kenji • 23 November 1903–1984 July 30 • Japan • Sauria
Namiye, Motokichi • 15 February 1854–1918 May 24 • Japan • (Anura), Sauria
Nankinov, D • ? • (?USSR) • *testudines
Narayan Rao, C. R. • ?–? • India • Anura, Serpentes
Nardo, Giovanni Domenico • 4 March 1802–1877 April 7 • Italy • Testudines
Narmandakh, P. • ? • USSR • *testudines
Nascetti, G. • ? • Italy • Anura, Caudata
Nascimento, Francisco Paiva do • ? • Brazil • Sauria, Serpentes
Natsuno, Tetsuya • ? • Japan • Serpentes
Natterer, Johann • 9 November 1787–1843 June 17 • Austria • (Anura), Crocodilia, (Sauria, Serpentes)
Navarro Barón, José • 5 May 1946 • Chile • Anura
Navás, Longinos • 1858–1938 • Spain • *caudata
Naylor, Bruce Gordon • 19 August 1950 • Canada • *caudata
Necker, Walter Ludwig • 27 December 1913–1979 December 17 • Germany / USA • (Anura, Sauria)
Neill, Wilfred T., Jr. • ?–? • USA • Anura, Caudata, Serpentes
Nelson, Craig Eugene • 21 May 1940 • USA • Anura
Nelson, Edward William • 8 May 1855–1934 May 19 • USA • (Anura, Sauria, Serpentes, Testudines)
Němec, J. • ? • Czechoslovakia • *anura
Nesov, L. A. • ? • USSR • *anura, *sauria, *testudines
Nesterov, P. V. • ? • Russia • Caudata
Netting, Morris Graham • 3 October 1904 • USA • Anura, Caudata, Serpentes
Neumann, Oscar Rudolph • 3 September 1867–1946 May 17 • Germany / USA • Sauria
Nevill, Hugh • 1848–1897 • Ceylon • Sauria
Nevo, Eviatar • 2 February 1929 • Israel • *anura
Newman, Walter B. • ? • USA • Caudata
Nguyen, Van Sang • ? • Vietnam • Sauria
Nicéforo María, (hermano, *formerly* Antoine Rouaire Siauzade) • 29 February 1888–1980 February 24 • France / Colombia • (Anura, Caudata, Gymnophiona, Sauria), Serpentes
Nicholls, George Edward • 1878–? • England / Australia • (Sauria)
Nicholls, Jesse C., Jr. • ? • USA • Caudata
Nicholls, Lucius • 22 January 1885–1969 September 3 • England / Ceylon / Singapore • Sauria, Serpentes

Nichols, John Treadwell • 11 June 1883–1958 November 10 • USA • (Sauria, Serpentes)
Nieden, Fritz • 1883–? • Germany • Anura, Gymnophiona, Sauria
Nielsen, Eigil Hans Aage • 1910 • Denmark • *testudines
Nikoloff, Ivan • ? • Bulgaria • *crocodilia
Nikolsky, Aleksander Mikhailovich • "18 February" (=3 March) 1858–1942 December 8 • Russia • Anura, Caudata, Sauria, Serpentes, Testudines
Nilson, Göran (=Goeran) • ? • Sweden • Anura, Sauria, Serpentes
Nilsson, Sven • 8 March 1787–1883 November 30 • Sweden • Anura
Ninni, Alessandro Pericle (conte) • 1837–1892 • Italy • Anura, Serpentes
Nitsche, Heinrich • 1845–1902 • Germany • (Serpentes)
Nixon, Charles William • 28 September 1925 • USA • Serpentes
Noble, Gladwyn Kingsley (*husband of* R. Noble) • 20 September 1894–1940 December 9 • USA • *Anura, *Caudata, (Gymnophiona), Sauria
Noble, Ruth Crosby (*wife of* G. K. Noble) • 15 March 1897 • USA • (Anura)
Noll, Friedrich Carl • 22 September 1832–1893 January 14 • Germany • (Sauria)
Nodder, Frederick Polydore • (*about* 1773)–1800 • England • Anura, Caudata, Serpentes
Nopsca, Ferencz (báró) • 3 May 1877–1933 April 25 • Romania / Austria • *crocodilia, *sauria, *testudines
Nordmann, Alexander Danilevsky von • 1805–1866 • Russia • Sauria, Serpentes
Norris, Kenneth Stafford • 11 August 1924 • USA • Sauria, Serpentes
Nöth, Lothar • ? • Germany • *sauria
Noulet, Jean Baptiste • 1802–1890 • France • *crocodilia, *testudines
Novo Rodríguez, Julio • ? • Cuba • Anura
Núñez Cepeda, Herman Antonio • 6 September 1953 • Chile • Sauria
Nussbaum, Ronald Archie • 9 February 1942 • USA • Caudata, Gymnophiona

- O -

Oates, Frank (=Francis) • 6 April 1840–1875 February 5 • England / South Africa • (Anura, [Serpentes])
Obst, Fritz Jürgen • 3 April 1939 • Germany • Caudata, Sauria
Ochoterena, Isaac • 1885–1950 • Mexico • (Sauria)
Oelrich, Thomas Mann • 27 May 1924 • USA • *testud.
Oeser, Richard • 6 July 1891–1974 July 4 • Germany • (Anura)
Oftedal, Olav T. • ? • USA • Sauria
Ogilby, James Douglas • 16 February 1853–1925 August 11 • Ireland / USA / Australia • Anura, Sauria, Serpentes, Testudines
Ogren, Larry H. • ? • USA • Sauria, Serpentes
Okada, Yaichirō • 24 June 1892–1976 April 28 • Japan • *Anura, Sauria, Serpentes
Oken (=Ockenfuss), **Lorenz** • 1 August 1779–1851 August 11 • Germany • Anura, Sauria, Serpentes
Okofuji, G. • ? • Japan • *testudines

Oldham, Robert S. • ? • (?England) / Nigeria / England • Anura
Oliver, James Arthur • 1 January 1914 • USA • Sauria, Serpentes
Oliver, Walter Reginald Brook • 7 September 1883–1957 May 16 • Tasmania / New Zealand • (Sauria)
Olivier, Guillaume-Antoine • 19 January 1756–1814 October 1 • France • Sauria, Serpentes
Olson, Everett Claire • 6 November 1910 • USA • *sauria
Olson, Rupert Earl • 30 June 1934 • USA / Haiti • Sauria
Onderstall, David • ? • South Africa • Sauria
O'Neill, F. Michael • 25 August 1940 • USA • *crocodilia
Oppel, Nikolaus Michael • 7 December 1782–1820 February 16 • Germany / France / Germany • Crocodilia, Sauria, Serpentes
Orbigny, Alcide Charles Victor Dessalines de (*brother of* A. d'Orbigny) • 1806–1876 • France • [Serpentes]
Orbigny, Alcide Dessalines de (*brother of* C. d'Orbigny) • 9 September 1802–1857 June 30 • France • (Anura), Sauria, (Serpentes, Testudines)
Orcés Villagomez, Gustavo • ? • Ecuador • (Anura), Sauria, (Serpentes)
Orcutt, Charles Russell • 1864–? • USA • (Sauria, Serpentes)
Orejas Miranda, Braulio R. • 19 February 1933–1985 May 8 • Uruguay • (Anura), Serpentes
Orlov, Nikolai L. • ? • USSR • Serpentes
Ortenburger, Arthur Irving • 8 June 1898 • USA • Serpentes
Ortiz Zapata, Juan Carlos • ? • Chile • Sauria
Orton, Grace Louise • 13 March 1916 • USA • (Serpentes)
Osbeck, Pehr (=Peter) • 9 May 1723–1805 December 23 • Sweden • Anura
Osborn, Henry Fairfield • 8 August 1857–1935 November 6 • USA • (Crocodilia), *crocodilia, (Serpentes)
O'Shaughnessy, Arthur William Edgar • 14 March 1844–1881 January 30 • England • Sauria
Oshima, Masamitsu • 21 June 1884–1965 June 26 • Japan • Sauria, Serpentes
Osman Hill, W. C. • ? • England • Anura
Ota, Hidetoshi • 25 September 1959 • Japan • Sauria
Otsu, T. • ? • (?Japan) / Taiwan • Anura
Otth, Carl Adolf (=Adolphe) • 2 April 1803–1839 May 16 • Switzerland • Anura
Ottley, John Randall • 10 May 1949 • USA • Sauria, Serpentes
Ouboter, Paul E. • ? • Holland • Sauria
Oudart (=Oudard), Paul-Louis • 1796–? • France • Sauria
Oudemans, Johannes Theodorus • 1862–1934 • Holland • Sauria, Testudines
Ouwens, P. A. (major) • ?–? • Holland / East Indies • (Anura), Sauria, Testudines
Owen, Richard (sir) • 20 July 1804–1892 December 18 • England • *crocodilia, Rhynchocephalia, (Sauria), *sauria, (Serpentes), *serpentes, *Testudines
Owens, David W. • ? • USA • Testudines
Özeti, Neclâ • 22 December 1932 • Turkey • Anura

- P -

Pacheco, Joviano A. A. • ? • Brazil • *testudines
Pack, Lloyd E., Jr. • ? • USA • Sauria
Packard, Alpheus Spring, Jr. • 19 February 1839–1905 February 14 • USA • Caudata
Packard, Earl Leroy • 10 April 1885–1983 January 20 • USA • *testudines
Pakenham, R. H. W. • ? • (?England / Zanzibar) • (Anura, Sauria)
Palacios Arribas, Fernando • ? • Spain • Sauria
Palisot, Ambroise-Marie-François-Joseph (baron de Beauvois) • 27 July 1752–1820 January 21 • France • Serpentes
Pallas, Peter Simon • 22 September 1741–1811 September 8 • Germany / Russia / Germany • Anura, Sauria, Serpentes, Testudines
Palmer, William M. • ? • USA • Sauria
Pan, Junhua • 11 November 1919 • China • (Testudines)
Paolillo O., Alfredo • ? • Venezuela • Testudines
Papenfuss, Theodore Johnstone • 22 July 1941 • USA • Anura, Caudata, Sauria
Parenzan, Pietro • ? • Italy • Testudines
Parker, Fred (=Frederick Stanley) • 18 December 1941 • Australia • Anura, Sauria, (Testudines)
Parker, Hampton Wildman • 5 July 1897–1968 September 2 • England • *Anura, Gymnophiona, Sauria, Serpentes
Parry, C. R. • ? • South Africa • Anura
Parsons, James • March 1705–1770 April 4 • England • (Sauria)
Pasquier: *see* Du Pasquier, L.
Passmore, Neville Ian • 11 September 1947 • South Africa • Anura
Pasteur, Georges • ? • France • Anura, Sauria
Patrizi, F. • ? • Italy • (Sauria)
Pattabiraman, R. • ? • India • Anura
Patterson, Bryan • 10 March 1909–1979 December 1 • England / USA • *crocodilia
Paulian, Renaud • ? • Madagascar • Sauria
Paulson, Dennis Roy • 29 Nov. 1937 • USA • Sauria
Pavlov (=Pavloff), P. A. • ? • USSR / China • Caudata
Peabody, Frank Elmer • 28 August 1914–1958 June 27 • USA • *caudata
Peale, Titian Ramsay • 17 November 1799–1885 March 13 • USA • Sauria
Peers, B. • ? • South Africa • (Sauria)
Péfaur, Jaime E. • 1944 • Venezuela • Anura, Sauria
Pehani, Hubert • 1900 • Yugoslavia • Caudata
Peixoto, Oswaldo Luiz • ? • Brazil • Anura
Pellegrin, Jacques • 1873–1944 • France • Sauria, Serpent.
Penna Varela, Mario • 1 December 1951 • Chile • Anura
Peracca, Mario Giacinto (conte) • 21 Nov. 1861–1923 May 23 • Italy • Anura, Caudata, Sauria, Serpentes
Pérez-Higareda, Gonzalo • 3 November 1932 • Mexico • Sauria, Serpentes
Pérez Mellado, Valentín • ? • Spain • Sauria
Peris, S. V. • ? • Spain • Anura
Perkins, Clarence Basil • 11 November 1888–1955 April 28 • USA • (Serpentes)
Pernetta, John C. • ? • Fiji / Papua New Guinea • Sauria
Péron, François • 22 August 1775–1810 December 14 • France • Anura, Sauria, (Serpentes)
Perret, Jean-Luc • ? • Switzerland • Anura, Sauria, Serpent.
Persson, Per Ove • ? • Sweden • *crocodilia
Peters, Günther • ? • Germany • (Anura), Sauria, Serpentes

Peters, James Arthur • 13 July 1922–1972 December 18 • USA • Anura, Sauria, *Serpentes, (Testudines)
Peters, Karl Ferdinand • 13 August 1825–1881 November 7 • Germany • *testudines
Peters, Wilhelm Karl Hartwig (=Guilielmus Carl Hartwig) • 22 April 1815–1883 April 20 • Germany • Anura, Caudata, Gymnophiona, Sauria, Serpentes, Testudines
Peterson, Harold W. • ? • USA • Caudata
Peterson, Magnus • ? • Australia • (Sauria)
Petit, Gabriel • 21 February 1870–? • France • (Sauria)
Petranka, James Walker • 31 Dec. 1951 • USA • Caudata
Pfeffer, Georg Johann • 1854–1931 • Germany • Anura, Sauria, Serpentes
Phelan, Robert L. • ? • USA • Sauria
Philippi, Rudolph Amandus • 14 September 1808–1904 July 23 • Germany / Chile • Anura, Sauria, Serpentes, Testudines
Phipson, H. M. • ?–? • British India • (Serpentes)
Pianka, Eric Rodger (*husband of* H. D. Pianka, 1965–1980) • 23 January 1939 • USA • (Sauria)
Pianka, H. D.: *see* Dunlap, H. L.
Piatt, Jean Barnett • 22 September 1909 • USA • Sauria
Picado Twight, Clodomiro • 17 January 1887–1944 May 16 • Costa Rica • (Anura, Caudata, Serpentes)
Picard, J.-J. • ? • Belgium • Anura
Pickersgill, Martin • ? • South Africa • Anura
Pictet de la Rive, François Jules • 27 Sept. 1809–1872 May 15 • Switzerland • (*sauria, Serpentes), *testudines
Pidoplichko, Ivan Grigorevich • 1905 • USSR • *testudines
Pienaar, Uys de Villiers • ? • South Africa • Sauria
Pieper, Harald • ? • Germany • Caudata
Pifano Capdeville, Félix • 1912 • Venezuela • (Serpentes)
Pilgrim, Guy Ellcock • 1875–1943 September 15 • Barbados / England / British India • *crocodilia
Pillai, R. S. • ? • India • Anura
Ping, Chih • 6 April 1886–1965 February 21 • China • (Anura), *testudines
Pintak, Thomas • ? • Germany • Anura
Pisanetz, Eh. M. • ? • USSR • Anura
Pitman, Charles Robert Senhouse (captain) • 19 March 1890–1975 September 18 • India / Kenya / Uganda / England • Serpentes
Piton, Louis E. • ? • France • *crocodilia
Piveteau, Jean • 23 September 1899 • France • *sauria
Pizarro, João Joaquim • ?–(1876)–? • Brazil • Anura
Plancy: *see* Collin de Plancy, V. É. M. J.
Plane, Michael • 1933 • Australia • *serpentes
Platz, James Ernest • 4 September 1943 • USA • Anura
Poche, Franz • ?–1945 • Austria • Serpentes
Poey y Aloy, Felipe • 1799–1891 • Cuba • [Testudines]
Poggesi, Marta • ? • Italy • Sauria
Poglayen-Neuwall, Ivo • ? • USA • Sauria
Poiret, Jean-Louis-Marie • 1755–1834 April 7 • France • (Caudata)
Pölz, Friedrich • 20 September 1894–1973 June 19 • Germany • (Caudata)
Pombal, José P., Jr. • ? • Brazil • Anura
Pomel, Nicolas Auguste • 1821–1898 • France / Algeria • *caudata, *crocodilia, *sauria, *serpentes, *testudines

Pope, Clifford Hillhouse (*husband of* S. H. Pope) • 11 April 1899–1974 June 3 • USA / China / USA • Anura, Caudata, Sauria, Serpentes, Testudines
Pope, Sarah Haydock Davis (*wife of* C. H. Pope) • ? • USA • Caudata, Serpentes
Porras, Louis William • 21 January 1948 • Costa Rica / USA • Anura
Porter, Ann P. • ? • USA • Sauria, Serpentes
Porter, Kenneth Raymond • 20 Oct. 1931 • USA • Anura
Portis, Alessandro • 1853–? • Italy • *anura, *serpentes, *testudines
Posada Arango, Andrés • 11 February 1839–1923 March 13 • Colombia • Caudata, Serpentes
Positano-Spada, Domenico • ?–(1892)–? • Italy • Sauria
Potter, Floyd E, Jr. • 6 December 1925–1987 July 9 • USA • Caudata, Serpentes
Power, John Hyacinth • ? • South Africa • Anura, Sauria
Powers, Arnold L. • ? • USA • Serpentes
Poynton, John Charles • ? • South Africa • Anura
Prado, Alcides • ? • Brazil • Serpentes
Prangner, Pater Engelbert • 12 September 1812–? • Austria • *crocodilia
Prasad, K. N. • ? • India • *testudines
Prater, Stanley Henry • 12 March 1890–1960 October 12 • British India / England • Serpentes
Prato, Alberto del • ? • Italy • Sauria
Pražák, J. Prokop • ?–(1898)–? • Czechoslovakia • Caudata
Pregill, Gregory Kent • 27 Nov. 1946 • USA • *sauria
Preston, Robert E. • ? • USA • *testudines
Preud'homme de Borre: *see* Borre, A. P. de
Price, Llewellyn Ivor • 9 Oct. 1905–1980 Mar. 14 • Brazil / USA / Brazil • (*gymnophiona), *sauria, *testudines
Price, Robert • 14 June 1951 • USA • Serpentes
Prigioni, Carlos Ma. • ? • Uruguay • Anura
Pringle, John Adams • 31 October 1910 • South Africa • Serpentes
Pritchard, Peter Charles Howard • 26 June 1943 • England / USA • Testudines
Procter, Joan Beauchamp • 5 Aug. 1897–1931 September 20 • England • Anura, Sauria, Serpentes, (Testudines)
Provancher, Léon Abel (l'abbé) • 10 March 1820–1892 March 23 • Canada • Caudata
Przhevalsky (=Prževalski), Nikolai Mikhailovich (general) • 1839–1888 "October 20" (=November 1) • Russia • [(Anura, Sauria), Serpentes]
Purschke, Carl Arthur • ?–(1885)–? • Austria • *testudines
Putnam, Frederick Ward • 16 April 1839–1915 August 14 • USA • (Serpentes, *testudines)
Pyburn, William (=Billy) Frank • 22 March 1927 • USA • Anura, (Serpentes)
Pyles, Rebecca A. • 22 July 1954 • USA • Anura

- Q -

Qu, Wenyuan • 4 September 1936 • China • Caudata
Queiroz: *see* de Queiroz, K.
Quelch, John Joseph • 1854–? • (?England) / British Guiana • (Anura)
Quinn, Hugh R. • 1 October 1946 • USA • Serpentes
Quoy, Jean-René-Constant • 10 November 1790–1869 July 4 • France • Sauria

- R -

Rabb, George Bernard • 2 January 1930 • USA • Caudata, Sauria, Serpentes
Rabor, Dioscoro S. • 18 May 1911 • Philippines • Sauria
Radcliffe, Charles W. • ? • USA • Serpentes
Raddi, Giuseppe • 1770–1829 • Italy • Anura
Radovanović, Milutin • 1900 • Yugoslavia • Caudata, Sauria
Raffles, Thomas Stamford (sir) • 5 July 1781–1826 July 5 • England / Malaya / Java / Sumatra / Singapore / England • Serpentes
Rafinesque Schmaltz, Constantine Samuel • 22 October 1783–1840 September 18 • Turkey / France / Italy / USA • Anura, Caudata, Sauria, Serpentes, Testudines
Rage, Jean-Claude • 1 March 1943 • France • *serpentes
Ramanantsoa, G. A. • ? • Madagascar • Sauria
Ramanna, B. S. • ?–? • India • Anura
Ramón de la Sagra: see Sagra, R. de la
Ramos Costa, Ana Maria M. • ? • Brazil • Sauria
Ramsay, Edward Pierson • 1842–1916 December 16 • Australia • Anura, Sauria, Serpentes, Testudines
Ramsden, Charles Theodore • 11 February 1876–1951 August 24 • Cuba • Sauria
Ramsey, Louis W. • ? • USA • Serpentes, Testudines
Rand, Austin Stanley • 19 September 1932 • USA • Caudata, Sauria, Serpentes
Rankin, Peter Robert • ?–? • Australia • Sauria
Rao: see Narayan Rao, C. R.
Rapp, Wilhelm Ludwig von • 3 June 1794–1868 November 11 • Germany • Anura, (Serpentes)
Rasmussen, Jens Bødtker • ? • Denmark • Serpentes
Rasmussen, Thomas Earl • ? • USA • *rhynchocephalia
Rathke, Martin Heinrich • 25 August 1793–1860 Sept. 3 • Poland / Estonia / Kaliningrad • Caudata, (Sauria)
Rau, Charles C. • ? • USA • Sauria
Raun, Gerald George • 14 July 1932 • USA • (Sauria)
Raw, Lynn R. G. • ? • South Africa • Anura, Sauria, Serpentes
Rawlinson, Peter Alan • ? • Australia • Sauria
Ray, Harish Chandra • 1906 • India • Serpentes
Razumovsky, Grigorii (=Grégoire de Razoumowsky, graf) • ?–1837 • Russia / Switzerland • Anura, Caudata, Sauria, Serpentes
Rebouças Spieker, Regina • ? • Brazil • Sauria
Reddell, James R. • ? • USA • Caudata
Redi, Francesco • 19 February 1626–1667 March 1 • Italy • (Serpentes)
Redkozubov, O. I. • ? • USSR • *testudines
Reese, Robert William • 3 August 1917 • USA • Caudata
Regal, Philip Joe • 2 December 1939 • USA • Caudata
Regenass, Urs • ? • USA • Serpentes
Rehák, Ivan • ? • Czechoslovakia • Serpentes
Reichenbach, Heinrich Gottlieb Ludwig • 8 June 1793–1879 March 17 • Germany • Caudata
Reichenow, Anton • 1 August 1847–1941 July 6 • Germany • Anura, Sauria, Serpentes
Reig, Osvaldo A. • 1929 • Argentina • *Anura
Reinach, Albert (baron) von • 7 November 1832–1905 January 12 • Germany • *testudines
Reinhardt, Johan Christopher Hagemann (*father of* J. T. Reinhardt) • 19 December 1776–1845 October 31 • Norway / Denmark • Serpentes
Reinhardt, Johannes Theodor (*son of* J. C. H. Reinhardt) • 3 December 1816–1882 October 23 • Denmark • Anura, Gymnophiona, Sauria, Serpentes
Reinwardt, Caspar Georg Carl • 3 July 1773–1854 • Holland • (Anura), Serpentes
Remacle, G. • ? • Belgium • (Sauria)
Rendahl, Carl Hialmar • 26 December 1891–1969 May 2 • Sweden • Gymnophiona, Sauria, Serpentes
Renous, Sabine • 2 July 1942 • France • Gymnophiona
Resnikova: see Reznikova, M.
Restrepo T., Jorge H. • ? • Colombia • Serpentes
Retzius, Anders Johan • 3 October 1742–1821 October 6 • Sweden • Caudata
Reuss, Adolph • 1804–1878 • Germany • Anura, Sauria, Serpentes
Reuss, T. (?=F. A. T.) • ? • Germany • Serpentes
Reznikova (=Resnikova), M. • ? • USSR • Sauria
Rhodin, Anders G. J. • ? • USA • Testudines
Riabinin, Anatoly Nikolaevich • ? • USSR • *testudines
Ribeiro: see Mirando Ribeiro, A. do
Rice, Frank Leon • ? • USA • Anura
Richards, Christina Macgregor • 5 August 1929 • USA • Anura
Richardson, Charles Howard, II • 7 July 1887–1977 January 9 • USA • Sauria
Richardson, John (sir) • 5 November 1787–1865 June 5 • Scotland / England • Sauria, (Serpentes)
Richmond, Neil Dwight • 17 November 1912 • USA • (Caudata), Serpentes
Richmond, Rollin Charles • 31 May 1944 • USA • Anura
Rickart, Eric Allan • 7 February 1950 • USA • *sauria
Ridley, Henry Nicholas • 10 December 1855–1956 October 24 • England • (Sauria)
Riemer, William John • 25 January 1924 • USA • Caudata, Serpentes
Rieppel, Olivier • 21 September 1951 • Switzerland • (Sauria), *sauria, *testudines
Risch, Jean-Paul • ? • France / Luxemburg • Anura, Caudata
Risso, Antoine • 8 April 1777–1845 August 25 • France • Anura, Sauria, Serpentes, Testudines
Ristori, Giuseppe • 1856–1905 • Italy • *crocodilia
Ritgen, Ferdinand August Max Franz von • 1787–1867 • Germany • Serpentes, Testudines
Rivero, Juan Arturo • 5 March 1923 • Venezuela / Puerto Rico • Anura, Sauria
Rivero Blanco, Carlos V. • ? • Venezuela • Anura, Sauria
Rivers, James John • 1824–1913 • England / USA • (Sauria), Testudines
Rivière, Émile • 1835–1922 • France • *anura
Robb, Joan • 11 December 1921 • New Zealand • Sauria, Serpentes
Roberts, John Dale • ? • Australia • Anura
Robertson, William Beckwith • 22 August 1924 • USA • Serpentes
Robinson, Douglas Clark • ? • USA / Costa Rica • Caudata

Robinson, Herbert Christopher • 4 November 1874–1929 May 30 • England / British Malaya / England • (Anura, Sauria), [Testudines]
Robinson, Pamela Lamplugh • ? • England • (*rhynch.)
Robison, Wilbur Gerald, Jr. • 27 December 1933 • USA • Sauria
Roček, Zbyněk • 16 August 1945 • Czechoslovakia • *sauria, *serpentes
Rochebrune, Alphonse Amédée Trémeau de • 18 Sept. 1836–1912 April 23 • France • Sauria, *Serpentes
Rodeck, Hugo George • 19 Sept. 1902 • USA • (Sauria)
Rodgers, Thomas Luthan • 4 July 1909 • USA • Sauria
Rodhain, Alphonse-Hubert-Jérôme • 25 January 1876–1956 September 26 • Belgium / Belgian Congo / Belgium • (Serpentes)
Rodrigues, João Barbosa • 1842–1909 • (?Brazil) • *croc.
Rodrigues, Miguel Trefaut Urbano • ? • Brazil • Sauria, Serpentes
Rodrigues da Cunha: see Cunha, O. R. da
Roemer, Karl Ferdinand • 5 January 1818–1891 December 14 • Germany • *serpentes
Roesel von Rosenhof: see Rösel von Rosenhof, A. J.
Roger, Otto • ?–? • Germany • *sauria, *testudines
Rogers, Karel Louise Kokx • 28 June 1947 • USA • *caudata, *serpentes, *testudines
Roig, Virgilio German • ? • Argentina • Anura, Sauria
Román, Benigno • ? • Upper Volta • Serpentes
Romano Hoge, Silvia Alma R. W. de Lemos (wife of A. R. Hoge) • ? • Brazil • Serpentes
Romer, Alfred Sherwood • 28 December 1894–1973 November 5 • USA • (*anura)
Romer, John Dudley • 9 September 1920–1982 March 15 • England / Hong Kong / England • Anura
Römer, M. • ? • Venezuela • Serpentes
Romiti, Massimo • ? • Italy • Sauria
Romoli Sassi, Anna • ? • Italy • Sauria
Rooij, "Nelly" de (=Petronella Johanna, mrs. Breijer-de Rooij) • 30 July 1883–1964 June 10 • Holland • Sauria, Serpentes
Roonwal, Mithan Lal • 18 Sept. 1908 • India • Anura
Roosevelt, Cornelius Van Schaack • 23 October 1915 • USA • Sauria
Rose, Francis Lewis • 20 December 1935 • USA • Caudata, *testudines
Rose, Walter • 1884–1964 • South Africa • Anura, Sauria
Rösel von Rosenhof, August Johann • 30 March 1705–1759 March 27 • Germany • (Anura)
Rosén, Nils Valfrid • 11 January 1882–? • Sweden • Sauria, Serpentes
Rosenberg, Hans • ? • Germany • (Sauria)
Roshchin, V. V. • ? • USSR • Sauria
Rossman, Douglas Athon (husband of N. J. Rossman) • 4 July 1936 • USA • Serpentes
Rossman, Nita Jane Kuster (wife of D. A. Rossman) • 18 March 1936 • USA • (Serpentes)
Rostombekov, V. • ? • USSR • Serpentes
Rothschild, Lionel Walter (second baron) • 8 Feb. 1868–1937 Aug. 27 • England • (Anura), Sauria, *Testudines
Roubieu, Guillaume Joseph • 1757–1834 • France • Sauria
Roule, Louis • 1861–1942 • France • (Serpentes)

Roux, Jean • 5 March 1876–1939 December 1 • Switzerland • Anura, Sauria, Serpentes
Roux-Estève, Rolande • ? • France • (Anura), Sauria, Serpentes
Rovereto, Cayetano (=Gaetano, marchese) • 1870–? • Italy / Argentina / Italy • *crocodilia, *sauria, *testudines
Roxo, Mathias Gonsalves de Oliveira • ? • Brazil • *crocodilia
Roy, Pranjalendu • ? • India • Anura
Roze, Janis Arnold • 31 October 1926 • Latvia / Germany / Venezuela / USA • (Anura), Gymnophiona, Sauria, Serpentes
Rueda Almonacid, José Vicente • ? • Colombia • Anura, Sauria
Ruibal, Rodolfo • 27 October 1927 • USA • Anura, Sauria
Ruiz, I. R. G. • ? • Portugal • Anura
Ruíz Carranza, Pedro Miguel • 3 February 1932 • Colombia • Anura, (Sauria)
Ruiz García, F. • ? • Cuba • Anura
Rumph, William Thomas • 24 March 1947 • USA • Caudata
Rüppell, Eduard Wilhelm Peter Simon • 20 November 1794–1884 December 10 • Germany • Anura, Crocodilia, Sauria, Serpentes, Testudines
Rusconi, Carlos • 2 November 1898 • Argentina • *crocodilia, *sauria
Rusconi, Mauro • 11 November 1776–1849 March 27 • Italy • Caudata
Russell, Anthony Patrick • 10 September 1947 • England / Canada • Sauria
Russell, Patrick • 6 February 1727–1805 July 2 • England / Syria / British India / England • (Sauria), Serpentes
Russell, Richard Westphal • 10 February 1929 • USA / Canada • Caudata
Ruthe, Johann Friedrich • 16 April 1788–1859 August 24 • Germany • Caudata
Ruthven, Alexander Grant • 1 April 1882–1971 January 18 • USA • Anura, Sauria, Serpentes, Caudata
Rzepakovsky, V. T. • ? • USSR • Sauria

- S -

Sá: see de Sá, R. O.
Sabanejev, L. L. • ? • Russia • Serpentes
Sacco, Federico • 5 Feb. 1864–1948 Oct. 4 • Italy • *croc.
Sackett, J. Townsend • ? • USA • Sauria, Serpentes
Sadlier, Ross A. • ? • Australia • Sauria
Sager, Abram • 22 December 1810–1877 March 6 • USA • Caudata
Sagra, Ramón de la • 7 December 1801–1871 May 25 • Cuba • Anura, (Sauria), Serpentes
Sahi, D. N. • ? • India • Sauria
Sahlberg, John Reinhold • 1845–1920 • Finland • Sauria
Sahni, Ashok • 1941 • India • *crocodilia
Saint Girons, Hubert Jean-Marie Jacques • 1926 • France • Serpentes
Sallaberry, Michel • ? • Chile • Anura, Sauria
Salvado y Rotea, Lucas José Rosendo (bishop of Victoria) • 1814–1900 • (?Spain) / Australia • Serpentes
Salvador, Alfredo • ? • Spain • Anura, Gymnoph., Sauria
Salvin, Osbert • 25 February 1835–1898 June 1 • England • (Anura), Serpentes, (Testudines)

Sampelayo, Primitivo Hernández • 1880–? • Spain • *testudines
Sanchíz, Francisco de Borja • 1949 • Spain • *Anura, [*caudata, *serpentes]
Sanders, Ottys E. • 25 April 1903 • USA • Anura
Sandner Montilla, Fernando • ? • Venezuela • Serpentes
Sang: see Nguyen, V. S.
Sanyal, D. P. • ? • India • Serpentes
Sarasin, Fritz (=Carl Friedrich, *cousin of* P. Sarasin) • 3 December 1859–1942 March 23 • Switzerland • Sauria, (Serpentes)
Sarasin, Paul Benedict (*cousin of* F. Sarasin) • 11 Dec. 1856–1929 April 7 • Switzerland • (Serpentes)
Sarkar, A. K. • ? • India • Anura
Sato, Ikio • 22 Nov. 1903–1945 Aug. 11 • Japan • Caudata
Sauer de Avila-Pires: see Avila-Pires, T. C. S. de
Sauter, Hans • 1871–1948 • (?Austria) / Formosa • (Anura, Sauria, Serpentes)
Sauvage, Henri Émile • 22 September 1842–1917 January 3 • France • Anura, Caudata, Sauria, Serpentes
Sauzier, Théodore • ?–(1892)–? • France • *testudines
Savage, Jay Mathers • 26 August 1928 • USA • Anura, (Caudata), Sauria, Serpentes
Savi, Paolo • 1798–1871 • Italy • Caudata
Savigny, Marie-Jules-César le Lorgne de • 5 April 1777–1851 October 5 • France • (Anura, Sauria, [Serpentes])
Saville-Kent: see Kent, W. S.
Savitzky, Alan Howard • 23 June 1950 • USA • Sauria, Serpentes
Sawaya, Paulo • ? • Brazil • Gymnophiona
Saxe, Leroy Hallowell, Jr. • 8 Oct. 1917 • USA • Sauria
Say, Thomas • 27 June 1787–1834 October 10 • USA • Anura, Caudata, Sauria, Serpentes
Sazima, Ivan • ? • Brazil • Anura
Scalabrini, Pedro • 21 December 1848–1916 April 24 • Italy / Argentina • *crocodilia, *sauria
Ščerbak: see Shcherbak, N. N.
Schaefer, N. • ? • South Africa • Serpentes
Schaeffer, Bobb • 27 September 1913 • USA • *anura
Schaffer, Herbert • 14 February 1936 • Austria • *testudines
Schätti, Beat • ? • Switzerland • Serpentes
Schatzinger, Richard • ? • USA • *sauria
Schauensee: see Meyer de Schauensee, R.
Scheben, Leonard • ? • Germany / German Southwest Africa • (Sauria)
Schenkel, Ehrenfried • 1869–1953 • Switzerland • Sauria, Serpentes
Scheuchzer, Johann Jacob • 2 August 1672–1733 June 23 • Switzerland • (*caudata)
Schilthuis, L. • ?–(1889)–? • Holland • Anura
Schinz, Heinrich Rudolf • 30 March 1777–1861 March 8 • Switzerland • Anura, Sauria, Serpentes
Schiøtz (=Schioetz), Arne • ? • Denmark • Anura
Schlegel, Hermann • 10 June 1804–1884 January 17 • Germany / Holland • Anura, Caudata, Crocodilia, Sauria, Serpentes, Testudines
Schleich, Hans Hermann • ? • Germany • *Sauria, *Testudines
Schlosser, Johann Albert • ?–1769 • Holland • Sauria
Schlosser, Max • 5 February 1854–1932 October • Germany • *caudata, *serpentes
Schmacker, Bernhard • (circa 1852)–1896 March 26 • Germany • (Anura, Serpentes, Testudines)
Schmelcher, Doris • ? • Germany • Sauria
Schmidt, Eduard Oscar • 21 February 1823–1886 January 17 • Germany • Anura
Schmidt, Franklin James White (*brother of* K. P. Schmidt) • 25 July 1901–1935 August 8 • USA • (Caudata), Serpentes
Schmidt, Hermann • 3 Nov. 1892 • Germany • *testudines
Schmidt, Karl Patterson (*brother of* F. J. W. Schmidt) • 19 June 1890–1957 September 26 • USA • Anura, Caudata, *Crocodilia, *Sauria, Serpentes, *Testudines
Schmidt, Philipp (=Moses Paul Philipp Friedrich) • 18 December 1800–1873 • Germany • Anura, Serpentes
Schmidtler, Josef Friedrich (*son of* J. J. Schmidtler) • 6 May 1942 • Germany • Anura, Sauria, Serpentes
Schmidtler, Josef Johann (*father of* J. F. Schmidtler) • 15 Aug. 1910–1983 Oct. 24 • Germany • Sauria, Serpentes
Schneider, Bert • ? • Germany • Anura, Serpentes
Schneider, Hans • ? • Germany • Anura
Schneider, Johann Gottlob Theaenus • 18 January 1750–1822 January 13 • Germany • Anura, Caudata, Crocodilia, Sauria, Serpentes, Testudines
Schneider, Karl Camillo • 1867–? • (?Austria) • (Serpent.)
Schnitzer, Eduard (*alias* Emin Pasha) • 28 March 1840–1892 October 20 • Prussia / Ottoman Empire / Equatorial Africa • (Sauria, Serpentes)
Schnurrenberger, Hans • ?–1964 October 6 • Switzerland / Libya / Nepal • (Serpentes)
Schoepff, Ioannis Davidis: see Schöpf, J. D.
Schomburgk, Moritz Richard • 5 October 1811–1891 March 24 • Germany / Australia • [(Anura, Sauria), Serpentes, Testudines]
Schöpf, Johann David • 8 March 1752–1800 September 10 • Germany / USA / Germany • Testudines
Schouteden, Henri • 1881–1972 • Belgium • (Anura, Sauria, Serpentes)
Schrank, Franz von Paula von • 21 August 1747–1835 December 22 • Germany • Anura, Caudata, Sauria
Schreber, Johann Christian Daniel von • 17 January 1739–1810 December 10 • Germany • Anura
Schreiber, Egid • 1836–1913 • Austria • Anura, Caudata, Sauria, Serpentes, Testudines
Schreibers, Carl (=Karl Franz Anton, ritter von) • 15 Aug. 1775–1852 May 21 • Austria • (Caudata, Sauria)
Schreitmüller, Wilhelm • 1870–? • Germany • (Sauria)
Schubotz, Hermann • 1881–? • (?Germany) • (Anura), Sauria
Schulte, Rainer • ? • Germany / Peru • Anura
Schultze, Oskar Maximilian Sigismund • 1859–1920 • Germany • (Testudines)
Schulze, Erwin • 1861–? • Germany • Anura
Schüttler, B. • ? • (?Germany) • Sauria
Schüz, Ernst • 24 October 1901 • Germany • Sauria
Schwartz, Albert • 13 September 1923 • USA • Anura, Caudata, Sauria, Serpentes
Schweigger, August Friedrich • 8 September 1783–1821 June 28 • Germany • Testudines
Schweizer, Hans • 11 September 1891–1975 July 20 • Switzerland • (Sauria, Serpentes)
Schweizerbarth-Roth, Elise Melitta von • ?–? • Germany • Caudata

Schwenk, Kurt • ? • USA • Sauria
Sclater, William Lutley • 23 September 1863–1944 July • South Africa / England • Anura, Sauria, Serpentes
Scolaro, José Alejandro • 9 Jan. 1946 • Argentina • Sauria
Scopoli, Giovanni Antonio • 13 June 1723–1788 May 8 • Italy • (Serpentes)
Scortecci, Giuseppe • 2 November 1898–1973 October 18 • Italy • Anura, Sauria, Serpentes
Scott, E. O. G. • ? • Australia • Anura
Scott, William Berryman • 12 February 1858–1947 March 29 • USA • *crocodilia
Scudday, James Franklin • 16 Sept. 1929 • USA • Sauria
Seba, Albert (=Albertus) • "2" (=13) May 1665–1736 May "2" (=13) • Holland • (Sauria, Serpentes, Testudines)
Seeley, Harry Govier • 18 February 1839–1909 January 8 • England • *crocodilia, *testudines
Seeliger, Lillian M. • ? • USA • Testudines
Seidlitz, Wilfried von • 1880–? • Germany • *crocodilia
Seiffert, Jürgen • ? • Germany • *anura
Seigel, Ricahrd A. • ? • USA • Sauria
Seliškar, Albin • 1896 • Yugoslavia • Caudata
Sellards, Elias Howard • 2 May 1875–1961 February 4 • USA • *crocodilia, *testudines
Semenov, D. V. • ? • USSR • Sauria
Semon, Richard Wolfgang • 1859–1918 • Germany • [Anura, Sauria, Testudines]
Semper, Carl Gottfried • 1832–1893 • Germany • (Serpentes)
Senanayake, F. Ranil • ? • Ceylon • Anura
Sentzen, U. I. • ?–(1796)–? • (?Germany) • Serpentes
Seoane y Pardo-Montenegro, Victor López • 1834–1900 • Spain • Anura, Caudata, Sauria, Serpentes
Sequeira, Eduardo Enrique Vieira Coelho de • 1861–? • Portugal • (Caudata)
Serié, Pedro • ? • Argentina • Serpentes
Serna D., Marco Antonio (hermano) • 11 July 1936 • Colombia • Anura, Sauria
Serventy, Dominic Louis • 28 March 1904 • Australia • (Sauria)
Seshachar, B. R. • ?–? • India • Gymnophiona
Sever, David Michael • 21 February 1948 • USA • Caudata
Severtzov, Nikolai Alekseevich • 1827–1885 • Russia • Sauria
Shaffer, Howard Bradley • 26 October 1953 • USA • Anura
Shannon, Frederick Albert • 4 May 1921–1965 August 31 • USA • Anura, Caudata, Sauria, Serpentes
Sharma, Ramesh Chandra • ? • India • Sauria
Shaw, Charles Edward • 31 July 1918–1971 May 30 • USA • Sauria
Shaw, George • 10 Dec. 1751–1813 July 22 • England • Anura, Caudata, Gymnoph., Sauria, Serpent., *Testud.
Shcherbak (=Ščerbak, =Szczerbak), Nikolai Nikolaevich • 31 October 1927 • USSR • Anura, Sauria
Shea, Glenn Michael • 5 February 1961 • Australia • Anura, Sauria
Shelford, Robert Walter Campbell • 1872–1912 • British Malaya • Serpentes
Shen, Youhui • 16 January 1932 • China • Anura
Shenbrot, G. I. • ? • USSR • Sauria
Sheplan, Bruce R. • ? • USA • Serpentes
Sheppard, Revett • ?–(1804)–? • England • Sauria, Serpentes

Sherbrooke, Wade C. • ? • USA • Serpentes
Sherman, Franklin, Jr. • 2 November 1877–1947 June 23 • USA • (Caudata)
Shikama, Tokio • 1912 • Japan • *testudines
Shipachev, V. G. • ? • Russia • Anura
Shreve, Benjamin • 1908 • USA • Anura, Sauria, Serpent.
Shufeldt, Robert Wilson (major) • 1 December 1850–1934 January 21 • USA • Testudines
Shuvalov, V. F. • ? • USSR • *testudines
Siebenrock, Friedrich • 20 January 1853–1925 January 28 • Austria • (Sauria), Testudines
Siebold, Carl Theodor Ernst von (*cousin of* P. F. von Siebold) • 16 February 1804–1885 April 7 • Germany • (Serpentes)
Siebold, Philipp Franz Jonkheer von (*cousin of* C. T. E. von Siebold) • 17 February 1796–1866 October 18 • Germany / Holland / Japan / Germany • [Anura, (Caudata)], (Serpentes), [Testudines]
Sierra de Soriano, Blanca • ? • Uruguay • Sauria
Sill, William Dudley • ? • USA • *crocodilia
Silva Coutinho, João Martino (major) da • ?–1889 • Brazil • Testudines
Silverstone, Philip A. • ? • USA • Anura
Simmons, John Edward • 15 April 1951 • USA • Anura
Simmons, Robert Stanley • 30 May 1924–1985 September 28 • USA • (Sauria), Serpentes
Simon, Martin Patrick • 15 October 1951 • USA • Sauria
Simpson, George Gaylord • 16 June 1902 • USA • *crocodilia, *rhynchocephalia, *serpentes, *testudines
Singh, G. • ? • India • *crocodilia
Sinitsin, Demetrius Theodorovich • 23 February 1871–? • Russia / USA • Sauria
Sites, Jack Walter, Jr. • 6 August 1951 • USA • Sauria
Sjöstedt, Bror Yngve • 3 August 1866–1948 • Sweden • (Anura), Sauria, Serpentes
Skilton, Avery Judd • 1802–1858 • USA • Caudata, Sauria
Skinner, Morris Fredrick • 14 September 1906 • USA • (*sauria, *serpentes), [*testudines]
Skuk, Gabriel • ? • Uruguay • Sauria
Slack, John Hamilton • ?–1874 Aug. 27 • USA • [Sauria]
Slater, James Alexander • 10 January 1920 • USA • Sauria
Slater, James Rodenburg • 11 June 1890 • USA • Anura, Caudata
Slater, Kenneth R. • ? • Australia • (Anura), Serpentes
Slavens, Frank Leo • 19 August 1947 • USA • (Serpentes)
Slevin, Joseph Richard • 13 September 1881–1957 February 15 • USA • Sauria, Serpentes
Slipp, John Wesley • 22 July 1917 • USA • Caudata
Sloane, Hans (sir) • 16 April 1660–1753 January 11 • England • (Sauria)
Smedley, Norman • ? • Singapore • Anura, Serpentes
Smets, Gérard (l'abbé) • 1857–? • Belgium • *testudines
Smith, Andrew (sir) • 3 December 1797–1872 August 11 • Scotland / South Africa / England • Anura, Sauria, Serpentes, Testudines
Smith, Charles Clinton • 30 April 1910–1966 • USA • (Anura)
Smith, Hobart Muir (*husband of* R. B. Smith) • 26 Sept. 1912 • USA • Anura, Caudata, Sauria, Serpent., Testud.
Smith, Hugh McCormick • 21 November 1865–1941 September 28 • USA • (Sauria, Serpentes)

Smith, James Edward (sir) • 2 December 1759–1828 March 17 • England • [Caudata]
Smith, John Alexander • 1818–1883 • Scotland • Sauria
Smith, Lawrence Alec • 1944 • Australia • Serpentes
Smith, Malcolm Arthur • 1875–1958 July 22 • England / Siam / England • Anura, Caudata, Sauria, Serpentes, Testudines
Smith, Meredith J. • ? • Australia • *serpentes
Smith, Philip Wayne • 2 December 1921–1986 October 11 • USA • Anura, Sauria, Serpentes
Smith, Rozella Pearl Blood (*wife of* H. M. Smith) • 18 May 1911–1987 December 15 • USA • (Anura), Sauria, Serpentes
Smith, Tarleton Friend • ? • USA • Anura, Serpentes
Snyder, David Hilton • 24 June 1938 • USA • Anura
Sobolevsky, N. I. • ? • USSR • Sauria
Sochurek, Erich • 26 April 1923–1987 November 21 • Austria • Caudata, Sauria, Serpentes
Sofianidou, Theodora • ? • Greece • Anura
Solano, Haydeé • ? • Venezuela • Anura, Gymnophiona
Solórzano, Alejandro • ? • Costa Rica • Serpentes
Soman, P. W. • ? • India • Anura
Song, Mingtao • 3 July 1937 • China • Sauria, Serpentes, Testudines
Sonnini de Manoncourt, Charles Nicolas Sigisbert • 1 February 1751–1812 May 29 • France • Anura, Sauria, Serpentes, [Testudines]
Sordelli, Ferdinando • 12 December 1837–1916 January 17 • Italy • Serpentes
Souleyet, Louis François Auguste • 1811–1852 • France • Serpentes
Sowerby, Arthur de Carle • 8 July 1885–1954 August 16 • China • (Anura), Caudata, (Sauria)
Spada: *see* Positano-Spada, D.
Spanò, S. • ? • Italy • Sauria
Sparrman, Anders • 27 February 1748–1820 August 9 • Sweden • Sauria, Serpentes
Spawls, Stephen • 1953 • England • Serpentes
Speir, Francis, Jr. • 1856–? • USA • *crocodilia
Spence, J. M. • ? • South Africa • Sauria
Spencer, Walter Baldwin (sir) • 23 June 1860–1929 July 14 • Australia • Anura
Spengel, Johann Wilhelm • ?–? • Germany • (Anura)
Spieler, Richard Arno • 8 April 1932 • USA • Sauria
Špinar, Zdeněk V. • 4 April 1916 • Czechoslovakia • *anura, (*sauria)
Spirandelli Cruz, E. F. • ? • Brazil • Anura
Spiteri, David Emmanuel • 2 February 1946 • Malta / USA • Serpentes
Spix, Johann Baptist von • 9 February 1781–1826 May 15 • Germany • Anura, Croc., Sauria, ([Serpentes]), Testud.
Springer, Stewart • 5 June 1906 • USA • Sauria
Spurrell, Herbert George Flaxman • 1877–1918 • England • (Anura)
Stadelman, Raymond E. • 27 August 1907 • USA • (Anura, Serpentes)
Stadie, Christian • ? • Germany • Anura
Staesche, Karl • 24 August 1902 • Germany / Brazil • *testudines
Stanley, Arthur • ? • (?England) / China • Serpentes
Stansbury, Howard • 8 February 1806–1863 April 17 • USA • [(Sauria), Serpentes]

Starrett, Priscilla Hollister • 1929 • USA • Anura
Stebbins, Robert Cyril • 31 March 1915 • USA • Caudata, Sauria
Steel, Thomas • 1858–1925 • Australia • [Serpentes]
Steenstrup, Johann Japetus Smith • 8 March 1813–1897 June 20 • Norway / Denmark • Anura
Stefano, Giuseppe de • ?–? • Italy • *caudata, *crocodilia, *sauria, *serpentes, *testudines
Steindachner, Franz • 11 November 1834–1919 December 10 • Austria • Anura, Caudata, Sauria, Serpentes, (Testudines)
Steiner, Hans M. • ? • Austria • Caudata
Steinitz, Heinz • 26 April 1909–1971 April 28 • Germany / Israel • Anura
Stejneger, Leonhard Hess • 30 October 1851–1943 February 28 • Norway / USA • Anura, Caudata, Gymnophiona, Sauria, Serpentes, Testudines
Stemmler-Gyger, Othmar • 30 May 1934 • Switzerland • Serpentes
Stemmler-Morath, Carl • 4 April 1904 • Switzerland • (Serpentes)
Štěpánek, Otakar • ? • Czechoslovakia • Sauria, Serpentes
Stephens, Frank • 2 April 1849–1937 October 5 • USA • (Serpentes)
Sternberg, Charles Hazelius (*father of* C. M. Sternberg) • 15 June 1850–1943 July 20 • USA / Canada • (*croc.)
Sternberg, Charles Mortram (*son of* C. H. Sternberg) • 18 Sept. 1885–1981 Sept. 8 • USA / Canada • *crocodilia
Sternfeld, Richard • 8 February 1884–1943 • Germany • Anura, Sauria, Serpentes
Stevens, Margaret Skeels • ? • USA • *sauria
Stevens, R. Andrew • ? • Malawi / England • Anura, Serpentes
Stewart, Margaret McBride • 6 Feb. 1927 • USA • (Anura)
Steyn, Willem (=William) J. • ?–1968 June 26 • South Africa • Sauria
Stickel, William Henson • 8 November 1912 • USA • (Anura, Sauria), Serpentes
Stimson, Andrew Francis • 7 December 1940 • England • Sauria
Stipanicic, Pedro Nicolas • 6 June 1921 • Argentina • *anura
Stirling, Edward Charles (sir) • 8 September 1848–1919 March 20 • Australia • Sauria, Serpentes
Stocks, Rupert • ? • Papua New Guinea • Anura
Stoddart, ? (colonel) • ?–? • British India • (Sauria)
Stokes, John Lort (admiral) • 1812–1885 June 11 • England • (Sauria, [Serpentes])
Stoliczka, Ferdinand • "28 May" (=7 June) 1838–1874 "June 9" (=June 19) • Austria / British India • Anura, Sauria, Serpentes
Stone, Witmer • 22 September 1866–1939 May 24 • USA • (Sauria)
Storer, David Humphreys • 26 March 1804–1891 September 10 • USA • Anura, Caudata, Serpentes
Storer, Tracy Irwin • 17 August 1889–1973 June 25 • USA • Caudata
Storm, Robert MacLeod • 9 July 1918 • USA • (Caudata)
Storr, Glen Milton • 1921 • Australia • Sauria, Serpentes
Strand, Embrik • 1876–1947 • Germany • Anura, Sauria

Strauch, Alexander (=Aleksander Aleksandrovich) • 1 March 1832–1893 August 14 • Russia • Anura, Caudata, Sauria, Serpentes, Testudines
Straughan, Ian R. • ? • Australia / USA • Anura
Straus, Adolf • 1904 • Germany • (*testudines)
Strecker, John Kern, Jr. • 10 July 1875–1933 January 9 • USA • Anura, (Serpentes)
Stritzke, Rüdiger • ? • Germany • *sauria
Stromer von Reichenbach, Karl Heinrich Ernst Wolfgang (freiherr) • 12 June 1871–? • Germany • *testudines
Stuart, Laurence Cooper • 9 July 1907–1983 May 27 • USA / Guatemala • Anura, Caudata, Sauria, Serpentes
Stucki-Stirn, Martin Christian • 17 August 1935 • Switzerland • Serpentes
Stuhlmann, Franz L. • 29 October 1863–1928 November 19 • Germany • (Serpentes)
Stull, Olive Griffith (=mrs. Loy Erman Davis) • 10 February 1905 • USA • Serpentes
Sturm, Jacob • 21 March 1771–1848 November 28 • Germany • Sauria
Sturn, Ann (=Anna Marie Sturn Skalka) • 2 July 1938 • USA • *testudines
Stutchbury, Samuel • 1797–1859 February 12 • England / Australia • Gymnophiona, Sauria
Su, Chengye • 2 March 1943 • China • Anura, Sauria
Suarez, José Martin • ? • Brazil • *testudines
Suchov, G. F. • ? • USSR • Sauria
Suckow, Georg Adolph • 28 January 1751–1813 May 13 • Germany • Sauria, Serpentes, Testudines
Sukhanov, V. B. • ? • USSR • *testudines
Sulimski, Andrzej • 1926 • Poland • *sauria
Sullivan, Charles Donald • 19 September 1938 • USA • Caudata
Sullivan, Robert Michael • 4 Aug. 1951 • USA • *sauria
Sumichrast, Adrien-Louis-Jean-François • 15 October 1828–1882 September 26 • Switzerland • (Anura, Sauria, Serpentes)
Sun, Ailin • 25 November 1930 • China • *serpentes
Suneja, I. J. • ? • India • *crocodilia
Swain, Tom A. • ? • USA • Sauria
Swainson, William • 8 October 1789–1855 December 7 • England / New Zealand • Sauria, Serpentes
Swan, Lawrence Wesley • 9 March 1922 • USA • ([Anura])
Swinhoe, Robert (consul) • (?1835 or 1836)–1877 • England / China • (Anura, Sauria, Serpentes, Testudines)
Swinton, William Elgin • 30 September 1900 • England / Canada • *croc., *rhynch., *serpentes, *testudines
Switak, Karl H. • 30 June 1938 • Germany / USA • (Sauria)
Swynnerton, Charles Francis Massey • 3 December 1877–1938 June 8 • British East Africa • (Sauria)
Symeonidis, Nikolaus • 1930 • Greece • *testudines
Szalai, Tibor • ? • Hungary • *testudines
Szczerbak: *see* Shcherbak, N. N.
Szeliga-Mierzeyevski, W. • ? • (?Lithuania) • Caudata
Szyndlar, Zbigniew • 3 February 1952 • Poland • *serpentes, [*caudata]

- T -

Taddei, Antonio • ? • Italy • Sauria

Tagliaferro, Napoleone • 1843–? • Malta • *testudines
Tago, Katsuya • 5 May 1877–1943 December 19 • Japan • (Anura), Caudata
Takahashi, Seiichi • ? • Japan • Serpentes
Takara, Tetsuo • 25 June 1913 • Japan • Serpentes
Tamsitt, James Ray • 22 November 1928 • USA / Canada *and* Colombia / USA • (Anura)
Tandy, Jocelyn • ? • USA • Anura
Tandy, Robert Mills • ? • USA • Anura
Tang, Zhenjie • 28 August 1934 • China • Anura
Tang, Ziming • 6 August 1932 • China • Caudata
Tang, Ziyin • 12 February 1927 • China • Caudata
Tanner, Vasco Myron • 29 October 1892 • USA • Sauria, Serpentes
Tanner, Wilmer W. • 17 December 1909 • USA • Anura, Caudata, Sauria, Serpentes
Tanzer, Ernest Claude • ?–1971 January 23 • USA • (Serpentes)
Tao, Hsi-jen • ? • Taiwan • *testudines
Tarashchuk (=Taraščuk), Vladimir Ivanovich • 1922–1987 • USSR • *testudines
Tarlo, Lambert Beverly Halstead • 13 June 1933 • England • *testudines
Tasman, Kenneth Robert (father) • 13 June 1890–1968 May • England / Rhodesia • (Sauria)
Tatarinov, Leonid Petrovich • 1926 • USSR • *anura
Taylor, Edward Harrison • 23 April 1889–1978 June 16 • USA / Philippine Islands / USA • *Anura, *Caudata, Gymnophiona, *Sauria, Serpentes, Testudines
Taylor, W. Edgar • ?–? • USA • (Serpentes), Testudines
Taylor, Walter Penn • 31 October 1888–? • USA • (Sauria)
Tchikvadze: *see* Chkhikvadze, V. M.
Telfair, Charles • (*circa* 1777)–1833 July 14 • Ireland / Mauritius • (Sauria)
Telford, Sam Rountree, Jr. • 25 August 1932 • USA • Serpentes
Temminck, Coenraad Jacob • 31 March 1778–1858 January 30 • Holland • Anura, Caudata, [Crocodilia], (Sauria, Serpentes), Testudines
Tennent, James Emerson (sir) • 7 April 1804–1869 March 6 • Ireland / Ceylon / England • (Sauria)
Tepper, Johann Gottlieb Otto • 1841–1922 • Australia • Sauria
Terentjev, Paul (=Pavel Victorovich) • 23 December 1903–1970 December 30 • USSR • Anura, Sauria
Test, Frederick Cleveland • 14 June 1869–? • USA • Anura
Test, Frederick Harold • 16 April 1912 • USA • Anura
Theobald, William • 1829–1908 • British India • Anura, Crocodilia, Sauria, Serpentes
Thienemann, Friedrich August Ludwig • 22 December 1793–1858 June 24 • Germany • Sauria
Thireau, Michel • ? • France • Sauria
Thomas, Alcide • ?–(1855)–? • France • Anura
Thomas, Bruce William • 6 July 1947 • New Zealand • Sauria
Thomas, John Paul Richard • 2 May 1938 • USA / Puerto Rico • Anura, Caudata, Sauria, Serpentes
Thomas, Robert Allen • 10 April 1946 • USA • Serpentes
Thominot, Alexandre • ?–(1878-1889)–? • France • Anura, Sauria
Thompson, Fred Gilbert • 13 November 1934 • USA • Sauria, Serpentes

Thompson, Helen Beulah: *see* Gaige, H. B. T.
Thompson, Joseph Cheesman • 1874–? • USA • Caudata, Sauria, Serpentes
Thomson, Donald Ferguson • 1901–1970 • Australia • Serpentes
Thon, Theodor • 14 May 1792–1838 November 17 • Germany • Serpentes
Thorn, Robert • 24 August 1925 • Luxemburg • Caudata
Thornton, Wilmot Arnold • 27 Dec. 1921 • USA • Anura
Thunberg, Karl Peter • 11 November 1743–1828 August 8 • Sweden • Testudines
Thurow, Gordon Ray • 13 February 1929 • USA • Caudata
Tian, Wanshu • 15 January 1930 • China • Anura, Caudata
Tiedemann, Franz • ? • Austria • Sauria
Tiedemann, Friedrich • 23 August 1781–1861 January 22 • Germany • (Sauria)
Tihen, Joseph Anton • 20 November 1918 • USA • *anura, *caudata, Sauria, (*serpentes)
Tilak, Raj • 1935 • India • Anura
Tilley, Stephen George • 21 July 1943 • USA • Caudata
Ting: *see* Ding, H.-p.
Tinkle, Donald Ward • 3 December 1930–1980 February 21 • USA • Sauria, Serpentes, Testudines
Tinsley, Richard C. • ? • England • Anura
Tirant, Gilbert • ?–(1885)–? • French Indochina • Serpentes, Testudines
Tiwari, K. K. • ? • India • Sauria, Serpentes
Toft, Catherine Ann • 11 October 1950 • USA • Anura
Toit, Cornelius Albertus du • ? • South Africa • (Anura)
Tomasini, Otto (ritter) von • ?–(1889)–? • Bosnia (*now in* Yugoslavia) • (Caudata), Sauria
Toriba, Michihisa • ? • Japan • Serpentes
Tornier, Gustav • 1859–1938 • Germany • (Anura), Gymnophiona, Sauria, Serpentes, (Testudines)
Torres, Frank • ? • Puerto Rico • Sauria
Toula, Franz • 1845–1920 • Austria • *croc., *testudines
Tourneville, Albert • ?–? • France • Caudata, (Sauria)
Townsend, Charles Haskins • 29 September 1859–1944 January 28 • USA • Sauria
Traill, Thomas Stewart • 29 October 1781–1862 July 30 • England • Serpentes
Trapido, Harold • 10 December 1916 • USA • Anura, Caudata, Serpentes
Trebbau M., Pedro • 20 May 1929 • Germany / Venezuela • (Serpentes), Testudines
Trefaut Rodrigues: *see* Rodrigues, M. T.
Trémeau de Rochebrune: *see* Rochebrune, A. T. de
Treviño Saldaña, Carlos Humberto • ? • Mexico • Serpent.
Trevisan, Pierluigi • ? • Italy • Caudata
Tristram, Henry Baker (reverend) • 11 May 1822–1906 March 8 • England • (Testudines)
Troedsson, Gustaf Timoteus • 9 May 1891–1954 December 12 • Sweden • *crocodilia
Troost, Gerard • 15 March 1776–1850 August 14 • Holland / USA • Serpentes, Testudines
Troschel, Franz Hermann (*or* Herrmann, *but not* Hugo) • 10 October 1810–1882 November 4 • Germany • Anura, Sauria, Serpentes, Testudines
Trouessart, Édouard-Louis • 25 August 1842–1927 June 30 • France • (Caudata)
Troxell, Edward Leffingwell • 15 April 1884–1972 September 21 • USA • *crocodilia

True, Frederick William • 8 July 1858–1914 June 25 • USA • (Anura)
Trueb, Linda (*wife of* W. E. Duellman) • 13 January 1942 • USA • Anura
Ts'ai, Ming-chang • 25 October 1937 • China • Anura
Tsarevski: *see* Zarevsky, S. F.
Tschernov: *see* Chernov, S. A.
Tschudi, Johann Jakob von • 25 July 1818–1889 August (?*or* October) 8 • Switzerland / Austria • Anura, Caudata, Sauria, Serpentes
Tschumakov (=Chumakov), Ivan Sergeevich • ? • USSR • *sauria
Tuckey, James Kingston • August 1776–1816 October 4 • England • [Serpentes]
Tunijev (=Tuniév), B. S. • ? • USSR • Serpentes, Testud.
Tunner, Heinz G. • 24 October 1940 • Austria • Anura
Turbott, Evan Graham • 1914 • New Zealand • Anura, (Sauria)
Tweedie, Michael Willmer Forbes • 1907 • Malaya • (Anura), Serpentes
Twente, John Wesley, Jr. • 18 December 1926 • USA • *sauria, (*testudines)
Twitty, Victor Chandler • 5 November 1901–1967 March 22 • USA • Caudata
Tyler, Michael James • 27 March 1937 • England / Australia • *Anura
Tymowska, Janina • ? • Switzerland • Anura
Tytler, Robert Francis Christopher Alexander (major general) • ?–1916 October 19 • England / British India • Sauria, (Serpentes)

- U -

Uéno, Shun-ichi • 8 December 1930 • Japan • Sauria
Ulasievicz, W. • ? • (?Lithuania) • Caudata
Underwood, Garth Leon • 16 July 1919 • England • Sauria
Unterstein, Walter • ? • Germany • (Anura), Caudata
Urbano, R. F. • ? • USA • Sauria
Urbina B., Medardo • ? • Chile • Sauria
Urich, Frederic William • 1870–1937 • Trinidad • (Anura)
Uthmöller, Wolfgang • ? • Germany • (Sauria), Serpentes
Uzzell, Thomas Marshall, Jr. • 6 April 1932 • USA • Anura, Sauria

- V -

Vaillant, Léon-Louis • 11 November 1834–1914 November 24 • France • (Anura), Caudata, *crocodilia, Sauria, (Serpentes), *Testudines
Valdés de la Osa, A. • ? • Cuba • Anura
Valencia, Jose • ? • Chile • Anura
Valenciennes, Achille • 9 August 1794–1865 April 13 • France • Caudata, Sauria, *Testudines
Valentine, Barry Dean (="B. O.") • 6 June 1924 • USA • Caudata
Valverde Gómez, José A. • ? • Spain • Sauria
van: *see* Beneden, Dam, Hasselt, Kampen, Lidth de Jeude
van Beurden, Eric K. • ? • Australia • Anura
Vancea, Stefan • ? • Romania • Sauria, *testudines
Vandelli, Domenico • 1735–1816 • Italy / Portugal • Sauria, Testudines

Van Denburgh, John • 23 August 1872–1924 October 24 • USA • Anura, Caudata, Sauria, Serpentes, *Testudines
van der Hoeven: *see* Hoeven, J. van der
van Frank, Richard • ? • USA • *caudata
Van Gelder, Richard George • 17 December 1928 • USA • Sauria
Vanni, S. • ? • Italy • Caudata, Serpentes
Vannini, Jay P. • ? • Guatemala • Sauria
Vanzolini, Paulo Emílio • ? • Brazil • (Anura), *Sauria, *Serpentes
Varela J., Manuel • ? • Mexico • Serpentes
Varona, Luis S. • ? • Cuba • *crocodilia, Sauria
Vashetko, Emilia Viktorovna • 1942 • USSR • Sauria
Vasvári, Miklós • ? • Hungary • Sauria
Vaz Ferreira, Raúl • 1918 • Uruguay • Sauria
Vedmederya, Valery Iosiphovich • 1946 • USSR • Sauria, Serpentes
Veith, Georg (hauptmann) • 1875–1925 • Austria • (Sauria, Serpentes)
Velasco, José Maria • 1840–1912 • Mexico • Caudata
Vellard, Jehan Albert • 18 July 1900 • France / Peru / Argentina • Anura, Serpentes
Veloso Martínez, Alberto • 29 October 1940 • Chile • Anura, (Sauria)
Venning, Francis Esmond Wingate (brigadier) • 26 January 1882–1970 August 28 • British India • (Serpentes)
Venzmer, Gerhard • 1 June 1893 • Germany • Serpentes, Testudines
Versilin, N. N. • ? • USSR • *testudines
Vesey FitzGerald, L. D. E. F. • 1910–1974 May 3 • Arabia / East Africa • (Sauria)
Vestergren, Greta • ? • Sweden • Gymnophiona, Serpentes
Vial, James Leslie • 19 December 1924 • USA • Caudata
Vianna, A. • ? • Portugal • *crocodilia
Vicente, Luis A. • ? • Portugal • Sauria
Videla, Fernando • ? • Argentina • Sauria
Vigle, Gregory O. • ? • USA • Anura
Vigny: *see* Loumont-Vigny, C.
Vilaró Diaz, Juan • ?–? • Cuba • Testudines
Villada, Manuel María • 1841–1924 • Mexico • Serpentes
Villa Rivas, Jaime Dolan • 28 March 1944 • Nicaragua / USA • (Sauria), Serpentes
Villiers, André • 9 April 1915–1983 June 8 • France / French West Africa / France • (Anura), Serpentes
Vinciguerra, Decio • 23 May 1856–1934 October 5 • Italy • Sauria, Serpentes
Vinson, Jean-Michel (*son of* J. Vinson) • ? • Mauritius • Sauria
Vinson, Joseph Lucien Jean (*father of* J.-M. Vinson) • 1906–1966 • Mauritius • Sauria
Viosca, Paul Percy, Jr. • 24 June 1892–1961 August 27 • USA • Anura, Caudata, (Testudines)
Vis: *see* De Vis, C. W.
Visser, John • ? • South Africa • Sauria
Voeltzkow, Alfred • 1860–1947 • Germany • ([Anura, Sauria])
Vogt, Carl Christopher • 5 July 1817–1895 May 5 • Germany / Switzerland • Crocodilia
Vogt, Richard Carl • ? • USA / Mexico • Sauria
Vogt, Theodor • ? • Germany • Anura, Sauria, Serpentes

Voigt, Friedrich Siegmund • 1 October 1781–1850 December 10 • Germany • Sauria
Vosmaer, Arnout • 1720–1799 • Holland • (Sauria, *testudines)
Vuillemin, Simone • ? • Madagascar • Testudines

- W -

Waga, Antoni Stanisław • 1799–1890 • Poland • Caudata
Wager, Vincent A. • ? • South Africa • Anura
Wagler, Johann Georg • 28 March 1800–1832 August 23 • Germany • Anura, Caudata, Crocodilia, Sauria, Serpentes, Testudines
Waite, Edgar Ravenswood • 5 May 1866–1928 January 19 • England / Australia / New Zealand / Australia • Serpentes
Wake, David Burton (*husband of* M. H. Wake) • 8 June 1936 • USA • Caudata, (Sauria)
Wake, Marvalee Hendricks (*wife of* D. B. Wake) • 31 July 1939 • USA • *Gymnophiona
Walbaum, Johann Julius • 30 June 1724–1799 August 21 • Germany • Anura
Walker, Charles Frederic • 27 December 1904–1979 February 16 • USA • Anura, Caudata
Walker, Cyril Alexander • 8 February 1939 • England • *testudines
Walker, James Martin • 21 October 1938 • USA • Sauria
Walker, Myrl Vincent • 20 March 1903 • USA • *sauria
Walker, Warren Franklin, Jr. • 27 September 1918 • USA • Serpentes
Wall, Frank (colonel) • 21 April 1868–1950 May 19 • British India / England • Serpentes
Wallace, Howard Keefer • 7 May 1907 • USA • (Caudata)
Wallach, Van Stanley Bartholomew • 6 April 1947 • USA • *Serpentes
Walley, Harlan D. • 7 November 1932 • USA • Sauria
Wallis, Kurt • ? • Austria • Serpentes
Waltl, Joseph • ?–(1832–1849)–? • Germany • (Caudata)
Wandolleck, Benno Edward Max Julius • 18 April 1864–? • Germany • Anura, Sauria
Wang, Chin-hsiang • ? • Taiwan • Anura, Sauria
Ward, Joseph Patric • 1944 • USA • Testudines
Warren, Ernest • 1871–1946 January 29 • England / South Africa / England • (Sauria, Serpentes)
Warren, James Wolfe • ? • USA / Australia • Serpentes
Watling, Dick • ? • England • Sauria
Watson, Gillian • ? • Rhodesia • Serpentes
Watson, Graeme Field • 17 Dec. 1942 • Australia • Anura
Weaver, William Glenn, Jr. • 1936 • USA • *testudines
Webb, Philip Barker • 10 July 1793–1854 August 31 • England • [Sauria]
Webb, Robert Gravem • 18 February 1927 • USA • Anura, Sauria, Serpentes, Testudines
Webber, Paul Nagle • 25 August 1949 • Australia • Sauria
Weber, Max Carl Wilhelm • 5 December 1852–1937 February 7 • Germany / Holland • Sauria, Serpentes
Webster, Thomas Preston, III • 14 July 1947–1975 November 9 • USA • Anura, (Caudata), Sauria
Weed, Alfred Cleveland • 28 July 1881–? • USA • Anura
Wegner, Theodor Hubert • 9 September 1880–1934 November 15 • Germany • *testudines
Wei, Feng • ? • China • *testudines

Wei, Jinlai • 22 November 1940 • China • Anura
Weigel, Christian Ehrenfried von • 24 May 1748–1831 August 8 • Germany • (Serpentes)
Weigelt, Johannes • 1890–1948 • Germany • (*caudata), *sauria
Weijenbergh: *see* Weyenbergh, H.
Weinland, David Friedrich • 1829–1915 • Germany • Anura, Sauria
Weismann, August F. L. • 17 January 1834–1914 • Germany • (Caudata)
Weiss, Christian Samuel • 26 February 1780–1856 October 1 • Germany • *testudines
Weithofer, K. Anton • ?–(1888)–? • Austria • *sauria
Weitzel, Karl • ? • Germany • *anura, *crocodilia
Welch, Kenneth R. G. • ? • England • Sauria
Wellborn, Vera • 1909 • Germany • Caudata
Weller, Worth Hamilton • 28 May 1913–1931 June 22 (age 18!) • USA • Caudata, Serpentes
Welles, Samuel Paul • 9 November 1907 • USA • *rhynchocephalia
Wellman, John • 6 December 1937 • USA • Sauria
Wellstead, Carl Frederick • 1951 • USA • *sauria
Welter, Wilfred August • 29 March 1906–1939 December 20 • USA • (Caudata)
Wen, Yetang • 27 August 1924 • China • Caudata
Werler, John E. • ? • USA • Anura, Caudata, Sauria, Serpentes
Wermuth, Heinz • ? • Germany • Sauria, Testudines, (*testudines)
Werner, Franz Josef Maria • 15 August 1867–1939 February 28 • Austria • Anura, Caudata, Gymnophiona, Sauria, Serpentes, Testudines
Werner, Yehudah Leopold • 16 June 1931 • Germany / Israel • Sauria
Westersheimb: *see* Wettstein-Westersheimb, O.
Westphal, Frank • 27 June 1930 • Germany • *caudata
Wetmore, Frank Alexander • 18 June 1886–1978 December 7 • USA • (Anura, Sauria, Serpentes)
Wettstein-Westersheimb, Otto • 7 August 1892–1967 July 10 • Austria • *Anura, Gymnophiona, Sauria, Serpentes, Rhynchocephalia, (Testudines)
Weyenbergh (=Weijenbergh), Hendrik • 6 December 1842–1885 July 25 • Holland / Argentina / Holland • Anura
Weyer, Dora • ? • USA / Costa Rica / Liberia • (Anura), Serpentes
Weygoldt, Peter • ? • Germany • Anura
Wheeler, George Montague (major) • 9 October 1842–1905 May 3 • USA • [Anura], (*crocodilia), [Sauria, Serpentes]
Whetstone, Kenneth N. • ? • USA • *testudines
Whitaker, Anthony Hume (="Tony") • 5 September 1944 • England / New Zealand • Sauria
Whitaker, Romulus • ? • India • Serpentes
White, Gilbert (reverend) • 18 July 1720–1793 June 26 • England • [(Testudines)]
White, John • (*circa* 1756)–1832 February 20 • England / Australia • Anura, Sauria
White, Theodore Elmer • 8 December 1905 • USA • *crocodilia, *sauria
Whiteside, D. I. • ? • England • *rhynchocephalia

Whymper, Edward • 27 April 1840–1911 September 16 • England • (Anura, Serpentes)
Wiedersheim, Robert Ernst Eduard • 21 April 1848–1923 July 12 • Germany • Caudata, (Sauria)
Wied-Neuwied, Maximilian Alexander Philipp (prinz zu) • 23 September 1782–1867 February 3 • Germany • Anura, Sauria, Serpentes
Wiegmann, Arend Friedrich August Heinrich • 2 June 1802–1841 January 15 • Germany • Anura, Sauria, Serpentes, Testudines
Wieland, George Reber • 24 January 1865–1953 January 18 • USA • *testudines
Wiggins, Ira Loren • 1 January 1899–1987 November 28 • USA • (Sauria)
Wilder, Harris Hawthorne (*husband of* I. W. Wilder) • 7 April 1864–1928 February 27 • USA • (Anura, Serpentes)
Wilder, Inez Luanne Whipple (*wife of* H. H. Wilder) • 19 May 1871–1929 April 29 • USA • (Caudata)
Wiley, Grace Olive Koontz • 1883–1948 July 20 • USA • (Sauria)
Wilhoft, Daniel Charles • 16 Nov. 1930 • USA • Sauria
Wilkinson, Mark • ? • USA • Gymnophiona
Williams, Ernest Edward • 7 January 1914 • USA • (Anura), Sauria, (Serpentes), *Testudines
Williams, Jorge Daniel • 10 October 1953 • Argentina • Sauria
Williams, Kenneth Lee • 4 September 1934 • USA • Anura, Sauria, Serpentes
Williston, Samuel Wendell • 10 July 1852–1918 August 30 • USA • *testudines
Wilson, Larry David • 10 April 1940 • USA • Anura, Sauria, Serpentes
Wilson, Richard Leland • ? • USA • *anura, *sauria, *serpentes
Wilson, Vivian J. • ? • Rhodesia • (Serpentes)
Wiman, Carl Johan Josef Ernst • 10 March 1867–1944 June 15 • USA / Sweden • *testudines
Winterl, Jakob József • 15 April 1739–1809 • Hungary • (Caudata)
Witschi, Emil • 18 February 1890 • Switzerland / USA • Anura
Witte, Gaston-François de • 12 June 1897–1980 June 1 • Belgium / Belgian Congo / Belgium • Anura, Sauria, Serpentes
Witten, Geoffrey J. • ? • Australia • Sauria
Wolf, Johann • 26 May 1765–1824 February 16 • Germany • Sauria
Wolterstorff, Willy • 16 June 1864–1943 January 21 • Germany • Anura, Caudata, (Sauria)
Wombey, John C. • ? • Australia • Sauria
Wood, Gilbert Congdon • 14 Nov. 1915 • USA • Serpentes
Wood, John George • 21 July 1827–1889 March 3 • England • Caudata
Wood, Roger Conant • 20 May 1941 • USA • *testudines
Wood, Searles Valentine • 14 February 1798–1880 October 26 • England • *crocodilia
Wood, Wallace F. • 1910 • USA • Caudata
Wood, William W. • ?–(1825)–? • USA • Caudata
Woodall, Harold T. • ?–World War II • USA • Caudata
Woodbury, Angus Munn (*father of* D. M. Woodbury) • 11 July 1886–1964 August 1 • USA • Sauria, Serpentes

Woodbury, Dixon Miles (*son of* A. M. Woodbury) • 6 August 1921 • USA • Serpentes
Woodin, William Hartman, III • 16 December 1925 • USA • Serpentes
Woodley, Jeremy D. • ? • Jamaica • Anura
Worrell, Eric Frederic Arthur • 1924–1987 July 12 • Australia • Sauria, Serpentes
Worthington, Richard Dane • 20 September 1941 • USA • Caudata
Worthy, Trevor Henry • 3 January 1957 • New Zealand • *anura
Wouters, Georges Florent • 1935 • Belgium • (*crocodilia)
Wright, Albert Hazen (*husband of* A. A. Wright) • 15 August 1879–1970 July 4 • USA • Anura, (Caudata, Sauria, Serpentes)
Wright, Anna Allen (*wife of* A. H. Wright) • 4 March 1882–1964 December 5 • USA • Anura, (Serpentes)
Wright, John William • 9 July 1936 • USA • Anura, Caudata, Sauria, Serpentes
Wright, Margaret Ruth • 24 March 1913 • USA • Caudata
Wu, Guanfu • 12 May 1935 • China • Anura
Wu, Hsien-wen • March 1900–1985 April 3 • China • Anura
Wu, Lu • 15 October 1915–1983 Dec. 2 • China • [Anura]
Wu, Shuhui • 26 July 1936 • China • Caudata
Wu, Weitang • ? • China • *testudines
Wucherer, Otho (=Otto Eduard Heinrich) • 7 July 1820–1874 May 7 • Portugal / England / Brazil • (Anura, Sauria), Serpentes
Wurfbain, Johann Paul • 14 December 1655–1711 January 12 • Germany • (Caudata)
Wussow, W. • ? • (?Prussia) • Testudines
Wynn, Addison Hartwell • 9 Nov. 1955 • USA • Caudata

- X -

Xantus, John (=János Xántus, *alias* Louis de Vésey) • 5 October 1825–1894 December 13 • Hungary / USA / Hungary • (Sauria)
Xavier, Françoise • 13 July 1939 • France • Anura
Xing, Yiling • 8 April 1961 • China • Caudata
Xu, Runhua • 26 August 1931 • China • Anura
Xu, Yubin • ? • China • *testudines

- Y -

Yáñez Valenzuela, José Lautaro • 18 April 1951 • Chile • Sauria
Yang, Datong • 12 March 1937 • China • Anura, Caudata, Gymnophiona, Sauria
Yang, Fuhua • 24 August 1930 • China • Anura, Caudata
Yang, Jungjien • ? • China • *anura
Yang, Rongsheng • ? • China • Caudata, Sauria
Yang, Suh Yung • 1934 • Korea • Caudata
Yarrow, Henry Crècy (="Harry") • 19 November 1840–1929 • USA • [Anura, (Sauria), Serpentes]

Yatkola, Daniel A. • 1947–1976 March 13 • USA • *sauria, (*serpentes)
Yazdani, G. M. • ? • India • Anura
Ye, Changyuan • 14 November 1938 • China • Anura, Caudata
Ye, Xiangkui (=Yeh, Hsiang-k'uei) • 10 October 1926 • China • *crocodilia, *testudines
Yefimov: *see* Efimov, M. B.
Yeh, H.-k.: *see* Ye, X.-k.
Yeriomtschenko, Valery Konstantinovich • 1949 • USSR • Sauria
Yilmaz, Irfan • 24 September 1953 • Turkey • Anura
Young, Chungchien • 1 June 1897–1979 January 15 • China • *caudata, *crocodilia, *testudines
Youngson, William Kenneth • ? • Australia • (Sauria)
Yu, Zhiwei • 7 January 1937 • China • Caudata
Yuan, Hong • 16 January 1955 • China • Serpentes
Yuztis, Enrique • ? • Venezuela • Anura

- Z -

Zammit-Maempel, George • 1925 • Malta • *sauria
Zangerl, Rainer • 19 November 1912 • USA • *crocodilia, *Testudines
Zarevsky (=Čarevskij, =Tsarevski), S. F. (=S. Th.) • ? • USSR • Anura, Sauria, Serpentes
Zenneck, Jonathan Adolf Wilhelm • 1871–1959 • Germany • Serpentes
Zerova, G. A. • ? • USSR • *sauria
Zhang, Fuji • 3 January 1946 • China • Caudata
Zhang, Minghua • 19 January 1947 • China • *Testudines
Zhang, Yuxia • 27 September 1931 • China • Anura
Zhao (=Djao), **Ermi** • 30 January 1930 • China • Anura, Caudata, Sauria, Serpentes
Zhao (=Chao), **Kentang** • 1 Sept. 1932 • China • Sauria
Zheng (=Cheng), **Ji** • 3 February 1931 • China • Serpentes
Zhou, Kaiya • 8 October 1932 • China • Sauria
Zietz, Amandus Heinrich Christian • 1874–1922 • Australia • Sauria, Serpentes
Zietz, Frederick Robert • ? • Australia • Sauria
Zimmermann, Elke • 1958 • Germany • Anura
Zimmermann, Helmut • ? • Germany • Anura
Zingo, Achille (barone) **de** • 14 January 1813–1892 January 15 • Italy • *serpentes
Zinniker, Walter • 29 May 1930 • Switzerland • Sauria, (Serpentes)
Zittel, Karl Alfred (ritter) **von** • 25 September 1839–1904 January 5 • Germany • *caudata, *rhynch., *serpentes
Zong, Yu • 29 December 1936 • China • Serpentes
Zonkov, J. • ? • Bulgaria • Caudata, Serpentes
Zug, George Robert • 16 November 1938 • USA • (Anura), Sauria, Serpentes, (Testudines)
Zugmayer, Erich Johann Georg • 16 May 1879–1938 • Austria / Germany • Sauria
Zuñiga U., Oscar • ? • Chile • Sauria
Zweifel, Richard George • 5 November 1926 • USA • *Anura, (Gymnophiona), Sauria, Serpentes

ACADEMIC LINEAGES OF DOCTORAL DEGREES IN HERPETOLOGY
by Ronald Altig

Professors have students who eventually have students, who themselves have students *ad infinitum*, and this sequence produces academic lineages. The First World Congress of Herpetology at Canterbury, England, seemed like the appropriate occasion to document as much herpetological academic history as possible. As a criterion for including persons in this tabulation, I looked for herpetological projects completed for doctoral requirements or extensive publication or graduation of students in herpetology after the doctorate. Once a lineage appeared, it was carried as far forward as possible, even if all persons were not herpetologists. The absence of a person's name indicates lack of information and not purposeful exclusion. Realize that the lineage of a person who went to another country for the doctorate will appear under that country and not under his or her eventual location. I had no intention of ignoring persons not holding the doctoral degree, but this list is not a directory—I used the doctoral degree as a limiter to keep the project manageable within the confines of the intended historical perspective. Likewise, I did not intend to label a person's expertise—many of those included may prefer not to call themselves herpetologists. Even so, the inclusion of persons from many biological and non-biological fields suggests that the central theme of herpetology is an interesting and highly unifying force!

Within the Lineage section, countries are arranged alphabetically, and under each country the known lineage sequences are given letter designations (e.g., Series A, Series B, etc.). Within each series, major professors are given numerical designations (1., 2., etc.). Within the Index section, people are listed alphabetically by their last name, and to the right in each entry is a two-letter code for country, a dash, a one- or two-letter code for series, a second dash, and a number for major professor. To locate a particular person in the Lineage section, first find him or her in the Index and then use the codes for country, series and major professor. Only one professor per student is given, even though co-advisors commonly exist. Names in parentheses are persons who, although not a major professor in the usual sense, were highly influential in the guidance of a student.

I extend a general note of thanks and appreciation to the numerous persons who tolerated my repeated requests and who provided information. The information is true to the extent of the information that I was able to obtain, and all errors are mine. I intend to continue this project, making corrections, filling in older parts of series, and keeping the tabulation current. Therefore I eagerly solicit additional information: Ronald Altig, Department of Biological Sciences, P.O. Drawer GY, Mississippi State University, Mississippi State, MS 38762 USA.

ACADEMIC LINEAGES

ARGENTINA (AR)

SERIES A
1. BECCARI, N.
    Cei, J. M. .... 1930 ... Pisa (IT) 2
2. CEI, J. M.
    Barbieri, F. D. .... 1954 ... Tucuman 3
    Bertini, F. .... 1961 ... Tucuman
    Cohen, R. .... 1963 ... U. Nac. Cuyo
    Crespo, E. G. .... 1980 ... Lisboa 4
    Rengel, D. .... 1953 ... Tucuman
3. BARBIERI, F. D.
    Cabada, M. O. .... 1973 ... La Plata
    Fernandez, M. R. .... 1967 ... Tucuman
    Legname, A. H. .... 1964 ... Tucuman
    Manes, M. E. .... 1976 ... Cordoba
    Miceli, D. C. .... 1979 ... Tucuman
    Raisman, J. S. .... 1975 ... Tucuman
    Salomon, H. L. .... 1968 ... Tucuman
    Sanchez, S. O. .... 1983 ... Tucuman
4. CRESPO, E. G.
    Vincente, L. A. .... 1988 ... Lisboa

SERIES B
1. BONPARTE, J. F.
    Powell, J. E. .... 1986 ... Tucuman

SERIES C
1. BRIEN, P.
    Laurent, R. F. .... 1940 ... Libre Brussel 2
2. LAURENT, R. F.
    La Villa, E. O. .... 1983 ... Tucuman
    Scrocchi, G. J. .... 1987 ... Tucuman

AUSTRALIA (AU)

SERIES A
1. BARWICK, R. E.
    Bustard, H. R. .... 1966 ... Aust. Nat. U.

SERIES B
1. BURNSTOCK, G.
    Bell, C. .... 1965 ... Melbourne

SERIES C
1. CAUGHLEY, G. J.
    Magnusson, W. .... 19?? ..... Sydney

SERIES D
1. DAVIDSON, J.
    Andrewartha, H. G. .... 19?? ..... Adelaide 2
2. ANDREWARTHA, H. G.
    Green, B. .... 1969 ... Adelaide
    King, R. D. .... 1973 ... Adelaide

SERIES E
1. DAWBIN, W. H.
    Clarke, C. J. .... 19?? .... Sydney
    Hill, L. G. .... 19?? .... Sydney
SERIES F
1. ? ? ? ?
    Edmonds, S. J. .... 1958 ... Adelaide 2
2. EDMONDS, S. J.
    Bennett, L. J. .... 1971 ... Adelaide
SERIES G
1. HAYMAN, D. L.
    King, M. .... 19?? ... Adelaide
SERIES H
1. KIKKAWA, J. T.
    Ingram, G. J. .... 1986 ... Queensland
SERIES I
1. NELSON, J.
    Gibbons, J. R. .... 1976 ... Monash
SERIES J
1. TYLER, M. J.
    Burton, T. C. .... 1983 ... Adelaide
    Davies, M. .... 1987 ... Adelaide
    Horton, P. .... 1986 ... Adelaide
SERIES K
1. WARING, H. W.
    Main, A. R. .... 1956 ... W. Aust. 2
2. MAIN, A. R.
    Blackwell, J. M. .... 1974 ... W. Aust.
    Bradshaw, S. D. .... 1965 ... W. Aust. 3
    Bull, C. M. .... 1973 ... W. Aust. 4
    Burbidge, A. A. .... 1968 ... W. Aust.
    Lee, A. K. .... 1965 ... W. Aust. 5
    Littlejohn, M. J. .... 1958 ... W. Aust. 6
    Storr, G. M. .... 1962 ... W. Aust.
3. BRADSHAW, S. D.
    Baverstock, P. R. .... 19?? .... W. Aust.
4. BULL, C. M.
    Dubas, G. .... 1987 ... Flinders
    Odendaal, F. .... 1981 ... Flinders
    Williamson, I. .... 1982 ... Flinders
    Yeatman, E. .... 1980 ... Flinders
5. LEE, A. K.
    Straughn, I. R. .... 1967 ... Queensland
    Wilson, K. J. .... 1971 ... Monash
6. LITTLEJOHN, M. J.
    Gartside, D. F. .... 19?? .... Melbourne
    Humphries, R. B. ...... 1979 ... Aust. Nat. U.
    Loftus-Hills, J. J. ...... 19?? .... Melbourne
    Martin, A. A. .... 1968 ... Melbourne
    Robertson, J. G. .... 1982 ... Aust. Nat. U.
    Watson, G. F. .... 1974 ... Melbourne

AUSTRIA   (AS)
SERIES A
1. VERSLUYS, J.
    Haas, G. .... 1928 ... Vienna 2
2. HAAS, G.
    Kochva, E. .... 19?? .... Hebrew U.
    Nevo, E. .... 1964 ... Hebrew U.
    Werner, Y. L. .... 19?? .... Hebrew U.

BRAZIL   (BR)
SERIES A
1. GOMES, N. M.
    Pexioto, O. L. .... 19?? .... Sao Paulo

CANADA   (CN)
SERIES A
1. STEWART, K. W.
    Cook, F. .... 1978 ... Manitoba

DENMARK   (DE)
SERIES A
1. BOHR, C.
    Krogh, A. .... 19?? .... ? ? ? ? 2
2. KROGH, A.
    Jorgensen, C. B. .... 19?? .... A. Krogh Inst. .. 3
3. JORGENSEN, C. B.
    Budtz, P. .... 19?? .... A. Krogh Inst.
    Larsen, E. H. .... 19?? .... ? ? ? ?
    Larsen, L. O. .... 19?? .... ? ? ? ?
    Rosenkilde, P. .... 19?? .... ? ? ? ?
    Spies, I. .... 19?? .... ? ? ? ?
SERIES B
1. STEENBERG, C. M.
    Baestrup, F. W. .... 19?? .... ? ? ? ? 2
    Volsøe, H. .... 1944 ... Copenhagen 3
2. BAESTRUP, F. W.
    Rasmussen, J. B. .... 1977 ... Copenhagen
3. VOLSØE, H.
    Schiøtz, A. .... 1967 ... Copenhagen

FRANCE   (FR)
SERIES A
1. CUVIER, G.
    Milne-Edwards, H. ...... 18?? .... ? ? ? ? 2
2. MILNE-EDWARDS, H.
    Vaillant, L. .... 1865 ... Paris
SERIES B
1. RICQLES, A. de
    Castanet, J. .... 1982 ... Paris 2
2. CASTANET, J.
    Caetano, M. E. .... 1988 ... Lisboa

GERMANY   (GE)
SERIES A
1. CHUN, C.
    Mertens, R. .... 1915 ... Leipzig
SERIES B
1. KRAMER, G.
    Medem, F. .... 1942 ... Tubingen
SERIES C
1. MÜLLER, J.
    Haeckel, E. .... 18?? ... Berlin 2
    Peters, W. .... 1838 ... Berlin
2. HAECKEL, E.
    Bedriaga, J. von .... 1875 ... Jena
    Goeldi, E. A. .... 1883 ... Jena

HUNGARY (HU)
SERIES A
1. MÉHELŸ, L.
   Fejérváry, G. von .... 1917 ... Budapest

ITALY (IT)
SERIES A
1. LESSONA, M.
   Peracca, M. G. .... 1886 ... Turin

JAPAN (JA)
SERIES A
1. KOMAI, T.
   Maki, M. .... 1932 ... Kyoto
   Nakamura, K. .... 1935 ... Kyoto 2
2. NAKAMURA, K.
   Fukada, H. .... 1961 ... Kyoto
SERIES B
1. TASUMI, M.
   Matsui, M. .... 1984 ... Kyoto

MEXICO (ME)
SERIES A
1. FERRUSQUIA, I.
   Casas-Andreu, G. .... 1982 ... Nac. U. Mex. ... 2
2. CASAS-ANDREU, G.
   Gadseden-Esparza, H. .. 1987 ... Nac. U. Mex.

NETHERLANDS (NE)
SERIES A
1. ? ? ? ?
   Boschma, H. .... 1920 ... Amsterdam 2
2. BOSCHMA, H.
   Parker, H. W. .... 1949 ... Leiden
SERIES B
1. DE JONG, J. K.
   Brongersma, L. D. ..... 1934 ... Amsterdam 2
2. BRONGERSMA, L. D.
   Hoogmoed, M. S. ...... 1973 ... Leiden
SERIES C
1. DENUCE, Y. M.
   Hemclaar, A. S. .... 1986 ... K. U. Nijmegen
SERIES D
1. ? ? ? ?
   Dullemeijer, P. .... 1956 ... Leiden 2
2. DULLEMEIJER, P.
   de Jongh, H. J. .... 1968 ... Leiden
SERIES E
1. OOMEN, H. C.
   Wynands, H. E. .... 1979 ... K. U. Nijmegen
SERIES F
1. STOCK, Y. H.
   Blommers-
     Schlosser, R. ...... 1980 ... Amsterdam

POLAND (PO)
SERIES A
1. HOYER, H.
   Grodzinski, Z. .... 1922 ... Jagellonian 2
2. GRODZINSKI, Z.
   Juszczyk, W. .... 1945 ... Jagellonian 3
   Młynarski, M. .... 1956 ... Jagellonian 4
   Szarski, H. .... 1937 ... Jagellonian 5
3. JUSZCZYK, W.
   Guzik, M. .... 1983 ... Pedagog. Acad.
   Schimscheiner, L. ...... 1983 ... Pedagog. Acad.
   Swierad, J. .... 1977 ... Pedagog. Acad.
   Zakrzewski, M. .... 1974 ... Pedagog. Acad.
   Zamachowski, W. ...... 1967 ... Pedagog. Acad.
   Zysk, A. .... 1973 ... Pedagog. Acad.
4. MŁYNARSKI, M.
   Kowalewski, L. .... 1969 ... Jagellonian
   Szyndlar, Z. .... 1982 ... Acad. Sci. Krakow
5. SZARSKI, H.
   Czopek, G. .... 1963 ... Copernicus
   Czopek, J. .... 1954 ... Copernicus 6
   Plytycz, B. .... 1975 ... Jagellonian
   Poczopko, P. .... 1959 ... Copernicus
   Rafinski, J. N. .... 1973 ... Jagellonian
   Strawinski, S. .... 1956 ... Copernicus
   Szymura, J. M. .... 1975 ... Jagellonian
   Witalinska, L. .... 1972 ... Jagellonian
6. CZOPEK, J.
   Andrzejewski, H. .... 1969 ... Copernicus
   Francio-
     Krassowska, A. ... 1975 ... Copernicus
   Przystalska, A. .... 1978 ... Copernicus
   Wilczynskja, B. .... 1977 ... Copernicus
SERIES B
1. KILARSKI, W.
   Kordylewski, L. .... 1975 ... Jagellonian
SERIES C
1. KOSCIELSKI, B.
   Ogielska, M. .... 1982 ... Krakow
SERIES D
1. NUSBAUM-HILAROWICZ, J.
   Hirschler, J. .... 19?? .... Jan Casimir 2
2. HIRSCHLER, J.
   Sembrat, K. .... 1925 ... Jan Casimir 3
3. SEMBRAT, K.
   Kielbowna, L. .... 1969 ... Krakow
   Orska, J. .... 1948 ... Wroclaw 4
4. ORSKA, J.
   Lecyk, M. .... 1961 ... Krakow
   Pawlowska-Indyk, A. .. 1975 ... Krakow
SERIES E
1. RAABE, Z.
   Dobrowolska, H. .... 1969 ... Warszawa
SERIES F
1. SIEDLECKI, M.
   Skowdon, S. .... 1922 ... Jagellonian 2
2. SKOWDON, S.
   Srebro, Z. .... 1961 ... Medic Acad. 3
3. SREBRO, Z.
   Sliwa, L. .... 1982 ... Medic Acad.
   Sura, P. .... 1985 ... Medic Acad.
SERIES G
1. SMRECZYNSKI, S.
   Rzechak, K. .... 1964 ... Jagellonian

SERIES H
1. URBANSKI, J.
    Berger, L. .... 1961 ... Poznan

SOUTH AFRICA (SA)
SERIES A
1. DE VILLIERS, C. G.
    van Dijk, D. E. .... 19?? .... Stellenbosch 2
2. VAN DIJK, D. E.
    Channing, A. .... 19?? .... Natal
SERIES B
1. PATTERSON, H.
    Passmore, N. .... 1978 ... Witwaters 2
2. PASSMORE, N.
    Bishop, P. .... 19?? .... Witwaters
    Dyson, M. .... 1989 ... Witwaters
    Pides, M. .... 1987 ... Witwaters
    Telford, S. .... 1983 ... Witwaters

SPAIN (SP)
SERIES A
1. BERNIS, F.
    Salvador, A. .... 1979 ... Complutense 2
2. SALVADOR, A.
    Valentin, P. M. .... 1981 ... Salamanca

SWEDEN (SE)
SERIES A
1. LINNAEUS, C.
    Forskål, P. .... 1760 ... Uppsala
    Hasselquist, F. .... 1750 ... Uppsala
    Kalm, P. .... 1745 ... Uppsala
    Lofling, P. .... 1749 ... Uppsala
    Osbeck, P. .... 1754 ... Uppsala
    Schreber, J. C. .... 1760 ... Uppsala
    Thunberg, C. P. .... 1770 ... Uppsala

SWITZERLAND (SW)
SERIES A
1. FISCHBERG, M.
    Loumont, C. .... 19?? .... Geneva
SERIES B
1. KRAMER, E.
    Regenass, U. .... 19?? .... Basel
SERIES C
1. ZISWILER, V.
    Dusej, G. .... 19?? .... Zurich
    Schätti, B. .... 19?? .... Zurich
SERIES D
1. SEMPER, C.
    Sarasin, C. F. .... 1883 ... Basel

TURKEY (TU)
SERIES A
1. BAUR, E.
    Kosswig, C. .... 1927 ... Hamburg (GE) .. 2
2. KOSSWIG, C.
    Başoğlu, M. .... 1942 ... Istanbul 3
3. BAŞOĞLU, M.

Atatür, M. K. .... 19?? .... Aegean
Baran, I. .... 1969 ... Aegean
Özeti, N. .... 1963 ... Aegean

UNITED KINGDOM (UK)
SERIES A
1. ? ? ? ?
    Bellairs, A. d'A. .... 1942 ... Univ. College ... 2
2. BELLAIRS, A. D'A.
    Bryant, S. V. .... 19?? .... London
    Maderson, P. F. .... 1962 ... London 3
3. MADERSON, P. F.
    Contard, P. C. .... 1987 ... C. U. New York
SERIES B
1. CHITTY, D.
    Smyth, M. B. .... 1964 ... Oxford 2
2. SMYTH, M. B.
    Braysher, M. L. .... 1972 ... Adelaide (AU)
    Henzell, R. P. .... 1972 ... Adelaide (AU)
    Roberts, J. D. .... 1976 ... Adelaide (AU)
SERIES C
1. DE BEER, G. R.
    Ford, P. .... 1955 ... London
    Fox, H. .... 1951 ... London 2
2. FOX, H.
    Turner, S. C. .... 1970 ... London 3
3. TURNER, S. C.
    Touey, J. .... 1987 ... Portsmouth
SERIES D
1. GRUNEBERG, H.
    Truslove, G. .... 1951 ... London
SERIES E
1. SIMNETT, J.
    Oates, C. .... 19?? .... Newcastle
SERIES F
1. TINBERGEN, N.
    Arnold, E. N. .... 19?? .... Oxford
SERIES G
1. WATSON, D. M.
    Newth, D. .... 19?? .... London 2
    Whitear, M. .... 1951 ... London
2. NEWTH, D.
    Blackler, A. .... 1965 ... London
    Hamilton, L. .... 1965 ... London
SERIES H
1. CULLEN, J. M.
    Halliday, T. R. .... 1972 ... Oxford 2
2. HALLIDAY, T. R.
    Verrell, P. A. .... 1983 ... Open
SERIES I
1. CLOUDSLEY-THOMPSON, J. L.
    Griffths, R. A. .... 1983 ... Birkbeck
SERIES J
1. DAVIES, P. M.
    Hailey, A. .... 19?? .... Nottingham

UNION OF SOVIET SOCIALIST REPUBLICS (SU)

Codes: SPIU = St. Petersburg Imperial University; ZIAS Len. = Zoological Institute, Academy of Sciences, Leningrad; ZIAS Kiev = Zoological Institute, Academy of Sciences, Kiev

SERIES A
1. BOGDANOV, M.
    Nikolsky, A. M. .... 1887 ... SPIU 2
2. NIKOLSKY, A. M.
    Chernov, S. A. .... 1926 ... Kharkov 3
    Laister, A. F. .... 19?? .... Kharkov
3. CHERNOV, S. A.
    Alekperov, A. M. ...... 19?? .... ZIAS Len.
    Darevsky, I. S. .... 1967 ... ZIAS Len. 4
    Emelianov, A. A. ...... 1944 ... ZIAS Len.
4. DAREVSKY, I. S.
    Agasyan, A. .... 1989 ... ZIAS Len.
    Ananjeva, N. B. .... 1972 ... ZIAS Len.
    Bakradze, M. .... 1977 ... ZIAS Len.
    Borkin, L. .... 1986 ... ZIAS Len.
    Danieljan, F. .... 1968 ... ZIAS Len.
    Kupriyanova, L. .... 19?? .... ZIAS Len.
    Negmedzyanov, V. ..... 19?? .... ZIAS Len.
    Orlov, N. L. .... 19?? .... ZIAS Len.
    Pizchelawri, V. .... 19?? .... ZIAS Len.
    Tran, K. .... 1984 ... ZIAS Len.
    Tuniyev, B. .... 1988 ... ZIAS Len.
    Zaune, L. .... 19?? .... ZIAS Len.
    Zellarius, A. .... 1986 ... ZIAS Len.

SERIES B
1. ? ? ? ?
    Shcherbak, N. N. .... 1962 ... ZIAS Kiev 2
2. SHCHERBAK, N. N.
    Badmajeva, V. .... 1983 ... ZIAS Kiev
    Dotsenko, I. .... 1986 ... ZIAS Kiev
    Erjomtschenko, V. ..... 1984 ... ZIAS Kiev
    Golubev, M. .... 1982 ... ZIAS Kiev
    Gontscharenko, A. ..... 1980 ... ZIAS Kiev
    Kotenko, T. .... 1983 ... ZIAS Kiev
    Pisanets, E. .... 1978 ... ZIAS Kiev
    Sattorov, T. .... 1981 ... ZIAS Kiev
    Shcherban, M. .... 1976 ... ZIAS Kiev
    Tarashchuk, S. .... 1987 ... ZIAS Kiev
    Tertishinkov, M. .... 1972 ... ZIAS Kiev

UNITED STATES OF AMERICA (US)

Codes: AU = Australia; BS = Bachelor of Science degree; BR = Brazil; CA = California; FL = Florida; GE = Germany; MS = Master of Science degree; OH = Ohio

SERIES A
1. OKEN, L.
    Agassiz, L. .... 1829 ... Munich (GE) 2
2. AGASSIZ, L.
    (Jordan, D. S.) ..... (MS)1872 ... Cornell 3
    Whitman, C. O. .... 19?? .... Harvard 4
    Wilder, B. G. .... 1868 ... Harvard 5
3. JORDAN, D. S.
    Grinnell, J. .... 1913 ... Stanford 6
    Myers, G. S. .... 1933 ... Stanford 7
4. WHITMAN, C. O.
    Adams, C. C. .... 1908 ... Chicago 8
5. WILDER, B. G.
    Gage, S. H. .... (BS)1877 ... Cornell 9
    Reed, H. G. .... 1899 ... Cornell 10
    Wright, A. H. .... 1908 ... Cornell 11
6. GRINNELL, J.
    Dice, L. R. .... 1915 ... CA Berkeley ... 12
    Fitch, H. S. .... 1937 ... CA Berkeley ... 13
    Storer, T. I. .... 1925 ... CA Berkeley
7. MYERS, G. S.
    Alcala, A. C. .... 1966 ... Stanford
    Anderson, S. C. .... 19?? .... Stanford
    Banta, B. H. .... 1961 ... Stanford
    Brown, W. C. .... 1950 ... Stanford
    Cliff, F. S. .... 1954 ... Stanford
    Funkhouser, J. W. ..... 1951 ... Stanford
    Leviton, A. E. .... 1960 ... Stanford
    Maslin, T. P. .... 1945 ... Stanford 14
    Savage, J. M. .... 1955 ... Stanford 15
    Schuierer, F. W. .... 1966 ... Stanford
8. ADAMS, C. C.
    Ruthven, A. G. .... 1906 ... Michigan 16
9. GAGE, S. H.
    Kingsbury, B. F. .... 1895 ... Cornell 17
10. REED, H. G.
    McMullen, E. C. .... 1936 ... Cornell
11. WRIGHT, A. H.
    Axtell, H. .... 1947 ... Cornell
    Bishop, S. C. .... 1925 ... Cornell 18
    Cowles, R. B. .... 1929 ... Cornell 19
    Funkhouser, W. D. .... 1916 ... Cornell
    Gilbert, P. W. .... 1940 ... Cornell 20
    Gordon, K. L. .... 1936 ... Cornell 21
    Hamilton, W. J., Jr. ... 1930 ... Cornell 22
    Harper, F. .... 1925 ... Cornell
    Kezer, L. J. .... 1948 ... Cornell
    Liu, C.-c. .... 1934 ... Cornell 23
    Livezey, R. L. .... 1946 ... Cornell
    McCauley, R. H. .... 1940 ... Cornell
    Schmidt, K. P ..... (BS)1917 ... Cornell 24
    Trapido, H. .... 1943 ... Cornell
12. DICE, L. R.
    Blair, W. F. .... 1938 ... Michigan 25
13. FITCH, H. S.
    Burkett, R. D. .... 1964 ... Kansas
    Clark, D. R. .... 1968 ... Kansas 26
    Hall, R. J. .... 1969 ... Kansas
    Legler, J. M. .... 1960 ... Kansas 27
    Loomis, R. B. .... 1956 ... Kansas
    Malaret, L. .... 1985 ... Kansas
    Platt, D. R. .... 1966 ... Kansas
    Plummer, M. V. .... 1975 ... Kansas
    Regan, G. T. .... 1972 ... Kansas
    Seigel, R. A. .... 1984 ... Kansas
    Smith, A. K. .... 1973 ... Kansas
    Waltner, R. C. .... 1977 ... Kansas
    Webb, R. G. .... 1962 ... Kansas
14. MASLIN, T. P.
    Beargie, K. M. .... 1971 ... Colorado
    Campbell, J. B. .... 1970 ... Colorado
    Cuellar, O. .... 1969 ... Colorado

Ferner, J. W. .... 1972 ... Colorado
Knopf, G. N. .... 1966 ... Colorado
McCoy, C. J. .... 1965 ... Colorado
Pennock, L. A. .... 1966 ... Colorado
Reese, R. W. .... 1969 ... Colorado
Walker, J. M. .... 1966 ... Colorado 28
Werth, R. J. .... 1969 ... Colorado
15. SAVAGE, J. M.
    Deweese, J. E. .... 1976 ... South. CA
    Donnelly, M. A .... 1987 ... Miami (FL)
    Emerson, S. B. .... 1971 ... South. CA
    Gaudin, A. J. .... 1969 ... South. CA
    Guyer, C. .... 1986 ... Miami (FL)
    Harris, R. T. .... 1973 ... South. CA
    Heyer, W. R. .... 1969 ... South. CA
    Kluge, A. G. .... 1964 ... South. CA 29
    Lieberman, S. S. .... 1982 ... South. CA
    McDiarmid, R. W. ..... 1969 ... South. CA
    Meyer, J. R. .... 1969 ... South. CA
    Miyamoto, M. M. ..... 1982 ... South. CA
    Morafka, D. J. .... 1973 ... South. CA
    Presch, W. F. .... 1970 ... South. CA
    Scott, N. J., Jr. .... 1969 ... South. CA 30
    Silverstone, P. A. ...... 1971 ... South. CA
    Talbot, J. J. .... 1976 ... South. CA
    Vial, J. A. .... 1965 ... South. CA
    Wake, D. B. .... 1964 ... South. CA 31
    Wake, M. H. .... 1968 ... South. CA 32
16. RUTHVEN, A. G.
    Bailey, J. R. .... 1940 ... Michigan 33
    Blanchard, F. N. .... 1919 ... Michigan 34
    Burt, C. E. .... 1930 ... Michigan
    Hartweg, N. E. .... 1934 ... Michigan 35
    Oliver, J. A. .... 1941 ... Michigan
    Ortenburger, A. I. ...... 1925 ... Michigan 36
    Stuart, L. C. .... 1933 ... Michigan
    Stull, O. G. .... 1929 ... Michigan
    Van Tyne, J. .... 1928 ... Michigan 37
    Walker, C. F. .... 1935 ... Michigan 38
17. KINGSBURY, B. F.
    Adelmann, H. B. .... 1912 ... Cornell
18. BISHOP, S. C.
    Grobman, A. B. .... 1944 ... Rochester 39
    Tihen, J. A. .... 1945 ... Rochester 40
19. COWLES, R. B.
    Brattstrom, B. H. .... 1959 ... CA Los Ang.
    Lowe, C. H., Jr. .... 1951 ... CA Los Ang. . 41
    Norris, K. S. .... 1959 ... CA Los Ang. . 42
    Stebbins, R. C. .... 1945 ... CA Los Ang. . 43
20. GILBERT, P. W.
    Evans, H. E. .... 1950 ... Cornell 44
    Snyder, R. C. .... 1948 ... Cornell 45
21. GORDON, K. L.
    Storm, R. M. .... 1948 ... Oregon St. 46
22. HAMILTON, W. J., Jr.
    Barbour, R. W. .... 1949 ... Cornell 47
    Harris, L. E., Jr. .... 1954 ... Loma Linda
    Jopson, H. G. .... 1938 ... Cornell
    Oplinger, C. S. .... 1953 ... Cornell
    Stewart, M. M. .... 1956 ... Cornell 48
    Tilger, G. M. .... 1963 ... Cornell

Whitaker, J. O. .... 1962 ... Cornell
23. LIU, C.-C.
    Hu, S.-q. .... 1938 ... Soochow
    Zhao, E.-m. .... 1953 ... W. China
24. SCHMIDT, K. P.
    Inger, R. F. .... 1954 ... Chicago 49
25. BLAIR, W. F.
    Awbrey, F. T. .... 1965 ... Texas 50
    Axtell, R. W. .... 1958 ... Texas
    Bogart, J. P. .... 1969 ... Texas 51
    Brown, L. E. .... 1967 ... Texas 52
    Brownell, J. A. .... 1971 ... Texas
    Follansbee, L. A. .... 1954 ... Texas
    Fouquette, M. J., Jr. ... 1959 ... Texas 53
    Gerhardt, H. C. .... 1970 ... Texas 54
    Greding, E. J. .... 1968 ... Texas
    Grubb, J. C. .... 1971 ... Texas
    Guttman, S. I. .... 1967 ... Texas 55
    Heredia-Cano, J. F. .... 1973 ... Texas
    Hunsaker, D., II .... 1960 ... Texas
    Hunt, H. W. .... 1971 ... Texas
    Jameson, D. L. .... 1952 ... Texas 56
    Kennedy, J. P. .... 1958 ... Texas
    Laughlin, H. E. .... 1966 ... Texas
    Lindsay, H. L. .... 1958 ... Texas
    Low, B. S. .... 1967 ... Texas
    Mares, M. A. .... 19?? .... Texas
    Martin, R. F. .... 1969 ... Texas
    Martin, W. F. .... 1970 ... Texas
    Matthews, T. .... 19?? .... Texas
    McAllister, W. .... 1961 ... Texas
    McKown, R. R. .... 1972 ... Texas
    Meacham, W. R. .... 1958 ... Texas
    Mecham, J. S. .... 1955 ... Texas 57
    Michaud, T. C. .... 1959 ... Texas
    Milstead, W. W. .... 1953 ... Texas
    Nelson, C. E. .... 1966 ... Texas 58
    Pettus, D. .... 1956 ... Texas 59
    Pierce, J. R. .... 1968 ... Texas
    Porter, K. R. .... 1962 ... Texas
    Pyburn, W. F. .... 1956 ... Texas
    Ralin, D. B. .... 1970 ... Texas
    Rogers, J. S. .... 1971 ... Texas
    Sage, R. D. .... 1974 ... Texas
    Tandy, R. M. .... 1972 ... Texas
    Thompson, W. L. ...... 19?? .... Texas
    Thornton, W. A. .... 1954 ... Texas
    Wasserman, A. O. ...... 1956 ... Texas
26. CLARK, D. R.
    Ballinger, R. E. .... 1971 ... Texas A&M ... 60
    Fleet, R. R. .... 19?? .... Texas A&M
27. LEGLER, J. M.
    Berry, J. F. .... 1978 ... Utah
    Bull, J. J. .... 1977 ... Utah
    Feuer, R. C. .... 1966 ... Utah
    Moll, E. O. .... 1969 ... Utah
    Winokur, R. M. .... 1972 ... Utah
28. WALKER, J. M.
    Arnett, J. T. .... 1974 ... Arkansas
    Crowell, R. L. .... 1981 ... Colorado
    Cundall, D. L. .... 1974 ... Arkansas 61
    Durham, M. S. .... 1980 ... Arkansas

Ireland, P. H. .... 1971 ... Arkansas
Killibrew, F. C. .... 1976 ... Arkansas
Kuss, B. D. .... 1986 ... Arkansas
Reagan, D. P. .... 1972 ... Arkansas
29. KLUGE, A. G.
Arnold, S. J. .... 1972 ... Michigan 62
Bookstein, F. L. .... 1977 ... Michigan
Jaslow, A. P. .... 1982 ... Michigan
Kerfoot, W. C. .... 1972 ... Michigan
Kraus, E. F. .... 1987 ... Michigan
Moody, S. M. .... 1980 ... Michigan
Tolson, P. J. .... 1983 ... Michigan
30. SCOTT, N. J., Jr.
Lotter, C. F. .... 1975 ... Connecticut
Mitchell, S. L. .... 1988 ... New Mexico
Reynolds, R. P. .... 1978 ... New Mexico
Woodward, B. D. .... 1981 ... New Mexico
31. WAKE, D. B.
Alberch, P. .... 1980 ... CA Berkeley ... 63
Busack, S. D. .... 1985 ... CA Berkeley
Cadle, J. E. .... 1982 ... CA Berkeley
Edwards, J. L. .... 1976 ... CA Berkeley ... 64
Good, D. A. .... 1985 ... CA Berkeley
Hanken, J. .... 1980 ... CA Berkeley
Houck, L. D. .... 1977 ... CA Berkeley
Larson, L. A. .... 1982 ... CA Berkeley
Lombard, R. E. .... 1971 ... Chicago
Lynch, J. F. .... 1974 ... CA Berkeley
Maiorana, V. C. .... 1974 ... CA Berkeley
Peterson, J. .... 1973 ... Chicago
Sessions, S. K. .... 1984 ... CA Berkeley
Shafland, J. L. .... 1968 ... Chicago
Stewart, J. R. .... 1976 ... CA Berkeley
Sweet, S. S. .... 1978 ... CA Berkeley
Wassersug, R. J. .... 1972 ... Chicago 65
Yanev, K. P. .... 1978 ... CA Berkeley
32. WAKE, M. H.
Bauer, A. M. .... 1986 ... CA Berkeley
Hetherington, T. E. .... 1979 ... CA Berkeley
Klosterman, L. L. ...... 1983 ... CA Berkeley
Schwenk, K. .... 1984 ... CA Berkeley
33. BAILEY, J. R.
Bruce, R. C. .... 1968 ... Duke
Davis, D. M. .... 1969 ... Duke
Hutchison, V. H. .... 1959 ... Duke 66
Murphy, T. D. .... 1963 ... Duke
Osgood, D. W. .... 1968 ... Duke
34. BLANCHARD, F. N.
Clark, H. D. .... 1941 ... Michigan
Clay, W. M. .... 1937 ... Michigan
George, I. D. .... 1940 ... Michigan
Gloyd, H. K. .... 1937 ... Michigan
Stickel, W. H. .... 19?? .... Michigan
35. HARTWEG, N. E.
Brown, B. C. .... 1955 ... Michigan
Cagle, F. R. .... 1943 ... Michigan 67
Damman, A. E. .... 1960 ... Michigan
Dowling, H. G. .... 1951 ... Michigan 68
Downs, F. L. .... 1965 ... Michigan
Duellman, W. E. .... 1956 ... Michigan 69
Etheridge, R. E. .... 1959 ... Michigan
Mosimann, J. E. .... 1956 ... Michigan

Orton, G. L. .... 1944 ... Michigan
Peters, J. A. .... 1952 ... Michigan
36. ORTENBURGER, A. I.
Holzapfel, R. A. .... 1935 ... Oklahoma
37. VAN TYNE, J.
Brodkorb, P. .... 19?? .... Michigan 70
38. WALKER, C. F.
Adler, K. K. .... 1968 ... Michigan 71
Allen, A. C. .... 1970 ... Michigan
Healy, W. R. .... 1966 ... Michigan
Martin, P. S. .... 1956 ... Michigan
Rabb, G. B. .... 1957 ... Michigan
Starrett, P. H. .... 1968 ... Michigan
Uzzell, T. M., Jr. .... 1962 ... Michigan
Wilbur, H. M. .... 1971 ... Michigan 72
Zug, G. R. .... 1969 ... Michigan
39. GROBMAN, A. B.
Auffenberg, W. .... 1956 ... Florida 73
Highton, R. T. .... 1956 ... Florida 74
40. TIHEN, J. A.
Chantell, C. J .... 1965 ... Notre Dame
41. LOWE, C. H., Jr.
Bezy, R. L. .... 1970 ... Arizona
Cole, C. J. .... 1969 ... Arizona
Cross, J. K. .... 1979 ... Arizona
Ferguson, J. H. .... 1965 ... Arizona
Gates, G. O. .... 1963 ... Arizona
Goldberg, S. R. .... 1970 ... Arizona
Halpern, E. A. .... 1970 ... Arizona
Lardner, P. J. .... 1969 ... Arizona
Robinson, M. D. .... 1972 ... Arizona
Stropes, S. L. .... 1975 ... Arizona
Tremor, J. W. .... 1962 ... Arizona
Wright, J. W. .... 1965 ... Arizona
42. NORRIS, K. S.
Asplund, K. K. .... 1968 ... CA Los Ang.
Barrett, R. .... 1969 ... CA Los Ang.
Brown, T. W. .... 1970 ... CA Los Ang.
Campbell, H. W. .... 1967 ... CA Los Ang.
Evans, W. .... 19?? .... CA Los Ang.
Heath, J. E. .... 1962 ... CA Los Ang.
Porter, W. P. .... 1966 ... CA Los Ang. . 75
Pough, F. H. .... 1968 ... CA Los Ang. . 76
Shallenberger, E. W. .. 1970 ... CA Los Ang.
Warren, P. P. .... 1966 ... CA Los Ang.
43. STEBBINS, R. C.
Anderson, J. D. .... 1960 ... CA Berkeley ... 77
Anderson, P. K. .... 1958 ... CA Berkeley
Berry, K. H. .... 1972 ... CA Berkeley
Brown, A. G. .... 1971 ... CA Berkeley
Brown, C. W. .... 1970 ... CA Berkeley
Bury, R. B. .... 1971 ... CA Berkeley
Fox, R. W. .... 1950 ... CA Berkeley
Glaser, R. .... 1960 ... CA Berkeley
Gorman, J. B., Jr. ...... 1954 ... CA Berkeley
Hendrickson, J. R. ..... 1952 ... CA Berkeley
Karlstrom, E. L. .... 1956 ... CA Berkeley
LaPointe, J. L. .... 1966 ... CA Berkeley ... 78
Marlow, R. W. .... 1979 ... CA Berkeley
McGinnis, S. .... 1964 ... CA Berkeley
Papenfuss, T. J. .... 1979 ... CA Berkeley
Reimer, W. J. .... 1956 ... CA Berkeley

Rodgers, T. L. .... 1953 ... CA Berkeley
Rosenthal, G. M., Jr. .. 1954 ... CA Berkeley
Ruth, S. B. .... 1977 ... CA Berkeley
Tollestrup, K. .... 1978 ... CA Berkeley
Turner, F. B. .... 1957 ... CA Berkeley
Wilhoft, D. C. .... 1962 ... CA Berkeley ... 79
Zweifel, R. G. .... 1954 ... CA Berkeley
44. EVANS, H. E.
    Blackburn, D. G. .... 1985 ... Cornell
45. SNYDER, R. C.
    Pianka, E. R. .... 1965 ... Washington 80
    Frazzeta, T. H. .... 1963 ... Washington 81
    Larsen, J. H., Jr. .... 1963 ... Washington 82
    Waters, J. F. .... 1969 ... Washington
46. STORM, R. M.
    Altig, R. G. .... 1969 ... Oregon St. 83
    Beatty, J. J. .... 1979 ... Oregon St.
    Briggs, J. L. .... 1971 ... Oregon St.
    Brodie, E. D., Jr. .... 1969 ... Oregon St. 84
    Clothier, G. W. .... 1971 ... Oregon St.
    Clover, R. C. .... 1975 ... Oregon St.
    Cohen, N. W. .... 1956 ... Oregon St.
    Darrow, T. D. .... 1967 ... Oregon St.
    Dumas, P. C. .... 1953 ... Oregon St. 85
    Ferguson, D. E. .... 1957 ... Oregon St. 86
    Freiberg, R. E. .... 1954 ... Oregon St.
    Haertel, J. D. .... 1970 ... Oregon St.
    Hanlin, H. G. .... 1980 ... Oregon St.
    Johnson, O. W. .... 1965 ... Oregon St.
    McKenzie, D. S. .... 1970 ... Oregon St.
    Nussbaum, R. A. .... 1972 ... Oregon St. 87
    Roberts, J. O. .... 1970 ... Oregon St.
    Smith, J. M. .... 1967 ... Oregon St.
    Stewart, G. R. .... 1964 ... Oregon St.
    Sype, W. E. .... 1975 ... Oregon St.
    Taylor, J. T. .... 1977 ... Oregon St.
47. BARBOUR, R. W.
    Batson, J. D. .... 1965 ... Kentucky
    Ernst, C. H. .... 1969 ... Kentucky
    Petranka, J. W. .... 1982 ... Kentucky
48. STEWART, M. M.
    Townsend, D. S. .... 1984 ... St. U. New York
    Woolbright, L. L. ...... 1985 ... St. U. New York
49. INGER, R. F.
    Bacon, J. P., Jr. .... 1970 ... Chicago
    Frogner, K. .... 1975 ... Chicago
    Schwalm, P. A. .... 1981 ... Chicago
    Voris, H. K. .... 1969 ... Chicago
50. AWBREY, F. T.
    Maxson, L. R. .... 1973 ... San Diego 88
51. BOGART, J. P.
    Green, D. M. .... 1981 ... Guelph (CN)
52. BROWN, L. E.
    Gray, R. H. .... 1971 ... Illinois St.
    Moll, D. L. .... 1977 ... Illinois St.
    Thrall, J. H. .... 1972 ... Illinois St.
53. FOUQUETTE, M. J., Jr.
    Bowker, R. W. .... 1988 ... Arizona St.
    Hulse, A. C. .... 1974 ... Arizona St.
    Platz, J. E. .... 1974 ... Arizona St.
    Renaud, M. L. .... 1977 ... Arizona St.
    Sullivan, B. K. .... 1983 ... Arizona St.

Vitt, L. J. .... 1976 ... Arizona St.
Wylie, S. R. .... 1981 ... Arizona St.
54. GERHARDT, H. C.
    Doherty, J. A. .... 1982 ... Missouri
55. GUTTMAN, S. I.
    Karlin, A. A. .... 1978 ... Miami (OH)
    Merkle, D. A. .... 1975 ... Miami (OH)
    Romano, M. A. .... 1984 ... Miami (OH)
56. JAMESON, D. L.
    Mackey, J. P. .... 1958 ... Oregon
    Quinn, H. R. .... 1987 ... Houston
57. MECHAM, J. S.
    Flury, A. G. .... 1972 ... Texas Tech
58. NELSON, C. E.
    Lin, E. .... 1975 ... Indiana
    Morris, M. R. .... 1987 ... Indiana
    Williams, P. K. .... 1973 ... Indiana
59. PETTUS, D.
    Corn, P. S. .... 19?? .... Colorado St.
    Gillis, J. E. .... 1975 ... Colorado St.
    Hess, J. B. .... 1969 ... Colorado St.
    Matthews, T. .... 1968 ... Colorado St.
    Post, D. D. .... 1972 ... Colorado St.
    Spencer, A. W. .... 1964 ... Colorado St.
    Tordoff, W., III .... 1970 ... Colorado St.
60. BALLINGER, R. E.
    Jones, S. M. .... 1983 ... Nebraska
61. CUNDALL, D. L.
    Reinert, H. .... 19?? .... Lehigh
62. ARNOLD, S. J.
    Kephart, D. G. .... 1981 ... Chicago
    Shaffer, H. B. .... 1982 ... Chicago
63. ALBERCH, P.
    Burke, A. C. .... 1989 ... Harvard
64. EDWARDS, J. L.
    Jayne, B. C. .... 19?? .... Michigan St.
65. WASSERSUG, R. J.
    Breden, F. .... 19?? .... Chicago
66. HUTCHISON, V. H.
    Braun, B. A. .... 1984 ... Oklahoma
    deLuque, O. .... 1969 ... Rh. Island
    Engbretson, G. A. ...... 1975 ... Oklahoma
    Erskine, D. J. .... 1981 ... Oklahoma
    Graham, T. E. .... 1972 ... Rh. Island
    Gratz, R. K. .... 1976 ... Oklahoma
    Guimond, R. J. .... 1972 ... Rh. Island
    Hazard, E. S. .... 1981 ... Oklahoma
    Kosh, R. J. .... 19?? .... Rh. Island
    Maness, J. D. .... 1979 ... Oklahoma
    McIlvaine, P. M. .... 1971 ... Rh. Island
    Miller, K. .... 1978 ... Oklahoma
    Rawding, R. .... 1989 ... Oklahoma
    Sievert, L. .... 1989 ... Oklahoma
    Songdahl, J. H. .... 1971 ... Rh. Island
    Vinegar, A. .... 1968 ... Rh. Island
    Whitford, W. G. .... 1964 ... Rh. Island 89
67. CAGLE, F. R.
    Boyer, D. R. .... 1958 ... Tulane
    Chaney, A. H. .... 1958 ... Tulane
    Dobie, J. L. .... 1966 ... Tulane 90
    Gordon, R. E. .... 1956 ... Tulane 91
    Shoop, C. R. .... 1963 ... Tulane 92

Tinkle, D. W. .... 1956 ... Tulane 93
68. DOWLING, H. G.
    Brummer, B. J. .... 1980 ... New York
    Estol, C. O. .... 1981 ... New York
    Jenner, J. V. .... 1981 ... New York
    Lazell, J. D. .... 19?? .... Rh. Island
    Muehlbauer, E. I. .... 1987 ... New York
    Price, R. M. .... 1981 ... New York
    Watanabe, M. E. .... 1980 ... New York
69. DUELLMAN, W. E.
    Berger, T. H. .... 1985 ... Kansas
    Burger, W. L. .... 1971 ... Kansas
    Caldwell, J. P. .... 1974 ... Kansas
    Campbell, J. A. .... 1983 ... Kansas
    Crump, M. L. .... 1974 ... Kansas 94
    Dutta, S. K. .... 1985 ... Kansas
    Echternacht, A. C. ..... 1970 ... Kansas 95
    Edwards, S. R. .... 1974 ... Kansas
    Fritts, T. H. .... 1972 ... Kansas
    Frost, D. R. .... 1987 ... Kansas
    Gyi, K. K. .... 1965 ... Kansas
    Joglar, R. L. .... 1986 ... Kansas
    Lee, J. C. .... 1977 ... Kansas
    Lynch, J. D. .... 1969 ... Kansas 96
    Montanucci, R. R. ..... 1972 ... Kansas
    Myers, C. W. .... 1970 ... Kansas
    Péfaur, J. E. .... 1978 ... Kansas
    Roble, S. M. .... 1985 ... Kansas
    Savitzky, A. H. .... 1979 ... Kansas
    Schwaner, T. D. .... 1979 ... Kansas
    Trueb, L. .... 1968 ... Kansas 97
70. BRODKORB, P.
    Holman, J. A. .... 1961 ... Florida 98
71. ADLER, K. K.
    Phillips, J. B. .... 1985 ... Cornell
    Rodda, G. H. .... 1982 ... Cornell
    Ryan, M. J. .... 1982 ... Cornell 99
    Villa, J. D. .... 1978 ... Cornell
    Waldman, B. .... 1983 ... Cornell
72. WILBUR, H. M.
    Alford, R. A. .... 1986 ... Duke
    Chambers, R. C. .... 1984 ... Duke
    Dakin, S. K. .... 1978 ... Duke
    Harris, R. N. .... 1986 ... Duke
    Morin, P. J. .... 1982 ... Duke
    Steinwascher, K. F. .... 1978 ... Duke
    Travis, J. A. .... 1980 ... Duke
73. AUFFENBERG, W.
    Auth, D. L. .... 1980 ... Florida
    Iverson, J. B., III .... 1977 ... Florida
    Jackson, D. .... 1977 ... Florida
    Meylan, P. A. .... 1985 ... Florida
    Taylor, W. .... 1982 ... Florida
    Weaver, W. G. .... 1967 ... Florida
74. HIGHTON, R. T.
    Danstedt, R. T., Jr. .... 1972 ... Maryland
    Fellers, G. M. .... 1976 ... Maryland
    Fraser, D. F. .... 1974 ... Maryland
    Hurlbutt, H. .... 1963 ... Maryland
    Jaeger, R. G. .... 1969 ... Maryland
    Peabody, R. B. .... 1978 ... Maryland
    Schlitter, D. .... 1976 ... Maryland

    Worthington, R. D. .... 1968 ... Maryland
75. PORTER, W. P.
    Muth, F. A. .... 1980 ... Wisconsin
    Tracy, C. R. .... 1972 ... Wisconsin 100
76. POUGH, F. H.
    Beuchat, C. A. .... 1982 ... Cornell
    Mautz, W. J. .... 1979 ... Cornell
    Taigen, T. E. .... 1981 ... Cornell
    Wells, K. D. .... 1976 ... Cornell 101
    Wilson, R. E. .... 1976 ... Cornell
77. ANDERSON, J. D.
    Graham, R. E. .... 1971 ... Rutgers
    Hassinger, D. D. .... 1972 ... Rutgers
78. LaPOINTE, J. L.
    Price, A. H. .... 1986 ... New Mexico
79. WILHOFT, D. C.
    Garrick, L. D. .... 1973 ... Rutgers
    Greenberg, N. B. .... 1973 ... Rutgers
    Marchisin, A. .... 1980 ... Rutgers
    Mele, J. A. .... 1980 ... Rutgers
    Walker, R. F. .... 1971 ... Rutgers
80. PIANKA, E. R.
    Schall, J. J. .... 1976 ... Texas
81. FRAZZETA, T. H.
    Kardong, K. V. .... 1972 ... Illinois
82. LARSEN, J. H., Jr.
    Beneski, J. T. .... 1986 ... Wash. St.
    Davitt, C. M. .... 1987 ... Wash. St.
    Guthrie, D. J. .... 1975 ... Wash. St.
    Herrington, R. E. .... 1985 ... Wash. St.
    Wickham, M. G. .... 1970 ... Wash. St.
    Williams, T. A. .... 1984 ... Wash. St.
    Zalisko, E. J. .... 1987 ... Wash. St.
83. ALTIG, R. G.
    Bonner, L. A. .... 1988 ... Miss. St.
    Fortman, J. R. .... 1973 ... Miss. St.
    Smith, H. C. .... 1975 ... Miss. St.
    Tomson, O. H. .... 1974 ... Miss. St.
84. BRODIE, E. D., Jr.
    Coss, H. T. .... 1974 ... Clemson
    Cupp, P. V., Jr. .... 19?? .... Clemson
    Dodd, C. K., Jr. .... 1974 ... Clemson
    Mushinsky, H. R. ...... 1973 ... Clemson
85. DUMAS, P. C.
    Metter, D. E. .... 1963 ... Idaho 102
86. FERGUSON, D. E.
    Landreth, H. E. .... 1966 ... Miss. St.
    McKeown, J. P. .... 1968 ... Miss. St.
    Murphy, G. G. .... 1970 ... Miss. St.
    Taylor, D. H. .... 1970 ... Miss. St. 103
87. NUSSBAUM, R. A.
    Feaver, P. E. .... 1977 ... Michigan
88. MAXSON, L. R.
    Maha, G. C. .... 1982 ... Illinois
89. WHITFORD, W. G.
    Delson, J. H. .... 1974 ... New Mexico St.
90. DOBIE, J. L.
    Fahey, K. M. .... 1987 ... Auburn
    Scott, A. F. .... 1976 ... Auburn
91. GORDON, R. E.
    Harrison, J. R., III ..... 1964 ... Notre Dame
    MacMahon, J. A. .... 1964 ... Notre Dame

Snyder, D. H. .... 1971 ... Notre Dame
Werner, J. K. .... 1968 ... Notre Dame
92. SHOOP, C. R.
Agardy, M. T. .... 1986 ... Rh. Island
Doty, T. L. .... 1978 ... Rh. Island
Hudson, D. M. .... 1984 ... Rh. Island
Krebs, S. L. .... 1985 ... Rh. Island
Nyman, S. .... 1985 ... Rh. Island
Thompson, N. B. .... 1981 ... Rh. Island
93. TINKLE, D. W.
Collins, J. P. .... 1975 ... Michigan 104
Devine, M. C. .... 1977 ... Michigan
Dunham, A. E. .... 1978 ... Michigan 105
Ferguson, G. W. .... 1969 ... Michigan 106
Flaspohler, R. J. .... 19?? .... Michigan
Hirshfield, M. .... 1976 ... Michigan
Howe, H. F. .... 1977 ... Michigan
McKinney, C. .... 1969 ... Michigan
Ruby, D. E. .... 1976 ... Michigan
Smith, D. C. .... 1977 ... Michigan
Tilley, S. G. .... 1970 ... Michigan
Vinegar, M. E. .... 1972 ... Michigan
94. CRUMP, M. L.
Etheridge, K. .... 19?? .... Florida
Freed, A. .... 19?? .... Florida
Pound, A. .... 19?? .... Florida
95. ECHTERNACHT, A. C.
Jones, R. L. .... 1982 ... Tennessee
Mitchell, J. C. .... 1982 ... Tennessee
Thompson, J. F. .... 1984 ... Tennessee
Wade, J. K. .... 1981 ... Tennessee
96. LYNCH, J. D.
Hoger, C. E. .... 1976 ... Nebraska
Korky, J. K. .... 1977 ... Nebraska
Kruse, K. C. .... 1978 ... Nebraska
97. TRUEB, L.
Anderson, M. D. .... 1978 ... Kansas
Cannatella, D. C. ...... 1985 ... Kansas
Ford, L. S. .... 1989 ... Kansas
Pregill, G. K. .... 1979 ... Kansas
Pyles, R. A. .... 198? ... Kansas
98. HOLMAN, J. A.
Fay, L. P. .... 1984 ... Michigan St.
Rogers, K. L. .... 1975 ... Michigan St.
Sullivan, R. M. .... 1980 ... Michigan St.
Wilson, V. V. .... 1975 ... Michigan St.
99. RYAN, M. J.
Billy, A. J. .... 1986 ... Texas
Wagner, W. .... 1989 ... Texas
100. TRACY, C. R.
Christian, K. A. .... 1984 ... Colorado St.
Snell, H. L. .... 1984 ... Colorado St.
Turner, J. S. .... 1983 ... Colorado St.
101. WELLS, K. D.
Dawley, E. .... 1984 ... Connecticut
Given, M. F. .... 1987 ... Connecticut
102. METTER, D. E.
Bradford, J. .... 1973 ... Missouri
Huser, C. F. .... 1986 ... Missouri
Peterson, C. L. .... 1985 ... Missouri
Powell, R. .... 1984 ... Missouri
103. TAYLOR, D. H.

DeRosa, C. .... 1977 ... Miami (OH)
Ford, N. B. .... 1979 ... Miami (OH)
Nowak, R. T. .... 1983 ... Miami (OH)
104. COLLINS, J. P.
Holomuzski, J. P. ...... 1986 ... Arizona St.
Jones, T. J. .... 1989 ... Arizona St.
Zerba, K. E. .... 1989 ... Arizona St.
105. DUNHAM, A. E.
Newman, R. A. .... 1987 ... Pennsylvania
106. FERGUSON, G. W.
Bohlen, C. H. .... 1976 ... Kansas St.
Cooper, W. E. .... 1974 ... Kansas St.
Derickson, W. K. ...... 1974 ... Kansas St.
Yedlin, I. N. .... 19?? .... ? ? ? ?

SERIES B

1. REDFIELD, A. C.
Clarke, G. C. .... 19?? .... ? ? ? ? 2
2. CLARKE, G. C.
Bartholomew, G. A. ... 1947 ... Harvard 3
3. BARTHOLOMEW, G. A.
Dawson, W. R. .... 1953 ... CA Los Ang. ... 4
Lillywhite, H. B. .... 1970 ... CA Los Ang. ... 5
Lustick, S. .... 1960 ... CA Los Ang. ... 6
MacMillen, R. E. ...... 1960 ... CA Los Ang. ... 7
Seymour, R. S. .... 1972 ... CA Los Ang. ... 8
4. DAWSON, W. R.
Bennett, A. F. .... 1971 ... Michigan 9
Carey, C. .... 1976 ... Michigan
DeWitt, C. B. .... 1963 ... Michigan
Dunson, W. A. .... 1965 ... Michigan 10
Gatten, R. E. .... 19?? ... Michigan
Lasiewski, R. C. .... 1962 ... Michigan 11
Licht, P. .... 1964 ... Michigan 12
Minnich, J. E. .... 1968 ... Michigan
Moberly, W. R., III .... 1966 ... Michigan
Shoemaker, V. H. ..... 1964 ... Michigan 13
Templeton, J. R. .... 19?? .... Michigan 14
5. LILLYWHITE, H. B.
Roberts, J. .... 1980 ... Kansas
Schmid, A. W. .... 1984 ... Kansas
6. LUSTICK, S.
Butler, J. A. .... 1978 ... Ohio St.
Christopher, E. A. ..... 1975 ... CA Irvine
Hinko, A. .... 1980 ... Ohio St.
7. MACMILLEN, R. E.
Case, T. J. .... 1974 ... CA Irvine
8. SEYMOUR, R. S.
Thompson, M. B. ...... 1983 ... Adelaide (AU)
9. BENNETT, A. F.
Adams, B. A. .... 1987 ... CA Irvine
Gleeson, T. T. .... 19?? .... CA Irvine
John-Alder, H. B. ...... 1983 ... CA Irvine
10. DUNSON, W. A.
Freda, J. .... 1985 ... Penn St.
Mazzotti, F. J. .... 1983 ... Penn St.
Stokes, G. D. .... 1981 ... Penn St.
11. LASIEWSKI, R. C.
Weathers, W. W. .... 1969 ... CA Los Ang.
12. LICHT, P.
Feder, M. E. .... 1977 ... CA Berkeley
Porter, D. A. .... 1986 ... CA Berkeley
13. SHOEMAKER, V. H.

Jones, R. M. .... 1978 ... CA Riverside
McClanahan, L. L. ..... 1966 ... CA Riverside
Stinner, J. N. .... 1980 ... CA Riverside
14. TEMPLETON, J. R.
Claussen, D. L. .... 1971 ... Montana 15
15. CLAUSSEN, D. L.
Costanzo, J. .... 1987 ... Miami (OH)
Layne, J. R. .... 1983 ... Miami (OH)
Manis, M. L. .... 1985 ... Miami (OH)

SERIES C

1. OSBORN, H. F.
Gregory, W. K. .... 1910 ... Columbia 2
2. GREGORY, W. K.
Camp, C. L. .... 1923 ... Columbia 3
Noble, G. K. .... 1922 ... Columbia 4
Romer, A. S. .... 1921 ... Columbia 5
Williams, E. E. .... 1950 ... Columbia 6
3. CAMP, C. L.
Estes, R. D. .... 1960 ... CA Berkeley 7
Gregory, J. T. .... 1938 ... CA Berkeley 8
4. NOBLE, G. K.
Evans, L. T. .... 19?? .... ? ? ? ?
Greenberg, B. .... 19?? .... ? ? ? ?
5. ROMER, A. S.
Rivero, J. A. .... 1953 ... Harvard
Vanzolini, P. E. .... 1951 ... Harvard 9
6. WILLIAMS, E. E.
Blake, J. A. .... 19?? .... Harvard
Gans, C. .... 1957 ... Harvard 10
Gorman, G. C. .... 1967 ... Harvard 11
Greer, A. E. .... 19?? .... Harvard
Hall, W. P. .... 19?? .... Harvard
Hertz, P. E. .... 19?? .... Harvard
Holt, R. D. .... 19?? .... Harvard
Huey, R. B. .... 1976 ... Harvard 12
Kiester, A. R. .... 19?? .... Harvard
Mayer, G. C. .... 19?? .... Harvard
Miyata, K. .... 1980 ... Harvard
Moermond, T. C. .... 19?? .... Harvard
Rand, A. S. .... 1961 ... Harvard
Roughgarden, J. .... 19?? .... Harvard
Seger, J. .... 19?? .... Harvard
Wood, R. C. .... 19?? .... Harvard
7. ESTES, R. D.
Meszoely, C. A. .... 1967 ... Boston
8. GREGORY, J. T.
Bramble, D. M. .... 1972 ... CA Berkeley
9. VANZOLINI, P. E.
Rodrigues, M. T. .... 1986 ... Sao Paulo (BR)
Spieker, R. L. .... 1972 ... Sao Paulo (BR)
10. GANS, C.
Taub, A. M. .... 1965 ... St. U. New York
11. GORMAN, G. C.
Adest, G. A. .... 1978 ... CA Los Ang.
Drewes, R. C. .... 1981 ... CA Los Ang.
Harwood, R. H., Jr. .... 1978 ... CA Los Ang.
Lieb, C. S. .... 1981 ... CA Los Ang.
Murphy, R. W. .... 1982 ... CA Los Ang.
Wyles, J. S. .... 1980 ... CA Los Ang.
12. HUEY, R. B.
Adolph, S. C. .... 1987 ... Washington
Tsuji, J. S. .... 1986 ... Washington

Van Berkum, F. H. .... 1985 ... Washington

SERIES D

1. MILLER, A. H.
Eaton, T. H., Jr. .... 1939 ... CA Berkeley 2
Mayhew, W. W. .... 1953 ... CA Berkeley 3
Test, F. H. .... 1940 ... CA Berkeley 4
2. EATON, T. H., Jr.
Callison, G. L. .... 1969 ... Kansas
3. MAYHEW, W. W.
Bartlett, P. N. .... 1970 ... CA Riverside
Wright, S. J. .... 1968 ... CA Riverside
4. TEST, F. H.
Bachman, M. D. .... 1964 ... Michigan
Beiswenger, R. E. ...... 19?? .... Michigan
Carpenter, C. C. .... 1951 ... Michigan 5
Delzell, D. E. .... 1958 ... Michigan
Dole, J. W. .... 1963 ... Michigan
Dundee, H. A. .... 1958 ... Michigan 6
Heatwole, H. F. .... 1960 ... Michigan 7
Martof, B. S. .... 1951 ... Michigan 8
Sexton, O. J. .... 1956 ... Michigan 9
5. CARPENTER, C. C.
Black, J. H. .... 1973 ... Oklahoma
Brackin, M. F. .... 1975 ... Oklahoma
Bussjaeger, L. J. .... 1971 ... Oklahoma
Clarke, R. F. .... 1963 ... Oklahoma
Gillingham, J. C. ...... 1976 ... Oklahoma
Griffith, S. J. .... 1966 ... Oklahoma
Jenssen, T. J. .... 1969 ... Oklahoma
Krupa, S. S. .... 1987 ... Oklahoma
Leuck, B. E. .... 1980 ... Oklahoma
Lynn, R. T. .... 1963 ... Oklahoma
Mahmoud, I. Y. .... 1960 ... Oklahoma
Marcellini, D. L. .... 1970 ... Oklahoma
McDaniel, S. J. .... 1966 ... Oklahoma
Ogilvie, P. W. .... 1966 ... Oklahoma
Paulissen, M. A. .... 1986 ... Oklahoma
Preston, W. P. .... 1970 ... Oklahoma
Purdue, J. R. .... 1974 ... Oklahoma
Sassaman, J. F. .... 1974 ... Oklahoma
Snow, J. E. .... 1984 ... Oklahoma
Tyler, J. D. .... 1969 ... Oklahoma
6. DUNDEE, H. A.
Bayless, L. E. .... 1966 ... Tulane
Hart, D. R. .... 1979 ... Tulane
Rose, F. L. .... 1965 ... Tulane 10
Sever, D. M. .... 1974 ... Tulane
7. HEATWOLE, H. F.
Burns, G. .... 1984 ... New Eng. (AU)
Shine, R. .... 1976 ... New Eng. (AU)
8. MARTOF, B. S.
Bush, F. M. .... 1962 ... Georgia
Forester, D. C. .... 1974 ... N. Carolina St.
Thompson, E. F. .... 1959 ... Georgia
9. SEXTON, O. J.
Axtell, C. B. .... 1973 ... Washington
Bizer, J. R. .... 1977 ... Washington
Hathaway, L. M. .... 1969 ... Washington
Marion, K. R. .... 1970 ... Washington
Ocks, K. F. .... 1970 ... Washington
Seale, D. B. .... 1973 ... Washington
10. ROSE, F. L.

Armentrout, D. .... 19?? .... Texas Tech
Judd, F. W. .... 1973 ... Texas Tech
Long, D. R. .... 1986 ... Texas Tech
Selcer, K. W. .... 1986 ... Texas Tech

SERIES E
1. ? ? ? ?
    Rogers, J. S. .... 19?? .... Michigan 2
2. ROGERS, J. S.
    Carr, A. F., Jr. .... 1937 ... Florida 3
    Goin, C. J. .... 1946 ... Florida 4
3. CARR, A. F., Jr.
    Belkin, D. A. .... 1961 ... Florida
    Bjorndal, K. A. .... 1979 ... Florida
    Caldwell, D. K. .... 1957 ... Florida
    Christman, S. P. .... 1975 ... Florida
    Corn, M. J. .... 1981 ... Florida
    Crenshaw, J. W., Jr. ... 1955 ... Florida 5
    Dietz, D. C. .... 1979 ... Florida
    Ehrenfeld, D. W. .... 1966 ... Florida..............6
    Goodman, D. E. .... 1971 ... Florida
    Gourley, E. V. .... 1969 ... Florida
    Hirth, H. F. .... 1962 ... Florida 7
    Jackson, C. G. .... 1964 ... Florida
    Jackson, J. F. .... 1972 ... Florida
    Meylan, A. B. .... 1984 ... Florida
    Mortimer, J. A. .... 1981 ... Florida
    Mount, R. H. .... 1961 ... Florida 8
    Pritchard, P. C. H. ..... 1969 ... Florida
    Rossman, D. A. .... 1962 ... Florida 9
4. GOIN, C. J.
    Jones, D. A. .... 1967 ... Florida
    Wilder, C. D. .... 1962 ... Florida
5. CRENSHAW, J. W., Jr.
    Newcomer, R. J. .... 1968 ... Maryland
6. EHRENFELD, D. W.
    Bareziley, S. S. .... 1980 ... Rutgers
7. HIRTH, H. F.
    Brown, W. S. .... 1973 ... Utah
    Parker, W. S. .... 1974 ... Utah
8. MOUNT, R. H.
    Shealy, R. M. .... 1973 ... Auburn
    Trauth, S. E. .... 1980 ... Auburn
    Weatherby, C. .... 1982 ... Auburn
9. ROSSMAN, D. A.
    Blaney, R. M. .... 1971 ... Louisiana St.
    Keiser, E. D., Jr. .... 1967 ... Louisiana St. .. 10
    Lawson, R. .... 1985 ... Louisiana St.
    Thomas, J. P. .... 1976 ... Louisiana St.
    Vaeth, R. H. .... 1985 ... Louisiana St.
    Varkey, A. .... 1973 ... Louisiana St.
    Williams, K. L. .... 1970 ... Louisiana St.
    Wilson, L. D. .... 1968 ... Louisiana St.
10. KEISER, E. D., Jr.
    Morgan, E. C. .... 1973 ... SW Louisiana

SERIES F
1. SPEMANN, H.
    Barth, L. .... 19?? .... Freiburg (GE).... 2
2. BARTH, L.
    Moore, J. A. .... 1940 ... Columbia 3
3. MOORE, J. A.
    Carleton, R. .... 19?? .... Columbia
    Ruibal, R. .... 1956 ... Columbia 4

    Salthe, S. N. .... 1963 ... Columbia 5
    Volpe, E. P. .... 1953 ... Columbia 6
4. RUIBAL, R.
    Brown, H. A. .... 19?? .... CA Riverside
    Dimmitt, M. A. .... 1975 ... CA Riverside
    Evans, K. J. .... 1964 ... CA Riverside
    Hillyard, S. D. .... 1974 ... CA Riverside
    Philibosian, F. R. ...... 1972 ... CA Riverside
    Pomeroy, L. V. .... 1981 ... CA Riverside
    Rose, B. R. .... 1981 ... CA Riverside
    Simon, C. A. .... 1973 ... CA Riverside
    Weintraub, J. D. .... 1968 ... CA Riverside
5. SALTHE, S. N.
    Kaplan, R. H. .... 1978 ... C. U. New York
    Levy, P. L. .... 1972 ... C. U. New York
6. VOLPE, E. P.
    Curtis, S. K. .... 1968 ... Tulane
    Dasgupta, S. .... 1962 ... Tulane
    DeGruy, C. V. .... 1974 ... Tulane
    Gebhardt, B. M. .... 1967 ... Tulane
    Roux, K. H. .... 1974 ... Tulane
    Skinner, J. M. .... 1966 ... Tulane
    Turpen, J. B. .... 1973 ... Tulane
    Zettergren, L. D. .... 1972 ... Tulane

SERIES G
1. LANE, H. N.
    Taylor, E. H. .... 1927 ... Kansas 2
2. TAYLOR, E. H.
    Hibbard, C. W. .... 1941 ... Kansas
    Smith, H. M. .... 1936 ... Kansas 3
    Tanner, W. W. .... 1950 ... Kansas 4
3. SMITH, H. M.
    Bell, E. L., II .... 1954 ... Illinois
    Brandon, R. A. .... 1962 ... Illinois 5
    Chrapliwy, P. S. .... 1964 ... Illinois
    Cuellar, H. S. .... 1972 ... Colorado
    Guillette, L. J. .... 1981 ... Colorado
    Hammerson, G. .... 1981 ... Colorado
    Landy, M. J. .... 1970 ... Illinois
    Langebartel, D. A. ..... 1958 ... Illinois
    List, J. C. .... 1956 ... Illinois 6
    Northcutt, G. .... 1968 ... Illinois
    Oldham, J. .... 1975 ... Colorado
    Olsen, R. E. .... 1980 ... Colorado
    Scudder, K. M. .... 1982 ... Colorado
    Smith, P. W. .... 1953 ... Illinois 7
    Swain, T. A. .... 1977 ... Colorado
    Taylor, H. L. .... 1983 ... Colorado
    White, F. N. .... 1955 ... Illinois
4. TANNER, W. W.
    Avery, D. F. .... 1970 ... Brigham Young
    Burkholder, G. L. .... 1973 ... Brigham Young
    Cox, D. C. .... 1976 ... Brigham Young
    Larsen, K. R. .... 1973 ... Brigham Young
    Smith, N. M. .... 1972 ... Brigham Young
5. BRANDON, R. A.
    Besharse, J. C. .... 1973 ... S. Illinois
    Labanick, G. M. .... 1978 ... S. Illinois
    Morris, M. A. .... 1987 ... S. Illinois
    Reilly, S. M. .... 1986 ... S Illinois
    Williams, A. A. .... 1976 ... S. Illinois
    Wortham, J. W., Jr. ... 1975 ... S. Illinois

6. LIST, J. C.
    Atkinson, M. J. .... 1985 ... Ball St.
    Branham, A. E. .... 1977 ... Ball St.
    Kramer, D. C. .... 1971 ... Ball St.
    Miller, D. E. .... 1985 ... Ball St.
    Moshner, D. D. .... 1982 ... Ball St.
    Shepherd, J. L. .... 1973 ... Ball St.
7. SMITH, P. W.
    Liem, S. S. .... 1969 ... Illinois

SERIES H

1. ? ? ? ?
    Allen, A. A. .... 1910 ... Cornell 2
2. ALLEN, A. A.
    Dilger, W. C. .... 1955 ... Cornell 3
    Hensley, M. M. .... 1951 ... Cornell 4
    Welter, W. A. .... 1932 ... Cornell
3. DILGER, W. C.
    Wiewandt, T. .... 1977 ... Cornell
4. HENSLEY, M. M.
    Gibbons, J. W. .... 1967 ... Michigan St. 5
    Nelson, D. H. .... 1974 ... Michigan St.
    Premo, D. B. .... 1985 ... Michigan St.
    Wacasey, J. W. .... 1961 ... Michigan St.
5. GIBBONS, J. W.
    Camp, C. .... 1985 ... Georgia
    Lamb, T. .... 1985 ... Georgia
    Semlitsch, R. D. .... 1984 ... Georgia

SERIES I

1. ? ? ? ?
    Hess, E. H. .... 1948 ... Johns Hopkins .. 2
2. HESS, E. H.
    Burghardt, G. M. .... 1966 ... Chicago 3
3. BURGHARDT, G. M.
    Bock, B. C. .... 1984 ... Tennessee
    Davis, M. .... 1981 ... Tennessee
    Drummond, H. M. ..... 1980 ... Tennessee
    Dugan, B. A. .... 1980 ... Tennessee
    Froese, A. D. .... 1974 ... Tennessee
    Gove, D. .... 1978 ... Tennessee
    Greene, H. W. .... 1977 ... Tennessee 4
    Scudder, L. M. .... 1983 ... Tennessee
    Weldon, P. J. .... 1983 ... Tennessee
4. GREENE, H. W.
    Jaksic, F. M. .... 1982 ... CA Berkeley
    Seib, R. L. .... 1985 ... CA Berkeley

SERIES J

1. ? ? ? ?
    Tanner, J. T. .... 1940 ... Cornell 2
2. TANNER, J. T.
    Bickert, J. H. .... 1970 ... Tennessee
    Orr, L. P. .... 1962 ... Tennessee 3
3. ORR, L. P.
    Bobes, M. .... 1973 ... Kent St.
    Davic, R. D. .... 1983 ... Kent St. 4
    Fitzpatrick, L. C. .... 1970 ... Kent St.
    Keen, W. H. .... 1975 ... Kent St.
    Kleeberger, S. R. .... 1982 ... Kent St.
    Maple, W. T. .... 1974 ... Kent St.
    Shaefer, G. C. .... 1972 ... Kent St.
4. FITZPATRICK, L. C.
    McAllister, C. .... 1989 ... N. Texas St.

SERIES K

1. BORROR, D. J.
    Alexander, R. D. .... 1956 ... Ohio St. 2
2. ALEXANDER, R. D.
    Howard, R. D. .... 1977 ... Michigan 3
    Pace, A. E. .... 1972 ... Michigan
    Sherman, C. K. .... 1980 ... Michigan
3. HOWARD, R. D.
    King, R. B. .... 1985 ... Purdue

SERIES L

1. ? ? ? ?
    Dillon, L. S. .... 1954 ... Texas A&M 2
2. DILLON, L. S.
    Dixon, J. R. .... 1962 ... Texas A&M 3
3. DIXON, J. R.
    Dean, R. H. .... 1984 ... Texas A&M
    Gallagher, D. S., Jr. ... 1979 ... Texas A&M
    Hendricks, F. S. .... 1975 ... Texas A&M
    Johnson, J. D. .... 1984 ... Texas A&M
    Ketchersid, C. A. .... 1974 ... Texas A&M
    Kofron, C. P. .... 1980 ... Texas A&M
    Kroll, J. C. .... 1973 ... Texas A&M
    Mather, C. M. .... 1976 ... Texas A&M
    Rivero-Blanco, C. V. .. 1979 ... Texas A&M
    Saxon, J. G. .... 1970 ... Texas A&M
    Scuddy, J. F. .... 1971 ... Texas A&M
    Sites, J. W. .... 1980 ... Texas A&M
    Thomas, R. A. .... 1976 ... Texas A&M
    Wiest, J. A. .... 1978 ... Texas A&M

SERIES M

1. DOBZHANSKY, T.
    Wallace, B. .... 19?? ..... Columbia 2
2. WALLACE, B.
    Hecht, M. K. .... 1953 ... Cornell 3
3. HECHT, M. K.
    Hinderstein, B. .... 1969 ... C. U. New York
    Kropach, C. N. .... 1973 ... C. U. New York
    Madden, R. C. .... 1975 ... C. U. New York
    Malone, B. S. .... 1979 ... C. U. New York

SERIES N

1. ? ? ? ?
    Dusi, J. L. .... 1950 ... Ohio St. 2
2. DUSI, J. L.
    Folkerts, G. W. .... 1968 ... Auburn 3
    Fugler, C. M. .... 1967 ... Auburn
3. FOLKERTS, G. W.
    Caldwell, R. S. .... 1977 ... Auburn

SERIES O

1. ELLINGER, P.
    Coulson, R. A. .... 19?? ..... London 2
2. COULSON, R. A.
    Dessauer, H. C. .... 1952 ... Louisiana St. 3
    Herbert, J. .... 19?? ..... Louisiana St.
3. DESSAUER, H. C.
    Densmore, L., III .... 1981 ... Louisiana St.
    DiMaggio, A. .... 1961 ... Louisiana St.
    Jiminez, J. M. .... 1963 ... Louisiana St.
    Rosenquist, J. W. .... 1969 ... Tulane
    Shochat, D. .... 1975 ... Louisiana St.
    Sutton, D. E. .... 1969 ... Louisiana St.

SERIES P

1. LARSEN, ??
    Chermock, R. L. .... 1948 ... Cornell 2
2. CHERMOCK, R. L.
    Boyles, J. M. .... 1966 ... Alabama
    Cliburn, J. W. .... 19?? .... Alabama 3
3. CLIBURN, J. W.
    Brode, W. E. .... 1969 ... S. Miss.
    Crain, J. L. .... 1966 ... S. Miss.
    Donovan, L. A. .... 1980 ... S. Miss.
    Schwab, D. S. .... 1972 ... S. Miss.
    Woods, J. E. .... 1968 ... S. Miss.

SERIES Q

1. PARK, O.
    Hairston, N. G. .... 1948 ... Northwestern 2
2. HAIRSTON, N. G.
    Brockelman, W. Y. .... 1968 ... Michigan
    Gill, D. .... 1971 ... Michigan 3
    Nishikawa, K. C. .... 1985 ... N. Carolina
    Organ, J. A. .... 1960 ... Michigan
    Southerland, M. T. ..... 1985 ... N. Carolina
    Stenhouse, S. L. .... 1984 ... N. Carolina
    Van Devender, R. W. .. 1975 ... Michigan
3. GILL, D.
    Bervan, K. .... 19?? .... Maryland

SERIES R

1. PARK, T.
    Cole, L. C. .... 1944 ... Chicago 2
2. COLE, L. C.
    Hurlbert, S. H. .... 1968 ... Cornell
    Sheldon, A. L. .... 19?? .... Cornell 3
3. SHELDON, A. L.
    Daugherty, C. H. .... 1979 ... Montana

SERIES S

1. WEISS, P.
    Kollros, J. J. .... 1942 ... Chicago 2
2. KOLLROS, J. J.
    Baird, J. J. .... 1957 ... Iowa
    Bovbjerg, A. M. .... 1961 ... Iowa
    Decker, R. S. .... 1970 ... Iowa
    Eichler, V. B. .... 1969 ... Iowa
    Flanigen, N. S. .... 1958 ... Iowa
    Harrison, J. L. .... 1958 ... Iowa
    Just, J. J. .... 1968 ... Iowa 3
    Kaung, H. C. .... 1971 ... Iowa
    Martin, E. W. .... 1962 ... Iowa
    Naughton, S. C. .... 1971 ... Iowa
    Payne, W. F. .... 1958 ... Iowa
    Pollack, E. D. .... 1970 ... Iowa
    Race, J., Jr. .... 1960 ... Iowa
    Reynolds, W. A. .... 1962 ... Iowa
    Shell, L. C. .... 1959 ... Iowa
    Stephens, L. B. .... 1957 ... Iowa
3. JUST, J. J.
    Forman, L. J. .... 1979 ... Kentucky

SERIES T

1. BADER, R. S.
    Underhill, D. K. .... 1967 ... Illinois 2
2. UNDERHILL, D. K.
    Tabachnick, W. J. ...... 19?? .... Rutgers

SERIES U

1. ? ? ? ?
    Baldauf, R. J. .... 1956 ... Texas A&M 2
2. BALDAUF, R. J.
    Robinson, D. C. .... 1968 ... Texas A&M

SERIES V

1. ? ? ? ?
    Bothner, R. C. .... 1959 ... Fordham 2
2. BOTHNER, R. C.
    Cerrone, T. H. .... 1983 ... St. Bonaventure

SERIES W

1. BRECKENRIDGE, W. J.
    Tester, J. R. .... 1960 ... Minnesota 2
2. TESTER, J. R.
    Ewert, M. A. .... 19?? .... Minnesota

SERIES X

1. DAVIS, W. B.
    Degenhardt, W. G. ..... 1960 ... Texas A&M 2
2. DEGENHARDT, W. G.
    Aldridge, R. D. .... 1973 ... New Mexico 3
    Applegarth, J. S. .... 1979 ... New Mexico
    Christiansen, J. L. ..... 1969 ... New Mexico
    Hardy, L. M. .... 1969 ... New Mexico
    Jacob, J. S. .... 1977 ... New Mexico
    Jones, K. L. .... 1974 ... New Mexico
    Lucchino, R. V. .... 1973 ... New Mexico
    Seidel, M. E. .... 1973 ... New Mexico
    Sugarman, R. A. .... 1975 ... New Mexico
    Williams, S. R. .... 1976 ... New Mexico
3. ALDRIDGE, R. D.
    Krohmer, R. W. .... 1984 ... St. Louis
    Minsky, J. .... 1985 ... St. Louis
    Weil, M. .... 1979 ... St. Louis

SERIES Y

1. EDDY, S.
    Bellis, E. D. .... 1957 ... Minnesota 2
2. BELLIS, E. D.
    Barthalmus, G. T. .... 1971 ... Penn St.
    Ehlert, G. W. .... 1973 ... Penn St.
    Pluto, T. G. .... 1983 ... Penn St.
    Sherbine, K. B. .... 1969 ... Penn St.

SERIES Z

1. ? ? ? ?
    Frye, B. E. .... 1956 ... Emory 2
2. FRYE, B. E.
    Bartell, M. H. .... 1969 ... Michigan
    Brown, P. R. .... 1968 ... Michigan
    Farrar, E. S. .... 1972 ... Michigan
    Noonen, R. A. .... 1977 ... Michigan
    Snyder, B. W. .... 1970 ... Michigan

SERIES AA

1. HUDSON, G. E.
    Dunlap, D. G. .... 1955 ... Wash St. 2
2. DUNLAP, D. G.
    Anderson, D. L. .... 1967 ... S. Dakota
    Timken, R. L. .... 1968 ... S. Dakota

SERIES AB

1. KEETON, W. T.
    Howland, H. C. .... 1968 ... Cornell 2
2. HOWLAND, H. C.
    Ballarino, J. .... 1982 ... Cornell
    Fleischman, L. J. .... 1986 ... Cornell

SERIES A C
1. LEHRMAN, D. S.
    Crews, D. .... 1973 ... Rutgers 2
2. CREWS, D.
    Garstka, W. .... 19?? .... Texas
    Lindzey, J. .... 19?? .... Texas
    Mason, R. T. .... 1987 ... Texas
    Whittier, J. M. .... 19?? .... Texas

SERIES A D
1. LETTVIN, J.
    Capranica, R. R. .... 1965 ... Mass. I. Tech. .. 2
2. CAPRANICA, R. R.
    Constantine-
        Paton, M. .... 1976 ... Cornell
    Narins, P. .... 1976 ... Cornell
    Rose, G. .... 1983 ... Cornell
    Zakon, H. .... 1981 ... Cornell

SERIES A E
1. ? ? ? ?
    McCauley, W. J. .... 1955 ... South. CA 2
2. McCAULEY, W. J.
    Campbell, C. E. .... 1974 ... Arizona

SERIES A F
1. ? ? ? ?
    McKinnell, R. G. ...... 1959 ... Minnesota 2
2. McKINNELL, R. G.
    Pogany, G. C. .... 1966 ... Tulane

SERIES A G
1. ? ? ? ?
    McNab, B. K. .... 1962 ... Wisconsin 2
2. McNAB, B. K.
    Ackerman, R. .... 1975 ... Florida
    Krakauer, T. H. .... 1970 ... Florida
    Ultsch, G. .... 1972 ... Florida
    Wygoda, M. L. .... 1982 ... Florida

SERIES A H
1. MORRIS, R. W.
    Grigg, G. C. .... 1968 ... Oregon 2
2. GRIGG, G. C.
    Courtice, G. P. .... 19?? .... Sydney (AU)
    Hagon, B. .... 19?? .... Sydney (AU)
    Kirshner, D. .... 1985 ... Sydney (AU)
    Taplin, L. .... 19?? .... Sydney (AU)
    Thompson, J. .... 19?? .... Sydney (AU)
    van Beurden, E. K. ..... 19?? .... Sydney (AU)
    Wright, J. .... 1985 ... Sydney (AU)

SERIES A I
1. SCHECHTMAN, A. M.
    Nace, G. W. .... 1950 ... CA Los Ang. ... 2
2. NACE, G. W.
    Gill, S. J. .... 1972 ... Michigan
    Glass, L. E. .... 1958 ... Duke
    Hegennauer, J. C. ...... 1970 ... Michigan
    Levinson, J. W. .... 1979 ... Michigan
    Mohanty-Hejmadi, P. . 1970 ... Michigan
    Ostrovsky, D. S. .... 1970 ... Michigan
    Rittschof, D. .... 1975 ... Michigan

SERIES A J
1. ALLEE, W. C.
    Pearson, P. G. .... 1954 ... Florida 2
2. PEARSON, P. G.
    Merchant, H. C. .... 1970 ... Rutgers
    Russo, P. M. .... 1972 ... Rutgers
    Taub, F. B. .... 1959 ... Rutgers

SERIES A K
1. RICH, E. R.
    King, F. W. .... 1966 ... Miami (FL) 2
2. KING, F. W.
    Vliet, K. A. .... 1987 ... Florida

SERIES A L
1. ? ? ? ?
    Sealander, J. A., Jr. .... 1949 ... Illinois 2
2. SEALANDER, J. A., Jr.
    Elick, G. E. .... 1970 ... Arkansas
    Spotila, J. R. .... 1970 ... Arkansas

SERIES A M
1. WILSON, E. O.
    Simberloff, D. S. .... 1969 ... Harvard 2
2. SIMBERLOFF, D. S.
    Means, D. B. .... 1975 ... Florida St.

SERIES A N
1. ? ? ? ?
    Valentine, B. D. .... 1960 ... Cornell 2
2. VALENTINE, B. D.
    Handler, C. C. .... 1971 ... Ohio St.
    Juterbock, J. E. .... 1979 ... Ohio St.
    Rubenstein, N. M. ..... 1969 ... Ohio St.

SERIES A O
1. ? ? ? ?
    Weigert, R. C. .... 1962 ... Minnesota 2
2. WIEGERT, R. G.
    Frazer, M. B. .... 1983 ... Georgia
    Richardson, J. I. .... 1982 ... Georgia

SERIES A P
1. AYALA, F. J.
    Avise, J. .... 19?? ..... CA Davis 2
2. AVISE, J.
    Bermingham, E. L. .... 19?? .... Georgia

SERIES A Q
1. BALL, G. H.
    Telford, S. R., Jr. .... 1964 ... CA Los Ang.

SERIES A R
1. BANKS, E. M.
    Moehn, L. D. .... 1971 ... Illinois

SERIES A S
1. (BARBOUR, T.)
    Bailey, J. W. .... 1927 ... Harvard
    Dunn, E. R. .... 1921 ... Harvard

SERIES A T
1. BASKETT, T. S.
    Schroeder, E. E. .... 1966 ... Missouri

SERIES A U
1. BERTKE, E. M.
    Lemanski, L. F. .... 1971 ... Arizona St.
    Mays, C. E. .... 1968 ... Arizona St.
    Nickerson, M. A. .... 1969 ... Arizona St.

SERIES A V
1. BIESELE, J. J.
    Barker, K. R. .... 1966 ... Texas

Man-Li, S. Y. .... 1966 ... Texas
SERIES A W
1. BLACK, C. P.
   Teubner, V. A. .... 1986 ... Toledo
SERIES A X
1. BLAUSTEIN, A. R.
   O'Hara, R. K. .... 1981 ... Oregon St.
SERIES A Y
1. BRISBIN, I. L.
   Standora, E. .... 19?? .... Georgia
SERIES A Z
1. BURT, W. H.
   Schwartz, A. .... 1952 ... Michigan
SERIES B A
1. COLE, R.
   Weigmann, D. L. .... 1982 ... Michigan St.
SERIES B B
1. CONWAY, C. H.
   Wilkinson, R. F. .... 1965 ... Missouri
SERIES B C
1. DANIEL, J. F.
   Eakin, R. M. .... 1935 ... CA Berkeley
SERIES B D
1. DENT, J. N.
   Benson, D. J., Jr. .... 1965 ... Virginia
SERIES B E
1. DICKINSON, J. C.
   Wharton, C. .... 1958 ... Florida
SERIES B F
1. ENGEL, W. L.
   Alexander, D. G. .... 1966 ... N. Carolina
SERIES B G
1. EVANS, F. C.
   deBenedictus, P. A. .... 1970 ... Michigan
SERIES B H
1. EVANS, W. L.
   Woods, W. H. .... 1973 ... Arkansas
SERIES B I
1. FANKHOUSER, G.
   Bachmann, K. .... 19?? .... Princeton
SERIES B J
1. FINDLEY, J. S.
   Fleharty, E. D. .... 1963 ... New Mexico
SERIES B K
1. GEBHARDT, L. P.
   de St. Jeor, S. C. .... 1969 ... Utah
SERIES B L
1. GILBERT, C. H.
   Van Denburgh, J. .... 1897 ... Stanford
SERIES B M
1. GOLLEY, F. B.
   Smith, G. C. .... 19?? .... Georgia
SERIES B N
1. GORDON, M. S.
   Bishop, L. G. .... 1965 ... CA Los Ang.
   Hillman, S. .... 1975 ... CA Los Ang.
   Putnam, R. W. .... 1978 ... CA Los Ang.

SERIES B O
1. HADLEY, N.
   Burns, T. .... 1969 ... Arizona St
   Congdon, J. D. .... 1977 ... Arizona St.
SERIES B P
1. HALL, F. G.
   McCutheon, F. H. ..... 1936 ... Duke
SERIES B Q
1. HALLAM, T. G.
   Hom, C. L. .... 1986 ... Tennessee
SERIES B R
1. HARGITT, G. T.
   McManus, M. I. .... 1935 ... Duke
SERIES B S
1. HERRNKIND, W. F.
   Brock, O. G. .... 1980 ... Florida
SERIES B T
1. HILDEBRAND, M.
   Reiser, G. D. .... 1977 ... CA Davis
SERIES B U
1. HOLMES, S.
   Atsatt, S. R. .... 19?? ..... CA Berkeley
SERIES B V
1. HOWARD, W. E.
   Moore, R. G. .... 1976 ... CA Davis
SERIES B W
1. HUNSICKER, G. R.
   Bradley, G. L. .... 1987 ... Loma Linda
SERIES B X
1. JAMESON, E. W., Jr.
   Allison, A. .... 1979 ... CA Davis
   Benes, E. S. .... 1966 ... CA Davis
   Simon, M. P. .... 1982 ... CA Davis
SERIES B Y
1. JOHNSTON, R. F.
   Packard, G. C. .... 19?? ..... Kansas 2
2. PACKARD, G. C.
   Gutske, W. B. .... 1984 ... Colorado St.
SERIES B Z
1. JONES, A. W.
   Powders, V. N. .... 1967 ... Tennessee
SERIES C A
1. JONES, R. E.
   Duvall, D. .... 1980 ... Colorado
   Fawcett, J. D. .... 1975 ... Colorado
   Fitzgerald, K. T. .... 1982 ... Colorado
   Gerrard, A. M. .... 1974 ... Colorado
   La Greek, F. T. .... 1975 ... Colorado
   Roth, J. J. .... 1974 ... Colorado
   Sherman, C. K. .... 1980 ... Michigan
   Tokarz, R. R. .... 1977 ... Colorado
SERIES C B
1. KAUFMAN, J. H.
   Dalton, S. A. .... 1979 ... Florida
SERIES C C
1. KEMP, N. E.
   Lee, P. A. .... 1965 ... Michigan
SERIES C D

1. KENDEIGH, S. C.
    Bresler, J. B. .... 1957 ... Illinois
    Heckrotte, C. .... 1960 ... Illinois
        SERIES C E
1. KINGSLEY, ? ?
    Higgins, G. M. .... 1919 ... Illinois
        SERIES C F
1. KLOPFER, P.
    Hailman, J. P. .... 1964 ... Duke
        SERIES C G
1. LARIMER, J. L.
    Parker, R. B. .... 1975 ... Texas
        SERIES C H
1. LEONE, C. A.
    Pearson, D. D. .... 1965 ... Kansas
        SERIES C I
1. LUTZ, P. L.
    Lutcavage, M. E. .... 1987 ... Miami (FL)
        SERIES C J
1. MADISON, D. M.
    Petokas, P. J. .... 1987 ... St. U. New York
        SERIES C K
1. MAHER, M. H.
    Waddick, J. W. .... 1971 ... Kansas
        SERIES C L
1. MALVESTUTO, S. P.
    Henword, T. A. .... 1987 ... Auburn
        SERIES C M
1. MICHENER, C. M.
    Andrews, R. M. .... 1971 ... Kansas
        SERIES C N
1. MILLS, K. S.
    Funkhouser, A. B. ..... 1966 ... Oklahoma
        SERIES C O
1. MORGAN, T. H.
    Boring, A. M. .... 1910 ... Bryn Mawr
    King, H. D. .... 1899 ... Bryn Mawr
        SERIES C P
1. NAGY, K. A.
    Fusari-Asplund, M. .... 19?? .... CA Los Ang
        SERIES C Q
1. NESS, J. C.
    Meyer, D. E. .... 1959 ... Wisconsin
    Urban, E. K. .... 1964 ... Wisconsin
        SERIES C R
1. NORRIS, D. O.
    Gern, W. A. .... 1976 ... Colorado
    Morman, M. F. .... 1985 ... Colorado
    Platt, J. E. .... 1974 ... Colorado
    Snow, G. E. .... 1977 ... Colorado
        SERIES C S
1. OWEN, W.
    Comuzzie, D. K. .... 1987 ... Texas A&M
        SERIES C T
1. PADIAN, K.
    Gauthier, J. .... 1984 ... CA Berkeley
        SERIES C U
1. PATTON, J. L.
    Case, S. M. .... 1976 ... CA Berkeley
    Rainey, W. .... 1983 ... CA Berkeley
        SERIES C V
1. REEDER, W. G.
    Ream, C. H. .... 1967 ... Wisconsin
    Vogt, R. C. .... 1978 ... Wisconsin
        SERIES C W
1. REICHERT, S. E.
    Middendorf, G. A. ...... 1979 ... Tennessee
        SERIES C X
1. RICHARDS, A.
    Bragg, A. N. .... 1937 ... Oklahoma
    Johnson, V. O. .... 1940 ... Oklahoma
    Trowbridge, M. S. ..... 1939 ... Oklahoma
        SERIES C Y
1. RIEDESEL, M. L.
    Sturbaum, B. A. .... 1972 ... New Mexico
        SERIES C Z
1. ROSE, S. M.
    Richards, C. M. .... 1957 ... Illinois
        SERIES D A
1. SALT, G. W.
    Willard, D. E. .... 1966 ... CA Davis
        SERIES D B
1. SCHULTZE, H. P.
    Cloutier, R. .... 1989 ... Kansas
        SERIES D C
1. SENNIG, W. C.
    Bourque, J. E. .... 1939 ... Cornell
        SERIES D D
1. SHUMWAY, W.
    Sanders, J. M. .... 1931 ... Illinois
        SERIES D E
1. STAHNKE, H. L.
    Johnson, B. D. .... 1966 ... Arizona St.
        SERIES D F
1. BARLOW, G.
    Stamps, J. A. .... 19?? ..... CA Berkeley 2
2. STAMPS, J. A.
    Ono, K. I. .... 1981 ... CA Davis
    Troyer, K. E. .... 1982 ... CA Davis
        SERIES D G
1. TERBORGH, J. W.
    Toft, C. A. .... 1977 ... Princeton
        SERIES D H
1. TIPTON, S. R.
    Turner, J. E. .... 1970 ... Tennessee
        SERIES D I
1. THORSON, T. B.
    Suhr, J. D. .... 1976 ... Nebraska
        SERIES D J
1. TORREY, H. B.
    Johnson, M. E. .... 1912 ... CA Berkeley
        SERIES D K
1. TRAMA, F. B.
    Dorando, S. L. .... 19?? .... Rutgers
        SERIES D L
1. YOUNG, F. N.

Thurow, G. R. .... 1955 ... Indiana
SERIES D M
1. WALLACE, R. P.
Howard, J. H. .... 1979 ... Utah
SERIES D N
1. MEAD, R. A.
Highfill, D. R. .... 1975 ... Utah
SERIES D O
1. JOHNSON, D. M.
Diller, L. V. .... 1981 ... Utah
SERIES D P
1. MOORE, C.
Reynolds, A. E. .... 1941 ... Chicago
SERIES D Q
1. DU SHANE, G. P.
Stearner, S. P. .... 1946 ... Chicago
SERIES D R
1. EDDS, M. V.
Luckenbill, L. M. ...... 1958 ... Brown
SERIES D S
1. BAGNARA, J. T.
Frost, J. S. .... 1971 ... Arizona 2
2. FROST, J. S.
Hillis, D. M. .... 1985 ... Kansas
SERIES D T
1. UNDERHILL, J.
Hedeen, S. .... 1970 ... Minnesota
SERIES D U
1. STOCKARD, A. H.
Fitch, K. L. .... 1957 ... Michigan
SERIES D V
1. CASE, E. C.
White, T. .... 1936 ... Michigan

ZIMBABWE   (ZI)
SERIES A
1. LOVERIDGE, J. P.
Ganes, I. .... 1989 ... Zimbabwe
Hutton, J. .... 19?? .... Zimbabwe
St. James, D. .... 1988 ... Zimbabwe

INDEX TO LINEAGES

Country codes:

AR = Argentina
AS = Austria
AU = Australia
BR = Brazil
CN = Canada
DE = Denmark
FR = France
GE = Germany
HU = Hungary
IT = Italy
JA = Japan
ME = Mexico
NE = Netherlands
PO = Poland
SA = South Africa
SE = Sweden
SP = Spain
SU = Union of Soviet Socialist Republics
SW = Switzerland
TU = Turkey
UK = United Kingdom
US = United States of America
ZI = Zimbabwe

- A -

Ackerman, R. ... US-AG-2
Adams, B. A. US-B-9
Adams, C. C. US-A-4
Adelmann, H. B. US-A-17
Adest, G. A. US-C-11
Adler, K. K. US-A-38
Adolph, S. C. ... US-C-12
Agardy, M. T. ... US-A-92
Agassiz, L. US-A-1
Agasyan, A. SU-A-4
Alberch, P. US-A-31
Alcala, A. C. US-A-7
Aldridge, R. D. US-X-2
Alekperov, A. M. SU-A-3
Alexander, D. G. US-BF-1
Alexander, R. D. .. US-K-1
Alford, R. A. US-A-72
Allee, W. C. US-AJ-1
Allen, A. A. US-H-1
Allen, A. C. US-A-38
Allison, A. US-BX-1
Altig, R. G. US-A-46
Ananjeva, N. B. ... SU-A-4
Anderson, D. L. . US-AA-2
Anderson, J. D. . US-A-43
Anderson, M. D. US-A-97
Anderson, P. K. . US-A-43
Anderson, S. C. ... US-A-7
Andrewartha, H. ... AU-D-1
Andrews, R. M. US-CM-1
Andrzejewski, H. . PO-A-6
Applegarth, J. S. . US-X-2
Armentrout, D. .. US-D-10
Arnett, J. T. US-A-28
Arnold, E. N. UK-F-1
Arnold, S. J. US-A-29
Asplund, K. K. .. US-A-42
Atatür, M. K. TU-A-3
Atkinson, M. J. ... US-G-6
Atsatt, S. R. US-BU-1
Auffenberg, W. .. US-A-39
Auth, D. L. US-A-73
Avery, D. F. US-G-4
Avise, J. US-AP-1
Awbrey, F. T. US-A-25
Axtell, C. B. US-D-9
Axtell, H. US-A-11
Axtell, R. W. US-A-25
Ayala, F. J. US-AP-1

- B -

Bachman, M. D. .. US-D-4
Bachmann, K. US-BI-1
Bacon, J. P., Jr. . US-A-49
Bader, R. S. US-T-1
Badmajeva, V. SU-B-2
Baestrup, F. W. DE-B-1
Bagnara, J. T. ... US-DS-1
Bailey, J. R. US-A-16
Bailey, J. W. US-AS-1
Baird, J. J. US-S-2
Bakradze, M. SU-A-4
Baldauf, R. J. US-U-1
Ball, G. H. US-AQ-1
Ballarino, J. US-AB-2
Ballinger, R. E. .. US-A-26
Banks, E. M. US-AR-1
Banta, B. H. US-A-7
Baran, I. TU-A-3
Barbieri, F. D. AR-A-2
Barbour, R. W. .. US-A-22
Barbour, T. US-AS-1
Bareziley, S. S. US-E-6
Barker, K. R. US-AV-1
Barlow, G. US-DF-1
Barrett, R. US-A-42
Bartell, M. H. US-Z-1
Barth, L. US-F-1
Barthalamus, G. ... US-Y-2
Bartholomew, G. .. US-B-2
Bartlett, P. N. US-D-3
Barwick, R. E. AU-A-1
Baskett, T. S. US-AT-1
Başoğlu, M. TU-A-2
Batson, J. D. US-A-47
Bauer, A. M. US-A-32
Baur, E. TU-A-1
Baverstock, P. R. AU-K-3
Bayless, L. E. US-D-6
Beargie, K. M. ... US-A-14
Beatty, J. J. US-A-46
Beccari, N. AR-A-1
Bedriaga, J. von ... GE-C-2
Beiswenger, R. E. US-D-4
Belkin, D. A. US-E-3
Bell, C. AU-B-1
Bell, E. L., II US-G-3
Bellairs, A. d'A. .. UK-A-1
Bellis, E. D. US-Y-1
Benes, E. S. US-BX-1
Beneski, J. T. US-A-82
Bennett, A. F. US-B-4
Bennett, L. J. AU-F-2
Benson, D. J., Jr. US-BD-1
Berger, L. PO-H-1
Berger, T. H. US-A-69
Bermingham, E. . US-AP-2
Bernis, F. SP-A-1
Berry, J. F. US-A-27

Berry, K. H. US-A-43	Brock, O. G. US-BS-1	Carr, A. F., Jr. US-E-2	Crump, M. L. ... US-A-69
Bertini, F. AR-A-2	Brockelman, W. ... US-Q-2	Casas-Andreu, G. . ME-A-1	Cuellar, H. S. US-G-3
Bertke, E. M. ... US-AU-1	Brode, W. E. US-P-3	Case, E. C. US-DV-1	Cuellar, O. US-A-14
Bervan, K. US-Q-3	Brodie, E. D., Jr. US-A-46	Case, S. M. US-CU-1	Cullen, J. M. UK-H-1
Besharse, J. C. US-G-5	Brodkorb, P. US-A-37	Case, T. J. US-B-7	Cundall, D. L. ... US-A-28
Beuchat, C. A. ... US-A-76	Brongersma, L. D. NE-B-1	Castanet, J. FR-B-1	Cupp, P. V., Jr. . US-A-84
Bezy, R. L. US-A-41	Brown, A. G. US-A-43	Caughley, G. J. ... AU-C-1	Curtis, S. K. US-F-6
Bickert, J. H. US-J-2	Brown, B. C. US-A-35	Cei, J. M. AR-A-1	Cuvier, G. FR-A-1
Biesele, J. J. US-AV-1	Brown, C. W. ... US-A-43	Cerrone, T. US-V-2	Czopek, G. PO-A-5
Billy, A. J. US-A-99	Brown, H. A. US-F-4	Chambers, R. C. US-A-72	Czopek, J. PO-A-5
Bishop, L. G. US-BN-1	Brown, L. E. US-A-25	Chaney, A. H. ... US-A-67	
Bishop, P. SA-B-2	Brown, P. R. US-Z-2	Channing, A. SA-A-2	- D -
Bishop, S. C. US-A-11	Brown, T. W. US-A-42	Chantell, C. J. ... US-A-40	
Bizer, J. R. US-D-9	Brown, W. C. US-A-7	Chermock, R. L. .. US-P-1	Dakin, S. K. US-A-72
Bjorndal, K. A. US-E-3	Brown, W. S. US-E-7	Chernov, S. A. ... SU-A-2	Dalton, S. A. US-CB-1
Black, C. P. US-AW-1	Brownell, J. A. .. US-A-25	Chitty, D. UK-B-1	Damman, A. E. . US-A-35
Black, J. H. US-D-5	Bruce, R. C. US-A-33	Chrapliwy, P. S. . US-G-3	Daniel, J. F. US-BC-1
Blackburn, D. G. US-A-44	Brummer, B. J. .. US-A-68	Christian, K. ... US-A-100	Danieljan, F. SU-A-4
Blackler, A. UK-G-2	Bryant, S. V. UK-A-2	Christiansen, J. L. US-X-2	Danstedt, R., Jr. .. US-A-74
Blackwell, J. M. . AU-K-2	Budtz, P. DE-A-3	Christman, S. P. .. US-E-3	Darevsky, I. S. SU-A-3
Blair, W. F. US-A-12	Bull, C. M. AU-K-2	Christopher, E. A. US-B-6	Darrow, T. D. US-A-46
Blake, J. A. US-C-6	Bull, J. J. US-A-27	Chun, C. GE-A-1	Dasgupta, S. US-F-6
Blanchard, F. N. . US-A-16	Burbidge, A. A. ... AU-K-2	Clark, D. R. US-A-13	Daugherty, C. H. . US-R-3
Blaney, R. M. US-E-9	Burger, W. L. US-A-69	Clark, H. D. US-A-34	Davic, R. D. US-J-3
Blaustein, A. R. US-AX-1	Burghardt, G. M. .. US-I-2	Clarke, C. J. AU-E-1	Davidson, J. AU-D-1
Blommers-S., R. .. NE-F-1	Burke, A. C. US-A-63	Clarke, G. C. US-B-1	Davies, M. AU-J-1
Bobes, M. US-J-3	Burkett, R. D. ... US-A-13	Clarke, R. F. US-D-5	Davies, P. M. UK-J-1
Bock, B. C. US-I-3	Burkholder, G. L. . US-G-4	Claussen, D. L. .. US-B-14	Davis, D. M. US-A-33
Bogart, J. P. US-A-25	Burns, G. US-D-7	Clay, W. M. US-A-34	Davis, M. US-I-3
Bogdanov, M. SU-A-1	Burns, T. US-BO-1	Cliburn, J. W. US-P-2	Davis, W. B. US-X-1
Bohlen, C. H. .. US-A-106	Burnstock, G. AU-B-1	Cliff, F. S. US-A-7	Davitt, C. M. US-A-82
Bohr, C. DE-A-1	Burt, C. E. US-A-16	Clothier, G. W. .. US-A-46	Dawbin, W. H. AU-E-1
Bonner, L. A. US-A-83	Burt, W. H. US-AZ-1	Cloudsley-T., J. L. UK-I-1	Dawley, E. US-A-101
Bonparte, J. F. AR-B-1	Burton, T. C. AU-J-1	Cloutier, R. US-DB-1	Dawson, W. R. US-B-3
Bookstein, F. L. US-A-29	Bury, R. B. US-A-43	Clover, R. C. US-A-46	Dean, R. H. US-L-3
Boring, A. M. .. US-CO-1	Busack, S. D. US-A-31	Cohen, N. W. US-A-46	De Beer, G. R. UK-C-1
Borkin, L. SU-A-4	Bush, F. M. US-D-8	Cohen, R. AR-A-2	DeBenedictus, P.. US-BG-1
Borror, D. J. US-K-1	Bussjaeger, L. J. .. US-D-5	Cole, C. J. US-A-41	Decker, R. S. US-S-2
Boschma, H. NE-A-1	Bustard, H. R. AU-A-1	Cole, L. C. US-R-1	Degenhardt, W. G. US-X-1
Bothner, R. C. US-V-1	Butler, J. A. US-B-6	Cole, R. US-BA-1	DeGruy, C. V. US-F-6
Bourque, J. E. ... US-DC-1		Collins, J. P. US-A-93	De Jong, J. K. NE-B-1
Bovbjerg, A. M. .. US-S-2	- C -	Comuzzie, D. K. US-CS-1	De Jongh, H. J. ...NE-D-2
Bowker, R. W. .. US-A-53		Congdon, J. D. . US-BO-1	Delson, J. H. US-A-89
Boyer, D. R. US-A-67	Cabada, M. O. AR-A-3	Constantine-P. .. US-AD-2	DeLuque, O. US-A-66
Boyles, J. M. US-P-2	Cadle, J. E. US-A-31	Contard, P. C. UK-A-3	Delzell, D. E. US-D-4
Brackin, M. F. US-D-5	Caetano, M. E. FR-B-2	Conway, C. H. .. US-BB-1	Densmore, L., III . US-O-3
Bradford, J. US-A-102	Cagle, F. R. US-A-35	Cook, F. CN-A-1	Dent, J. N. US-BD-1
Bradley, G. L. ... US-BW-1	Caldwell, D. K. ... US-E-3	Cooper, W. E. . US-A-106	Denuce, Y. M. NE-C-1
Bradshaw, S. D. .. AU-K-2	Caldwell, J. P. .. US-A-69	Corn, M. J. US-E-3	Derickson, W. . US-A-106
Bragg, A. N. US-CX-1	Caldwell, R. S. ... US-N-3	Corn, P. S. US-A-59	DeRosa, C. US-A-103
Bramble, D. M. ... US-C-8	Callison, G. L. US-D-2	Coss, H. T. US-A-84	Dessauer, H. C. ... US-O-2
Brandon, R. A. US-G-3	Camp, C. US-H-5	Costanzo, J. US-B-15	De St. Jeor, S. ... US-BK-1
Branham, A. E. US-G-6	Camp, C. L. US-C-2	Coulson, R. A. ... US-O-1	De Villiers, C. G. SA-A-1
Brattstrom, B. H. US-A-19	Campbell, C. E. US-AE-2	Courtice, G. P. .. US-AH-2	Devine, M. C. ... US-A-93
Braun, B. A. US-A-66	Campbell, H. W. US-A-42	Cowles, R. B. US-A-11	DeWeese, J. E. .. US-A-15
Braysher, M. L. ... UK-B-2	Campbell, J. A. . US-A-69	Cox, D. C. US-G-4	DeWitt, C. B. US-B-4
Breckenridge, W. . US-W-1	Campbell, J. B. . US-A-14	Crain, J. L. US-P-3	Dice, L. R. US-A-6
Breden, F. US-A-65	Cannatella, D. ... US-A-97	Crenshaw, J., Jr. .. US-E-3	Dickinson, J. C. US-BE-1
Bresler, J. B. US-CD-1	Capranica, R. R. US-AD-1	Crespo, E. AR-A-2	Dietz, D. C. US-E-3
Brien, P. AR-C-1	Carey, C. US-B-4	Crews, D. US-AC-1	Diller, L. V. US-DO-1
Briggs, J. L. US-A-46	Carleton, R. US-F-3	Cross, J. K. US-A-41	Dilger, W. C. US-H-2
Brisbin, I. L. US-AY-1	Carpenter, C. C. .. US-D-4	Crowell, R. L. ... US-A-28	Dillon, L. S. US-L-1

DiMaggio, A. US-O-3
Dimmitt, M. A. ... US-F-4
Dixon, J. R. US-L-2
Dobie, J. L. US-A-67
Dobrowolska, H. .. PO-E-1
Dobzansky, T. US-M-1
Dodd, C. K., Jr. . US-A-84
Doherty, J. A. ... US-A-54
Dole, J. W. US-D-4
Donnelly, M. A. US-A-15
Donovan, L. A. ... US-P-3
Dorando, S. L. .. US-DK-1
Dotsenko, I. SU-B-2
Doty, T. L. US-A-92
Dowling, H. G. . US-A-35
Downs, F. L. US-A-35
Drewes, R. C. ... US-C-11
Drummond, H. M. US-I-3
Dubas, G. AU-K-4
Duellman, W. E. US-A-35
Dugan, B. A. US-I-3
Dullemeijer, P. NE-D-1
Dumas, P. C. US-A-46
Dundee, H. A. US-D-4
Dunham, A. E. .. US-A-93
Dunlap, D. D. .. US-AA-1
Dunn, E. R. US-AS-1
Dunson, W. A. US-B-4
Durham, M. S. .. US-A-28
Dusej, G. SW-C-1
DuShane, G. P. US-DQ-1
Dusi, J. L. US-N-1
Dutta, S. K. US-A-69
Duvall, D. US-CA-1
Dyson, M. SA-B-2

- E -

Eakin, R. M. US-BC-1
Eaton, T. H., Jr. .. US-D-1
Echternacht, A. .. US-A-69
Edds, M. V. US-DR-1
Eddy, S. US-Y-1
Edmonds, S. J. AU-F-1
Edwards, J. L. US-A-31
Edwards, S. R. ... US-A-69
Ehlert, G. W. US-Y-2
Ehrenfeld, D. W. .. US-E-3
Eichler, V. B. US-S-2
Elick, G. E. US-AL-2
Ellinger, P. US-O-1
Emelianov, A. A. . SU-A-3
Emerson, S. B. .. US-A-15
Engbretson, G. .. US-A-66
Engel, W. L. US-BF-1
Erjomtschenko, V. SU-B-2
Ernst, C. H. US-A-47
Erskine, D. J. US-A-66
Estes, R. D. US-C-3
Estol, C. O. US-A-68
Etheridge, K. US-A-94
Etheridge, R. E. . US-A-35

Evans, F. C. US-BG-1
Evans, H. E. US-A-20
Evans, K. J. US-F-4
Evans, L. T. US-C-4
Evans, W. US-A-42
Evans, W. L. US-BH-1
Ewert, M. A. US-W-2

- F -

Fahey, K. M. US-A-90
Fankhouser, G. ... US-BI-1
Farrar, E. S. US-Z-2
Fawcett, J. D. ... US-CA-1
Fay, L. P. US-A-98
Feaver, P. E. US-A-87
Feder, M. E. US-B-12
Fejérváry, G. von HU-A-1
Fellers, G. M. ... US-A-74
Ferguson, D. E. . US-A-46
Ferguson, G. W... US-A-93
Ferguson, J. H. . US-A-41
Fernandez, M. R. . AR-A-3
Ferner, J. W. US-A-14
Ferrusquia, I. ME-A-1
Feuer, R. C. US-A-27
Findley, J. S. US-BJ-1
Fischberg, M. SW-A-1
Fitch, H. S. US-A-6
Fitch, K. L. US-DU-1
Fitzgerald, K. T. US-CA-1
Fitzpatrick, L. C. .. US-J-3
Flanigen, N. S. ... US-S-2
Flaspohler, R. J. US-A-93
Fleet, R. R. US-A-26
Fleharty, E. D. ... US-BJ-1
Fleischman, L. J. US-AB-2
Flury, A. G. US-A-57
Folkerts, G. W. ... US-N-2
Follansbee, L. A. US-A-25
Ford, L. S. US-A-97
Ford, N. B. US-A-103
Ford, P. UK-C-1
Forester, D. C. US-D-8
Forman, L. J. US-S-3
Forskål, P. SE-A-1
Fortman, J. R. .. US-A-83
Fouquette, M., Jr. US-A-25
Fox, H. UK-C-1
Fox, R. W. US-A-43
Francio-K., A. PO-A-6
Fraser, D. F. US-A-74
Frazer, M. B. US-AO-2
Frazzeta, T. H. .. US-A-45
Freda, J. US-B-10
Freed, A. US-A-94
Freiberg, R. E. .. US-A-46
Fritts, T. H. US-A-69
Froese, A. D. US-I-3
Frogner, K. US-A-49
Frost, D. R. US-A-69
Frost, J. S. US-DS-1

Frye, B. E. US-Z-1
Fugler, C. M. US-N-2
Fukada, H. JA-A-2
Funkhouser, A. . US-CN-1
Funkhouser, J. W. US-A-7
Funkhouser, W. . US-A-11
Fusari-A., M. ... US-CP-1

- G -

Gadseden-E., H. ... ME-A-2
Gage, S. H. US-A-5
Gallagher, D., Jr. .. US-L-3
Ganes, I. ZI-A-1
Gans, C. US-C-6
Garrick, L. D. US-A-79
Garstka, W. US-AC-2
Gartside, D. F. AU-K-6
Gates, G. O. US-A-41
Gatten, R. E. US-B-4
Gaudin, A. J. US-A-15
Gauthier, J. US-CT-1
Gebhardt, B. M. ... US-F-6
Gebhardt, L. P. . US-BK-1
George, I. D. US-A-34
Gerhardt, H. C. .. US-A-25
Gern, W. A. US-CR-1
Gerrard, A. M. .. US-CA-1
Gibbons, J. R. AU-I-1
Gibbons, J. W. ... US-H-4
Gilbert, C. H. US-BL-1
Gilbert, P. W. US-A-11
Gill, D. US-Q-2
Gill, S. J. US-AI-2
Gillingham, J. C. US-D-5
Gillis, J. E. US-A-59
Given, M. F. ... US-A-101
Glaser, R. US-A-43
Glass, L. E. US-AI-2
Gleeson, T. T. US-B-9
Gloyd, H. K. US-A-34
Goeldi, E. A. GE-C-2
Goin, C. J. US-E-2
Goldberg, S. R. .. US-A-41
Golley, F. B. US-BM-1
Golubev, M. SU-B-2
Gomes, N. M. BR-A-1
Gontscharenko, A. SU-B-2
Good, D. A. US-A-31
Goodman, D. E. ... US-E-3
Gordon, K. L. US-A-11
Gordon, M. S. .. US-BN-1
Gordon, R. E. US-A-67
Gorman, G. C. US-C-6
Gorman, J., Jr. ... US-A-43
Gourley, E. V. US-E-3
Gove, D. US-I-3
Graham, R. E. ... US-A-77
Graham, T. E. US-A-66
Gratz, R. K. US-A-66
Gray, R. H. US-A-52
Greding, E. J. US-A-25

Green, B. AU-D-2
Green, D. M. US-A-51
Greenberg, B. US-C-4
Greenberg, N. B. US-A-79
Greene, H. W. US-I-3
Greer, A. E. US-C-6
Gregory, J. T. US-C-3
Gregory, W. K. US-C-1
Griffith, S. J. US-D-5
Griffiths, R. A. UK-I-1
Grigg, G. C. US-AH-1
Grinnell, J. US-A-3
Grobman, A. B. . US-A-18
Grodzinski, Z. PO-A-1
Grubb, J. C. US-A-25
Gruneberg, H. UK-D-1
Guillette, L. J. US-G-3
Guimond, R. J. .. US-A-66
Guthrie, D. J. ... US-A-82
Gutske, W. B. ... US-BY-2
Guttman, S. I. ... US-A-25
Guyer, C. US-A-15
Guzik, M. PO-A-3
Gyi, K. K. US-A-69

- H -

Haas, G. AS-A-1
Hadley, N. US-BO-1
Haeckel, E. GE-C-1
Haertel, J. D. US-A-46
Hagon, B. US-AH-2
Hailey, A. UK-J-1
Hailman, J. P. ... US-CF-1
Hairston, N. G. ... US-Q-1
Hall, F. G. US-BP-1
Hall, R. J. US-A-13
Hall, W. P. US-C-6
Hallam, T. G. US-BQ-1
Halliday, T. R. ... UK-H-1
Halpern, E. A. ... US-A-41
Hamilton, L. UK-G-2
Hamilton, W., Jr. US-A-11
Hammerson, G. ... US-G-3
Handler, C. C. .. US-AN-2
Hanken, J. US-A-31
Hanlin, H. G. US-A-46
Hardy, L. M. US-X-2
Hargitt, G. T. US-BR-1
Harper, F. US-A-11
Harris, L. E., Jr. US-A-22
Harris, R. N. US-A-72
Harris, R. T. US-A-15
Harrison, J. L. US-S-2
Harrison, J. R. ... US-A-91
Hart, D. R. US-D-6
Hartweg, N. US-A-16
Harwood, R., Jr. US-C-11
Hasselquist, F. SE-A-1
Hassinger, D. D. US-A-77
Hathaway, L. M. .. US-D-9
Hayman, D. L. ... AU-G-1

Hazard, E. S. US-A-66	Hurlbert, S. H. US-R-2	Keen, W. H. US-J-3	Lane, H. N. US-G-1
Healy, W. R. US-A-38	Hurlbutt, H. US-A-74	Keeton, W. T. ... US-AB-1	Langebartel, D. A. US-G-3
Heath, J. E. US-A-42	Huser, C. F. US-A-102	Keiser, E. D., Jr. .. US-E-9	LaPointe, J. L. .. US-A-43
Heatwole, H. F. ... US-D-4	Hutchison, V. H. US-A-33	Kemp, N. E. US-CC-1	Lardner, P. J. US-A-41
Hecht, M. K. US-M-2	Hutton, J. ZI-A-1	Kendeigh, S. C. US-CD-1	Larimer, J. L. ... US-CG-1
Heckrotte, C. US-CD-1		Kennedy, J. P. ... US-A-25	Larsen, E. H. DE-A-3
Hedeen, S. US-DT-1	- I -	Kephart, D. G. ... US-A-62	Larsen, J. H., Jr. US-A-45
Hegennauer, J. C. US-AI-2		Kerfoot, W. C. ... US-A-29	Larsen, K. R. US-G-4
Hemclaar, A. S. ... NE-C-1	Inger, R. F. US-A-24	Ketchersid, C. A. .. US-L-3	Larsen, L. O. DE-A-3
Hendricks, F. S. ... US-L-3	Ingram, G. J. AU-H-1	Kezer, L. J. US-A-11	Larsen, ? ? US-P-1
Hendrickson, J. .. US-A-43	Ireland, P. H. US-A-28	Kielbowna, L. PO-D-3	Larson, L. A. US-A-31
Hensley, M. M. ... US-H-2	Iverson, J. B., III US-A-73	Kiester, A. R. US-C-6	Lasiewski, R. C. .. US-B-4
Henword, T. A. .. US-CL-1		Kikkawa, J. T. AU-H-1	Laughlin, H. E. . US-A-25
Henzell, R. P. UK-B-2	- J -	Kilarski, W. PO-B-1	Laurent, R. F. AR-C-1
Herbert, J. US-O-2		Killibrew, F. C. . US-A-28	LaVilla, E. O. AR-C-2
Heredia-Cano, J. . US-A-25	Jackson, C. G. US-E-3	King, F. W. US-AK-1	Lawson, R. US-E-9
Herrington, R. E. US-A-82	Jackson, D. US-A-73	King, H. D. US-CO-1	Layne, J. R. US-B-15
Herrnkind, W. F. US-BS-1	Jackson, J. F. US-E-3	King, M. AU-G-1	Lazell, J. D. US-A-68
Hertz, P. E. US-C-6	Jacob, J. S. US-X-2	King, R. B. US-K-3	Lecyk, M. PO-D-4
Hess, E. H. US-I-1	Jaeger, R. G. US-A-74	King, R. D. AU-D-2	Lee, A. K. AU-K-2
Hess, J. B. US-A-59	Jaksic, F. M. US-I-4	Kingsbury, B. F. . US-A-9	Lee, J. C. US-A-69
Hetherington, T. US-A-32	Jameson, D. L. .. US-A-25	Kingsley, ? ? US-CE-1	Lee, P. A. US-CC-1
Heyer, W. R. US-A-14	Jameson, E., Jr. . US-BX-1	Kirschner, D. US-AH-2	Legler, J. M. US-A-13
Hibbard, C. W. US-G-2	Jaslow, A. P. US-A-29	Kleeberger, S. R. .. US-J-3	Legname, A. H. .. AR-A-3
Higgins, G. M. .. US-CE-1	Jayne, B. C. US-A-64	Klopfer, P. US-CF-1	Lehrman, D. S. . US-AC-1
Highfill, D. R. . US-DN-1	Jenner, J. V. US-A-68	Klosterman, L. L. US-A-32	Lemanski, L. F. US-AU-1
Highton, R. T. .. US-A-39	Jenssen, T. J. US-D-5	Kluge, A. G. US-A-15	Leone, C. A. US-CH-1
Hildebrand, M. ... US-BT-1	Jiminez, J. M. US-O-3	Knopf, G. N. US-A-14	Lessona, M. IT-A-1
Hill, L. G. AU-E-1	Joglar, R. US-A-69	Kochva, E. AS-A-2	Lettvin, J. US-AD-1
Hillis, D. M. US-DS-2	John-Alder, H. B. . US-B-9	Kofron, C. P. US-L-3	Leuck, B. E. US-D-5
Hillman, S. US-BN-1	Johnson, D. M. . US-DO-1	Kollros, J. J. US-S-1	Levinson, J. W. .. US-AI-2
Hillyard, S. D. US-F-4	Johnson, B. D. .. US-DE-1	Komai, T. JA-A-1	Leviton, A. E. US-A-7
Hinderstein, B. US-M-3	Johnson, J. D. US-L-3	Kordylewski, L. ... PO-B-1	Levy, P. L. US-F-5
Hinko, A. US-B-6	Johnson, M. E. .. US-DJ-1	Korky, J. K. US-A-96	Licht, P. US-B-4
Hirschler, J. PO-D-2	Johnson, O. W. . US-A-46	Koscielski, B. PO-C-1	Lieb, C. S. US-C-11
Hirshfield, M. US-A-93	Johnson, V. O. .. US-CX-1	Kosh, R. J. US-A-66	Lieberman, S. S. US-A-15
Hirth, H. F. US-E-3	Johnston, R. F. .. US-BY-1	Kosswig, C. TU-A-1	Liem, S. S. US-G-7
Hoger, C. E. US-A-96	Jones, A. W. US-BZ-1	Kotenko, T. SU-B-2	Lillywhite, H. B. . US-B-3
Holman, J. A. ... US-A-70	Jones, D. A. US-E-4	Kowalewski, L. ... PO-A-4	Lin, E. US-A-58
Holmes, S. US-BU-1	Jones, K. L. US-X-2	Krakauer, T. H. . US-AG-2	Lindsay, H. L. ... US-A-25
Holomuzki, J. . US-A-104	Jones, R. E. US-CA-1	Kramer, D. C. US-G-6	Lindzey, J. US-AC-2
Holt, R. D. US-C-6	Jones, R. L. US-A-95	Kramer, E. SW-B-1	Linnaeus, C. SE-A-1
Holzapfel, R. A. US-A-36	Jones, R. M. US-B-13	Kramer, G. GE-B-1	List, J. C. US-G-3
Hom, C. L. US-BQ-1	Jones, S. M. US-A-60	Kraus, E. F. US-A-29	Littlejohn, M. J. . AU-K-2
Hoogmoed, M. S. NE-B-2	Jones, T. J. US-A-104	Krebs, S. L. US-A-92	Liu, C.-c. US-A-11
Horton, P. AU-J-1	Jopson, H. G. ... US-A-22	Krogh, A. DE-A-1	Livezey, R. L. ... US-A-11
Houck, L. D. US-A-31	Jordan, D. S. US-A-2	Kroll, J. C. US-L-3	Lofling, P. SE-A-1
Howard, J. H. ... US-DM-1	Jorgensen, C. B. .. DE-A-2	Krohmer, R. W. .. US-X-3	Loftus-Hills, J. J. AU-K-6
Howard, R. D. US-K-2	Judd, F. W. US-D-10	Kropach, C. N. ... US-M-3	Lombard, R. E. .. US-A-31
Howard, W. E. ... US-BV-1	Just, J. J. US-S-2	Krupa, S. S. US-D-5	Long, D. R. US-D-10
Howe, H. F. US-A-93	Juszczyk, W. PO-A-2	Kruse, K. C. US-A-96	Loomis, R. B. ... US-A-13
Howland, H. C. . US-AB-1	Juterbock, J. E. . US-AN-2	Kupriyanova, L. .. SU-A-4	Lotter, C. F. US-A-30
Hoyer, H. PO-A-1		Kuss, B. D. US-A-28	Loumont, C. SW-A-1
Hu, S.-q.. .... US-A-23	- K -		Loveridge, J. P. ZI-A-1
Hudson, D. M. .. US-A-92		- L -	Low, B. S. US-A-25
Hudson, G. E. .. US-AA-1	Kalm, P. SE-A-1		Lowe, C. H., Jr. US-A-19
Huey, R. B. US-C-6	Kaplan, R. H. US-F-5	Labanick, G. M. .. US-G-5	Lucchino, R. US-X-2
Hulse, A. C. US-A-53	Kardong, K. V. .. US-A-81	La Greek, F. T. . US-CA-1	Luckenbill, L. .. US-DR-1
Humphries, R. B. AU-K-6	Karlin, A. A. US-A-55	Laister, A. F. SU-A-2	Lustick, S. US-B-3
Hunsaker, D., II . US-A-25	Karlstrom, E. L. US-A-43	Lamb, T. US-H-5	Lutcavage, M. E. US-CI-1
Hunsicker, G. R. US-BW-1	Kaufman, J. H. .. US-CB-1	Landreth, H. E. .. US-A-86	Lutz, P. L. US-CI-1
Hunt, H. W. US-A-25	Kaung, H. C. US-S-2	Landy, M. J. US-G-3	Lynch, J. D. US-A-69

Lynch, J. F. US-A-31
Lynn, R. T. US-D-5

- M -

Mackey, J. P. US-A-56
MacMahon, J. A. US-A-91
MacMillen, R. E. . US-B-3
Madden, R. C. US-M-3
Maderson, P. F. .. UK-A-2
Madison, D. M. . US-CJ-1
Magnusson, W. .. AU-C-1
Maha, G. C. US-A-88
Maher, M. H. ... US-CK-1
Mahmoud, I. Y. ...US-D-5
Main, A. R. AU-K-1
Maiorana, V. C. . US-A-31
Maki, M. JA-A-1
Malaret, L. US-A-13
Malone, B. S. US-M-3
Malvestuto, S. P. US-CL-1
Manes, M. E. AR-A-3
Maness, J. D. US-A-66
Manis, M. L. US-B-15
Man-Li, S. Y. .. US-AV-1
Maple, W. T. US-J-3
Marcellini, D. L. .. US-D-5
Marchisin, A. US-A-79
Mares, M. A. US-A-25
Marion, K. R. US-D-9
Marlow, R. W. .. US-A-43
Martin, A. A. AU-K-6
Martin, E. W. US-S-2
Martin, P. S. US-A-38
Martin, R. F. US-A-25
Martin, W. F. US-A-25
Martof, B. S. US-D-4
Maslin, T. P. US-A-7
Mason, R. T. ... US-AC-2
Mather, C. M. US-L-3
Matsui, M. JA-B-1
Matthews, T. US-A-25
Matthews, T. US-A-59
Mautz, W. J. US-A-76
Maxson, L. R. ... US-A-50
Mayer, G. C. US-C-6
Mayhew, W. W. ..US-D-1
Mays, C. E. US-AU-1
Mazzotti, F. J. ... US-B-10
McAllister, C. US-J-1
McAllister, W. .. US-A-25
McCauley, R. H. US-A-11
McCauley, W. J. US-AE-1
McClanahan, L. . US-B-13
McCoy, C. J. US-A-14
McCutheon, F. .. US-BP-1
McDaniel, S. J. ...US-D-5
McDiarmid, R. .. US-A-15
McGinnis, S. US-A-43
McIlvaine, P. M. US-A-66
McKenzie, D. S. US-A-46
McKeown, J. P. . US-A-86

McKinnell, R. G. US-AF-1
McKinney, C. ... US-A-93
McKown, R. R. US-A-25
McManus, M. I. US-BR-1
McMullen, E. C. US-A-10
McNab, B. K. ... US-AG-1
Meacham, W. R. US-A-25
Mead, R. A. US-DN-1
Means, D. B. ... US-AM-2
Mecham, J. S. ... US-A-25
Medem, F. GE-B-1
Méhelÿ, L. HU-A-1
Mele, J. A. US-A-79
Merchant, H. C. . US-AJ-2
Merkle, D. A. ... US-A-55
Mertens, R. GE-A-1
Meszoely, C. A. .. US-C-7
Metter, D. E. US-A-85
Meyer, D. E. US-CQ-1
Meyer, J. R. US-A-15
Meylan, A. B. US-E-3
Meylan, P. A. ... US-A-73
Miceli, D. C. AR-A-3
Michaud, T. C. .. US-A-25
Michener, C. M. US-CM-1
Middendorf, G. . US-CW-1
Miller, A. H. US-D-1
Miller, D. E. US-G-6
Miller, K. US-A-66
Mills, K. S. US-CN-1
Milne-Edwards, H. FR-A-1
Milstead, W. W. US-A-25
Minnich, J. E. US-B-4
Minsky, J. US-X-3
Mitchell, J. C. .. US-A-95
Mitchell, S. L. .. US-A-30
Miyamoto, M. .. US-A-15
Miyata, K. US-C-6
Młynarski, M. PO-A-2
Moberly, W., III .. US-B-4
Moehn, L. D. US-AR-1
Moermond, T. C. . US-C-6
Mohanty-H., P. .. US-AI-2
Moll, D. L. US-A-52
Moll, E. O. US-A-27
Montanucci, R. . US-A-69
Moody, S. M. ... US-A-29
Moore, C. US-DP-1
Moore, J. A. US-F-2
Moore, R. G. US-BV-1
Morafka, D. J. ... US-A-15
Morgan, E. C. US-E-10
Morgan, T. H. ... US-CO-1
Morin, P. J. US-A-72
Morman, M. F. . US-CR-1
Morris, M. A. US-G-5
Morris, M. R. ... US-A-58
Morris, R. W. ... US-AH-1
Mortimer, J. A. ... US-E-3
Moshner, D. D. ... US-G-6
Mosimann, J. E. US-A-35
Mount, R. H. US-E-3

Muehlbauer, E. I. US-A-68
Müller, J. GE-C-1
Murphy, G. G. ... US-A-86
Murphy, R. W. .. US-C-11
Murphy, T. D. ... US-A-33
Mushinsky, H. ... US-A-84
Muth, F. A. US-A-75
Myers, C. W. US-A-69
Myers, G. S. US-A-3

- N -

Nace, G. W. US-AI-1
Nagy, K. A. US-CP-1
Nakamura, K. JA-A-1
Narins, P. US-AD-2
Naughton, S. C. ... US-S-2
Negmedzyanov SU-A-4
Nelson, C. E. US-A-25
Nelson, D. H. US-H-4
Nelson, J. AU-I-1
Ness, J. C. US-CQ-1
Nevo, E. AS-A-2
Newcomer, R. J. .. US-E-5
Newman, R. A. US-A-105
Newth, D. UK-G-1
Nickerson, M. A. US-AU-1
Nikolsky, A. M. . SU-A-1
Nishikawa, K. C. US-Q-2
Noble, G. K. US-C-2
Noonen, R. A. US-Z-2
Norris, D. O. US-CR-1
Norris, K. S. US-A-19
Northcutt, G. US-G-3
Nowak, R. T. .. US-A-103
Nusbaum-H., J. ... PO-D-1
Nussbaum, R. A. US-A-46
Nyman, S. US-A-92

- O -

Oates, C. UK-E-1
Ocks, K. F. US-D-9
Odendaal, F. AU-K-4
Ogielska, M. PO-C-1
Ogilvie, P. W. US-D-5
O'Hara, R. K. ... US-AX-1
Oken, L. US-A-1
Oldham, J. US-G-3
Oliver, J. A. US-A-16
Olsen, R. E. US-G-3
Ono, K. I. US-DF-2
Oomen, H. C. NE-E-1
Oplinger, C. S. .. US-A-22
Organ, J. A. US-Q-2
Orlov, N. L. SU-A-4
Orr, L. P. US-J-2
Orska, J. PO-D-3
Ortenburger, A. I. US-A-16
Orton, G. L. US-A-35
Osbeck, P. SE-A-1
Osborn, H. F. US-C-1

Osgood, D. W. .. US-A-33
Ostrovsky, D. S. . US-AI-2
Owen, W. US-CS-1
Özeti, N. TU-A-3

- P -

Pace, A. E. US-K-2
Packard, G. C. ... US-BY-1
Padian, K. US-CT-1
Papenfuss, T. J. . US-A-43
Park, O. US-Q-1
Park, T. US-R-1
Parker, H. W. NE-A-2
Parker, R. B. US-CG-1
Parker, W. S. US-E-7
Passmore, N. SA-B-1
Patterson, H. SA-B-1
Patton, J. L. US-CU-1
Paulissen, M. A. ..US-D-5
Pawlowska-I., A. . PO-D-4
Payne, W. F. US-S-2
Peabody, R. B. .. US-A-74
Pearson, D. D. .. US-CH-1
Pearson, P. G. ... US-AJ-1
Péfaur, J. E. US-A-69
Pennock, L. A. .. US-A-14
Peracca, M. G. IT-A-1
Peters, J. A. US-A-35
Peters, W. GE-C-1
Peterson, C. L. US-A-102
Peterson, J. US-A-31
Petokas, P. J. US-CJ-1
Petranka, J. W. .. US-A-47
Pettus, D. US-A-25
Pexioto, O. L. BR-A-1
Philibosian, F. R. US-F-4
Phillips, J. B. US-A-71
Pianka, E. R. US-A-45
Pides, M. SA-B-2
Pierce, J. R. US-A-25
Pisanets, E. SU-B-2
Pizchelawri, V. SU-A-4
Platt, D. R. US-A-13
Platt, J. E. US-CR-1
Platz, J. E. US-A-53
Plummer, M. V. US-A-13
Pluto, T. G. US-Y-2
Plytycz, B. PO-A-5
Poczopko, P. PO-A-5
Pogany, G. C. ... US-AF-2
Pollack, E. D. US-S-2
Pomeroy, L. V. ... US-F-4
Porter, D. A. US-B-12
Porter, K. R. US-A-25
Porter, W. P. US-A-42
Post, D. D. US-A-59
Pough, F. H. US-A-42
Pound, A. US-A-94
Powders, V. N. .. US-BZ-1
Powell, J. E. AR-B-1
Powell, R. US-A-102

Pregill, G. K. US-A-97
Premo, D. B. US-H-4
Presch, W. F. US-A-15
Preston, W. P. US-D-5
Price, A. H. US-A-78
Price, R. M. US-A-68
Pritchard, P. C. US-E-3
Przystalska, A. PO-A-6
Purdue, J. R. US-D-5
Putnam, R. W. .. US-BN-1
Pyburn, W. F. ... US-A-25
Pyles, R. A. US-A-97

- Q -

Quinn, H. R. US-A-56

- R -

Raabe, Z. PO-E-1
Rabb, G. B. US-A-38
Race, J., Jr. US-S-2
Rafinski, J. N. PO-A-5
Rainey, W. US-CU-1
Raisman, J. S. AR-A-3
Ralin, D. B. US-A-25
Rand, A. S. US-C-6
Rasmussen, J. B. . DE-B-2
Rawding, R. US-A-66
Reagan, D. P. US-A-28
Ream, C. H. US-CV-1
Redfield, A. C. US-B-1
Reed, H. G. US-A-5
Reeder, W. G. ... US-CV-1
Reese, R. W. US-A-14
Regan, G. T. US-A-13
Regenass, U. SW-B-1
Reichert, S. E. .. US-CW-1
Reilly, S. M. US-G-5
Reimer, W. J. US-A-43
Reinert, H. US-A-61
Reiser, G. D. US-BT-1
Renaud, M. L. ... US-A-53
Rengel, D. AR-A-2
Reynolds, A. E. . US-DP-1
Reynolds, R. P. . US-A-30
Reynolds, W. A. .. US-S-2
Rich, E. R. US-AK-1
Richards, A. US-CX-1
Richards, C. M. . US-CZ-1
Richardson, J. I. US-AO-2
Ricqles, A. de FR-B-1
Riedesel, M. L. . US-CY-1
Rittschof, D. US-AI-2
Rivero, J. A. US-C-5
Rivero-B., C. V. .. US-L-3
Roberts, J. US-B-5
Roberts, J. D. UK-B-2
Roberts, J. O. US-A-46
Robinson, D. C. .. US-U-2
Robertson, J. G. . AU-K-6
Robinson, M. D. US-A-41

Roble, S. M. US-A-69
Rodda, G. H. US-A-71
Rodgers, T. L. ... US-A-43
Rodrigues, M. T. . US-C-9
Rogers, J. S. US-A-25
Rogers, J. S. US-E-1
Rogers, K. L. US-A-98
Romano, M. A. . US-A-55
Romer, A. S. US-C-2
Rose, B. R. US-F-4
Rose, F. L. US-D-6
Rose, G. US-AD-2
Rose, S. M. US-CZ-1
Rosenkilde, P. DE-A-3
Rosenquist, J. W. US-O-3
Rosenthal, G., Jr. US-A-43
Rossman, D. A. .. US-E-3
Roth, J. J. US-CA-1
Roughgarden, J. ... US-C-6
Roux, K. H. US-F-6
Rubenstein, N. .. US-AN-2
Ruby, D. E. US-A-93
Ruibal, R. US-F-3
Russo, P. M. ... US-AJ-2
Ruth, S. B. US-A-43
Ruthven, A. G. .. US-A-8
Ryan, M. J. US-A-71
Rzechak, K. PO-G-1

- S -

Sage, R. D. US-A-25
Salomon, H. L. . AR-A-3
Salt, G. W. US-DA-1
Salthe, S. N. US-F-3
Salvador, A. SP-A-1
Sanchez, S. O. .. AR-A-3
Sanders, J. M. . US-DD-1
Sarasin, C. F. ... SW-D-1
Sassaman, J. F. .. US-D-5
Sattorov, T. SU-B-2
Savage, J. M. US-A-7
Savitzky, A. H. US-A-69
Saxon, J. G. US-L-3
Sbrebo, Z. PO-F-3
Schall, J. J. US-A-80
Schätti, B. SW-C-1
Schimscheiner, L. PO-A-3
Schechtman, A. . US-AI-1
Schiøtz, A. DE-B-3
Schlitter, D. US-A-74
Schmid, A. W. ... US-B-5
Schmidt, K. P. . US-A-11
Schreber, J. C. ... SE-A-1
Schroeder, E. E. US-AT-1
Schuierer, F. W. . US-A-7
Schultze, H. P. . US-DB-1
Schwab, D. S. US-P-3
Schwalm, P. A. US-A-49
Schwaner, T. D. US-A-69
Schwartz, A. US-AZ-1
Schwenk, K. US-A-32

Scott, A. F. US-A-90
Scott, N. J., Jr. US-A-15
Scrocchi, G. J. ... AR-C-2
Scudder, K. M. ... US-G-3
Scudder, L. M. US-I-3
Scuddy, J. F. US-L-3
Sealander, J., Jr. US-AL-1
Seale, D. B. US-D-9
Seger, J. US-C-6
Seib, R. L. US-I-4
Seidel, M. E. US-X-2
Seigel, R. A. ... US-A-13
Selcer, K. W. ... US-D-10
Sembrat, K. PO-D-2
Semlitsch, R. D. US-H-5
Semper, C. SW-D-1
Sennig, W. C. . US-DC-1
Sessions, S. K. US-A-31
Sever, D. M. US-D-6
Sexton, O. J. US-D-4
Seymour, R. S. .. US-B-3
Shaefer, G. C. US-J-3
Shaffer, H. B. .. US-A-62
Shafland, J. L. . US-A-31
Shallenberger, E. US-A-42
Shcherbak, M. ... SU-B-1
Shcherban, K. SU-B-2
Shealy, R. M. US-E-8
Sheldon, A. L. ... US-R-2
Shell, L. C. US-S-2
Shepherd, J. L. ... US-G-6
Sherbine, K. B. .. US-Y-2
Sherman, C. K. ... US-K-2
Sherman, C. K. US-CA-1
Shine, R. US-D-7
Shochat, D. US-O-3
Shoemaker, V. ... US-B-4
Shoop, C. R. ... US-A-67
Shumway, W. .. US-DD-1
Sieb, R. L. US-I-4
Siedlecki, M. PO-F-1
Sievert, L. US-A-66
Silverstone, P. . US-A-15
Simberloff, D. US-AM-1
Simnett, J. UK-E-1
Simon, C. A. US-F-4
Simon, M. P. .. US-BX-1
Sites, J. W. US-L-3
Skinner, J. M. ... US-F-6
Skowdon, S. PO-F-1
Sliwa, L. PO-F-3
Smith, A. K. ... US-A-13
Smith, D. C. ... US-A-93
Smith, G. C. .. US-BM-1
Smith, H. C. ... US-A-83
Smith, H. M. US-G-2
Smith, J. M. ... US-A-46
Smith, N. M. US-G-4
Smith, P. W. US-G-3
Smreczynski, S. . PO-G-1
Smyth, M. B. UK-B-1
Snell, H. L. US-A-100

Snow, G. E. US-CR-1
Snow, J. E. US-D-5
Snyder, B. W. US-Z-2
Snyder, D. H. ... US-A-91
Snyder, R. C. ... US-A-20
Songdahl, J. H. . US-A-66
Southerland, M. . US-Q-2
Spemann, H. US-F-1
Spencer, A. W. . US-A-59
Spieker, R. L. ... US-C-9
Spies, I. DE-A-3
Spotila, J. R. .. US-AL-2
Srebro, Z. PO-F-2
Stamps, J. A. ... US-DF-1
Standora, E. US-AY-1
Stanke, H. L. .. US-DE-1
Starrett, P. H. ... US-A-38
Stearner, S. P. . US-DQ-1
Stebbins, R. C. US-A-19
Steenberg, C. M. DE-B-1
Steinwascher, K. US-A-72
Stenhouse, S. L. ... US-Q-2
Stephens, L. B. ... US-S-2
Stewart, G. R. .. US-A-46
Stewart, J. R. .. US-A-31
Stewart, K. W. .. CN-A-1
Stewart, M. M. . US-A-22
Stickel, W. H. .. US-A-34
Stinner, J. N. ... US-B-13
St. James, D. ZI-A-1
Stock, Y. H. NE-F-1
Stockard, A. H. US-DU-1
Stokes, G. D. ... US-B-10
Storer, T. I. US-A-6
Storm, R. M. ... US-A-21
Storr, G. M. AU-K-2
Straughn, I. R. .. AU-K-5
Strawinski, S. ... PO-A-5
Stropes, S. L. ... US-A-41
Stuart, L. C. US-A-16
Stull, O. G. US-A-16
Sturbaum, B. A. US-CY-1
Sugarman, R. US-X-2
Suhr, J. D. US-DI-1
Sullivan, B. K. .. US-A-53
Sullivan, R. M. US-A-98
Sura, P. PO-F-3
Sutton, D. E. US-O-3
Swain, T. A. US-G-3
Sweet, S. S. US-A-31
Swierad, J. PO-A-3
Sype, W. E. US-A-46
Szarski, H. PO-A-2
Szymura, J. M. .. PO-A-5
Szyndlar, Z. PO-A-4

- T -

Tabachnick, W. J. US-T-2
Taigen, T. E. US-A-76
Talbot, J. J. US-A-15
Tandy, R. M. ... US-A-25

Tanner, J. T. US–J–1
Tanner, W. W. ... US–G–2
Taplin, L. US–AH–2
Tarashchuk, S. SU–B–2
Tasumi, M. JA–B–1
Taub, A. M. US–C–10
Taub, F. B. US–AJ–2
Taylor, D. H. ... US–A–86
Taylor, E. H. US–G–1
Taylor, H. L. US–G–3
Taylor, J. T. US–A–46
Taylor, W. US–A–73
Telford, S. SA–B–2
Telford, S., Jr. . US–AQ–1
Templeton, J. R. ..US–B–4
Terborgh, J. W. US–DG–1
Tertishinkov, M. . SU–B–2
Test, F. H. US–D–1
Tester, J. R. US–W–1
Teubner, V. A. US–AW–1
Thomas, J. P. US–E–9
Thomas, R. A. ... US–L–3
Thompson, E. F. US–D–8
Thompson, J. ... US–AH–2
Thompson, J. F. US–A–95
Thompson, M. B. US–B–8
Thompson, N. .. US–A–92
Thompson, W. . US–A–25
Thornton, W. A. US–A–25
Thorson, T. B. .. US–DI–1
Thrall, J. H. US–A–52
Thunberg, C. P. .. SE–A–1
Thurow, G. R. . US–DL–1
Tihen, J. A. US–A–18
Tilger, G. M. ... US–A–22
Tilley, S. G. US–A–93
Timken, R. L. . US–AA–2
Tinbergen, N. UK–F–1
Tinkle, D. W. ... US–A–67
Tipton, S. R. .. US–DH–1
Toft, C. A. US–DG–1
Tokarz, R. R. .. US–CA–1
Tollestrup, K. ... US–A–43
Tolson, P. J. US–A–29
Tomson, O. H. . US–A–83
Tordoff, W., III . US–A–59
Torrey, H. B. US–DJ–1
Touey, J. UK–C–3
Townsend, D. S. US–A–48
Tracy, C. R. US–A–75
Trama, F. B. ... US–DK–1
Tran, K. SU–A–4
Trapido, H. US–A–11
Trauth, S. E. US–E–8

Travis, J. A. US–A–72
Tremor, J. W. .. US–A–41
Troyer, K. E. ... US–DF–2
Trowbridge, M. US–CX–1
Trueb, L. US–A–69
Truslove, G. UK–D–1
Tsuji, J. S. US–C–12
Tuniyev, B. SU–A–4
Turner, F. B. US–A–43
Turner, J. E. ... US–DH–1
Turner, J. S. .. US–A–100
Turner, S. C. UK–C–2
Turpen, J. B. US–F–6
Tyler, J. D. US–D–5
Tyler, M. J. AU–J–1

- U -

Ultsch, G. US–AG–2
Underhill, D. K. . US–T–1
Underhill, J. US–DT–1
Urban, E. K. ... US–CQ–1
Urbanski, J. PO–H–1
Uzzell, T., Jr. ... US–A–38

- V -

Vaeth, R. H. US–E–9
Vaillant, L. FR–A–2
Valentin, P. M. .. SP–A–2
Valentine, B. D. US–AN–1
Van Berkum, F. US–C–12
Van Beurden, E. . US–AH–2
Van Denburgh, J.US–BL–1
Van Devender, R. US–Q–2
Van Dijk, D. E. .. SA–A–1
Van Tyne, J. US–A–16
Vanzolini, P. E. . US–C–5
Varkey, A. US–E–9
Verrell, P. A. UK–H–2
Versluys, J. AS–A–1
Vial, J. A. US–A–15
Villa, J. D. US–A–71
Vincente, L. A. . AR–A–4
Vinegar, A. US–A–66
Vinegar, M. E. . US–A–93
Vitt, L. J. US–A–53
Vliet, K. A. US–AK–2
Vogt, R. C. US–CV–1
Volpe, E. P. US–F–3
Volsøe, H. DE–B–1
Voris, H. K. US–A–49

- W -

Wacasey, J. W. .. US–H–4
Waddick, J. W. . US–CK–1
Wade, J. K. US–A–95
Wagner, W. US–A–99
Wake, D. B. US–A–15
Wake, M. H. ... US–A–15
Waldman, B. US–A–71
Walker, C. F. .. US–A–16
Walker, J. M. .. US–A–14
Walker, R. F. .. US–A–79
Wallace, B. US–M–2
Wallace, R. P. US–DM–1
Waltner, R. C. . US–A–13
Waring, H. W. ... AU–K–1
Warren, P. P. ... US–A–42
Wasserman, A. . US–A–25
Wassersug, R. .. US–A–31
Watanabe, M. E. US–A–68
Waters, J. F. US–A–45
Watson, D. M. ... UK–G–1
Watson, G. F. AU–K–6
Weatherby, C. US–E–8
Weathers, W. W. US–B–11
Weaver, W. G. .. US–A–73
Webb, R. G. US–A–13
Weigert, R. C. . US–AO–1
Weigmann, D. . US–BA–1
Weil, M. US–X–3
Weintraub, J. D. . US–F–4
Weiss, P. US–S–1
Weldon, P. J. US–I–3
Wells, K. D. US–A–76
Welter, W. A. US–H–2
Werner, J. K. ... US–A–91
Werner, Y. L. AS–A–2
Werth, R. J. US–A–14
Wharton, C. US–BE–1
Whitaker, J. O. US–A–22
White, F. N. US–G–3
White, T. US–DV–1
Whitear, M. UK–G–1
Whitford, W. G. US–A–66
Whittier, J. M. . US–AC–2
Whitman, C. O. . US–A–2
Wickham, M. G. US–A–82
Wiest, J. A. US–L–3
Wiewandt, T. US–H–3
Wilbur, H. M. . US–A–38
Wilczynskja, B. .. PO–A–6
Wilder, B. G. US–A–2
Wilder, C. D. US–E–4
Wilhoft, D. C. . US–A–43
Wilkinson, R. .. US–BB–1
Willard, D. E. .. US–DA–1

Williams, A. A. . US–G–5
Williams, E. E. . US–C–2
Williams, K. L. .. US–E–9
Williams, P. K. US–A–58
Williams, S. R. . US–X–2
Williams, T. A. US–A–82
Williamson, I. ... AU–K–4
Wilson, K. J. AU–K–5
Wilson, L. D. US–E–9
Wilson, E. O. . US–AM–1
Wilson, R. E. ... US–A–76
Wilson, V. V. .. US–A–98
Winokur, R. M. US–A–27
Witalinska, L. ... PO–A–5
Wood, R. C. US–C–6
Woods, J. E. US–P–3
Woods, W. H. . US–BH–1
Woodward, B. D. US–A–30
Woolbright, L. . US–A–48
Wortham, J., Jr. . US–G–5
Worthington, R. US–A–74
Wright, A. H. US–A–5
Wright, J. US–AH–2
Wright, J. W. ... US–A–41
Wright, S. J. US–D–3
Wygoda, M. L. US–AG–2
Wyles, J. S. US–C–11
Wylie, S. R. US–A–53
Wynands, H. E. .. NE–E–1

- Y -

Yanev, K. P. US–A–31
Yeatman, E. AU–K–4
Yedlin, I. N. ... US–A–106
Young, F. N. .. US–DL–1

- Z -

Zakon, H. US–AD–2
Zakrzewski, M. .. PO–A–3
Zalisko, E. J. ... US–A–82
Zamachowski, W. PO–A–3
Zaune, L. SU–A–4
Zellarius, A. SU–A–4
Zerba, K. E. ... US–A–104
Zettergren, L. D. . US–F–6
Zhao, E.-m. US–A–23
Ziswiler, V. SW–C–1
Zug, G. R. US–A–38
Zweifel, R. G. .. US–A–43
Zysk, A. PO–A–3

PUBLICATIONS OF THE
SOCIETY FOR THE STUDY OF AMPHIBIANS AND REPTILES

PUBLICATIONS OF THE SOCIETY may be purchased from the Publications Secretary, Robert D. Aldridge, Department of Biology, St. Louis University, St. Louis, Missouri 63103 U.S.A. (telephone: area code 315, 658-3900). Prices are effective through December 1989. Please make checks payable to "SSAR." Prices include shipping charges (book rate) within the USA; all overseas orders will be billed only for the *additional* postage charges in excess of domestic rates. Publications sent at customer's risk; however, packages can be insured at customer's cost. Overseas customers must make payment in USA funds, by International Money Order, or by MasterCard or VISA (in which case account number and expiration date must be provided). For those paying by credit card, a 4% bank charge will be added to your credit card account. Items marked "out-of-print" are no longer available.

CONTRIBUTIONS TO HERPETOLOGY
Book-length monographs, comprising taxonomic revisions, results of symposia, and other major works. Issued irregularly by subscription.

No. 1. *Reproductive Biology and Diseases of Captive Reptiles*, by James B. Murphy and Joseph T. Collins (eds.). 1980. Results of a Society-sponsored symposium, including papers by 37 authors. 287 p., illus. Paperbound $26.00.

No. 2. *The Turtles of Venezuela*, by Peter C. H. Pritchard and Pedro Trebbau. 1984. An exhaustive natural history, covering half of the turtle species of South America. 414 p., 48 color plates (25 watercolor portraits and 165 photographs of turtles and habitats), 16 maps. Regular edition, clothbound $45.00; patron's edition, two leatherbound volumes in slipcase, signed and numbered by authors and artist $300.00; set of 25 watercolor prints of turtles, on heavy paper stock and in protective wrapper $30.00.

No. 3. *Introduction to the Herpetofauna of Costa Rica / Introducción a la Herpetofauna de Costa Rica*, by Jay M. Savage and Jaime Villa R. 1986. Bilingual edition, in English and Spanish, with distribution checklist, bibliographies, and extensive keys. 220 p., illus. (one color plate). Clothbound $30.00.

No. 4. *Studies on Chinese Salamanders*, by Ermi Zhao, Qixiong Hu, Yaoming Jiang, and Yuhua Yang. 1988. Evolutionary review with keys, distribution maps. 80 p., 7 plates (including 10 color photographs of salamanders and habitats). Clothbound $12.00.

No. 5. *Contributions to the History of Herpetology*, by Kraig Adler, John S. Applegarth, and Ronald Altig. 1989. Biographies of 152 leading herpetologists (with portraits and signatures), index to 2500 authors in taxonomic herpetology, and academic lineages of 1300 herpetologists. Worldwide coverage. 202 p., 149 photographs, 1 color plate. Clothbound $20.00.

No. 6. *Snakes of the* Agkistrodon *Complex: A Monographic Review*, by Howard K. Gloyd and Roger Conant. 1989. Comprehensive treatment of 33 taxa of poisonous snakes included in four genera: *Agkistrodon* of Asia and America, *Calloselasma* of Southeast Asia and Java, *Deinagkistrodon* of China, and *Hypnale* of India and Sri Lanka. Also includes nine supplementary chapters by leading specialists. About 550 p., 33 color plates (247 photographs of snakes and habitats), 20 uncolored plates, 60 text figures, 28 maps. Clothbound $75.00.

FACSIMILE REPRINTS IN HERPETOLOGY
Photolithographic reprints of classic and important books and papers. Most titles typically have extensive new introductions by leading authorities. Issued irregularly by subscription.

ANDERSON, J. 1896. *Contribution to the Herpetology of Arabia*. Introduction and new checklist of Arabian amphibians and reptiles by Alan E. Leviton and Michele L. Aldrich. 160 p., illus. (one plate in color). Clothbound $25.00.

BELL, T. 1842-1843. *Herpetology of the "Beagle."* Part 5 of Darwin's classic, "Zoology of the Voyage of H.M.S. Beagle," containing descriptions of amphibians and reptiles collected on the expedition. Introduction by Roberto Donoso-Barros. 100 p., 20 plates. Paperbound $13.00, clothbound $18.00.

BOJANUS, L. H. 1819-1821. *Anatome Testudinis Europaeae*. The standard atlas of turtle anatomy. Introduction by Alfred Sherwood Romer. 200 p., 40 foldout plates. Out-of-print.

SOCIETY PUBLICATIONS

BOULENGER, G. A. 1877-1920. *Contributions to American Herpetology.* A series of collected papers from various journals; complete in 18 parts totalling 900 p., numerous illustrations. Introduction by James C. Battersby. Paperbound.
   Table of Contents: Parts 1-10, 1877-1897, 20 p., $2.00.
   Part 1: 1877-1881, 6 papers, 32 p., $4.00.
        2: 1882-1883, 12 papers, 48 p., $4.00.
        3: 1884-1885, 10 papers, 44 p., $4.00.
        4: 1886-1887, 10 papers, 56 p., $4.00.
        5: 1888, 13 papers, 32 p., $4.00.
        6: 1889-1890, 10 papers, 48 p., $4.00.
        7: 1891-1893, 11 papers, 44 p., $4.00.
        8: 1894, 5 papers, 40 p., $4.00.
        9: 1895-1896, 11 papers, 52 p., $4.00.
        10: 1897, 2 papers, 56 p., $4.00.
   Table of Contents: Parts 11-18, 1898-1920, 12 p. $2.00.
   Part 11: 1898, 4 papers, 52 p., $4.00.
        12: 1899-1900, 9 papers, 48 p., $4.00.
        13: 1901-1902, 7 papers, 34 p., $4.00.
        14: 1903-1904, 8 papers, 28 p., $4.00.
        15: 1905-1911, 12 papers, 42 p., $4.00.
        16: 1912-1913, 6 papers, 46 p., $4.00.
        17: 1914-1919, 8 papers, 38 p., $4.00.
        18: 1920, 3 papers, 64 p., $4.00.
   Index, 44 p., $2.00.
   Complete set: 18 parts plus Index and Tables of Contents, in parts as issued, $55.00.

BULLETIN OF THE ANTIVENIN INSTITUTE OF AMERICA. Volumes 1-5, 1927-1932. Complete set of an important journal devoted to venomology and to the taxonomy, biology, and distribution of snakes, lizards, salamanders, and crocodilians. Introduction by Sherman A. Minton. 575 p., 163 photographs, maps, index. Out-of-print.

CAMP, C. L. 1923. *Classification of the Lizards.* The foundation of modern lizard systematics. New preface by the author and an introduction by Garth Underwood. 220 p., 112 figures, index. Out-of-print.

CHANG, M. L. Y. 1936. *Amphibiens Urodèles de la Chine.* The classic work on Chinese salamanders, with a new checklist by Arden H. Brame. 168 p., illus., 5 plates. Paperbound $8.00.

COPE, E. D. 1864. *Papers on the Higher Classification of Frogs.* Reprinted from Proceedings of the Academy of Natural Sciences of Philadelphia and Natural History Review. 32 p. Paperbound $3.00.

COPE, E. D. 1871. *Catalogue of Batrachia and Reptilia Obtained by McNiel in Nicaragua; Catalogue of Reptilia and Batrachia Obtained by Maynard in Florida.* 8 p. Paperbound $1.00.

COPE, E. D. 1892. *The Osteology of the Lacertilia.* An important contribution to lizard anatomy, reprinted from Proceedings of the American Philosophical Society. 44 p., 6 plates. Paperbound $4.00.

COWLES, R. B. and C. M. BOGERT. 1944. *A Preliminary Study of the Thermal Requirements of Desert Reptiles.* The foundation of thermoregulation biology, with extensive review of recent studies by F. Harvey Pough. Reprinted from Bulletin of American Museum of Natural History. 52 p., 11 plates. Paperbound $5.00.

DUNN, E. R. 1926. *Salamanders of the Family Plethodontidae.* A recognized classic treatment of the plethodontid salamanders, including tropical and European species. Introductions by David B. Wake and Arden H. Brame. 480 p., illus., 3 plates, 86 maps, index. Paperbound $20.00.

ESCHSCHOLTZ, F. 1829-1833. *Zoologischer Atlas* (herpetological sections). Descriptions of new reptiles and amphibians from California and the Pacific. Introduction by Kraig Adler. 32 p., 4 plates. Paperbound $3.00.

ESPADA, M. J. de la. 1875. *Vertebrados del Viaje al Pacifico: Batracios.* Espada's major work on South American frogs. Introduction by Jay M. Savage. 208 p., 6 plates. Clothbound $20.00.

FITZINGER, L. J. 1843. *Systema Reptilium.* An important nomenclatural landmark for herpetology, including Amphibia as well as reptiles; world-wide in scope. Introduction by Robert Mertens. 128 p., index. Paperbound $15.00, clothbound $20.00.

GLOYD, H. K. 1940. *The Rattlesnakes, Genera* Sistrurus *and* Crotalus. Introduction and new checklist by Hobart M. Smith and Herbert M. Harris. 300 p., plus 31 plates of photographs, index. Clothbound $28.00.

GRAY, J. E. 1825. *A Synopsis of the Genera of Reptiles and Amphibia.* Reprinted from Annals of Philosophy. 32 p. Paperbound $3.00.

GRAY, J. E. 1831-1844. *Zoological Miscellany.* A privately printed journal, devoted mostly to descriptions of amphibians, reptiles and birds from throughout the world. Introduction by Arnold G. Kluge. 86 p., 4 plates. Paperbound $6.00, clothbound $10.00.

GÜNTHER, A. 1885-1902. *Biologia Centrali-Americana. Reptilia and Batrachia.* The standard work on Middle America, with 76 full-page plates (12 in full color). Introductions by Hobart M. Smith, A. E. Gunther, and Kraig Adler. 575 p. clothbound $50; separate set of 12 color plates, in protective wrapper $18.

HOLBROOK, J. E. 1842. *North American Herpetology.* Five volumes bound in one. The classic work by the father of American herpetology. This is an exact facsimile of the second edition, including all 147 plates (20 reproduced in full color). Introduction and checklists by Richard and Patricia Worthington and by Kraig Adler. 1032 p. Regular edition, clothbound $60.00; patron's edition, out-of-print.

JUNIOR SOCIETY OF NATURAL SCIENCES (CINCINNATI, OHIO). 1930-1932. Herpetological papers from the society's Proceedings, with articles by Weller, Walker, Dury and others. 56 p. Paperbound $3.00.

KIRTLAND, J. P. 1838. *Zoology of Ohio* (herpetological portion). 8 p. Paperbound $1.00.

SOCIETY PUBLICATIONS

LECONTE, J. E. 1824-1828. *Three Papers on Amphibians*, from the Annals of the Lyceum of Natural History, New York. 16 p. Paperbound $2.00.

LINNAEUS, C. 1766-1771. *Systema Naturae* (ed. 12) and *Mantissa Plantarum* (herpetological portions from both). 56 p. Introduction by Kraig Adler. Out-of-print.

LOVERIDGE, A. 1946. *Reptiles [and Amphibians] of the Pacific*. The standard review of the herpetofauna of the Pacific region including Australia and extending from Indonesia to Hawaii and the Galápagos Islands. 271 p., 7 plates, 1 double-page map, index. Paperbound $18.00.

McILHENNY, E. A. 1935. *The Alligator's Life History*. The most complete natural history of the American alligator. Introduction by Archie Carr and a review of recent literature by Jeffrey W. Lang. 125 p., 18 photographs and a portrait. Clothbound $20.00.

McLAIN, R. B. 1899. *Contributions to North American Herpetology*. Three privately printed papers containing important distributional records and the description of a new form. 28 p., index. Paperbound $2.00.

ORBIGNY, A. d' [and G. BIBRON]. 1847. *Voyage dans l'Amérique Méridionale*. This extract comprises the complete section on reptiles and amphibians. 14 p., 9 plates. $3.00.

RAFINESQUE, C. S. 1820. *Annals of Nature* (herpetological and ichthyological sections), 4 p. Paperbound $1.00.

RAFINESQUE, C. S. 1822. *On Two New Salamanders of Kentucky*. 2 p. Paperbound $1.00.

RAFINESQUE, C. S. 1832-1833. *Five Herpetological Papers from the Atlantic Journal*. 4 p. Paperbound $1.00.

SOWERBY, J. DeC., E. LEAR and J. E. GRAY. 1872. *Tortoises, Terrapins, and Turtles Drawn From Life*. The finest atlas of turtle illustrations ever produced. Introduction by Ernest E. Williams. 26 p., 61 full-page plates. Clothbound $25.00.

SPIX, J. B. von and J. G. WAGLER. 1824-1825. *Herpetology of Brazil*. The most comprehensive and important early survey of Brazilian herpetology. Introduction by P. E. Vanzolini. 400 p., 98 plates (one in color). Clothbound $36.00.

TROSCHEL, F. H. 1850 (1852). Cophosaurus texanus, *neue Eidechsengattung aus Texas*. 8 p. Paperbound $1.00.

TSCHUDI, J. J. von. 1838. *Classification der Batrachier*. A major work in systematic herpetology, with introduction by Robert Mertens. 118 p., 6 plates. Paperbound $18.00.

TSCHUDI, J. J. von. 1845. *Reptilium Conspectus*. Descriptions of new reptiles and amphibians from Peru. 24 p. Paperbound $2.00.

VANDENBURGH, J. 1895-1896. *Herpetology of Lower California*. A review of the herpetology of Baja California (collected papers). 101 p., 11 plates, index. Paperbound $8.00.

WIEGMANN, A. F. A. 1834. *Herpetologia Mexicana*. Introduction by Edward H. Taylor. 66 p., 10 plates (folio, measuring 10 x 14 inches). Out-of-print.

WILCOX, E. V. 1891. *Notes on Ohio Batrachians*. 3 p. Paperbound $1.00.

WILLISTON, S. W. 1925. *Osteology of the Reptiles*. Covers living and extinct forms, with introduction by Claude W. Hibbard. 304 p., 191 text figures, index. Paperbound $18.00.

WRIGHT, A. H. and A. A. WRIGHT. 1962. *Handbook of Snakes of the United States and Canada, Vol. 3, Bibliography*. Out-of-print since about 1969, this bibliography is a necessary companion to Volumes 1 and 2. 187 p. Clothbound $18.00.

HERPETOLOGICAL REVIEW

The Society's official newsletter, international in coverage. Most issues contain extensive listings of current literature, a section that from 1983 became a separate serial, *Recent Herpetological Literature*. Issued quarterly as part of Society membership. All numbers paperbound as issued.

Volume 1, number 1 (1967), $4.00.
number 2 (1968), $4.00.
number 3 (1968), $4.00.
number 4 (1968), $4.00.
number 5 (1968), $4.00.
number 6 (1969), $4.00.
number 7 (1969), out-of-print.
number 8 (1969), $4.00.
number 9 (1969), $4.00.
Volume 2, number 1 (1970), $4.00.
number 2 (1970), $4.00.
number 3 (1970), $4.00.
number 4 (1970), $4.00.
Volume 3, number 1 (1971), $4.00.
number 2 (1971), out-of-print.
number 3 (1971), $4.00.
number 4 (1971), $4.00.
number 5 (1971), $4.00.
number 6 (1971), $4.00.
Volume 4, number 1 (1972), out-of-print.
number 2 (1972), $4.00.
number 3 (1972), $4.00.
number 4 (1972), $4.00.
number 5 (1972), $4.00.
number 6 (1972), $4.00.
Volume 5, number 1 (1974), out-of-print.
number 2 (1974), out-of-print.
number 3 (1974), $4.00.
number 4 (1974), $4.00.
Volume 6, number 1 (1975), out-of-print.
number 2 (1975), $4.00.
number 3 (1975), $4.00.
number 4 (1975), out-of-print.
Volume 7, number 1 (1976), $4.00.

SOCIETY PUBLICATIONS

        number 2 (1976), $4.00.
        number 3 (1976), out-of-print.
        number 4 (1976), out-of-print.
Cumulative Index for Volumes 1-7 (1967-1976), $4.00.
Volume 8, number 1 (1977), $4.00.
        number 2 (1977), $4.00.
        number 3 (1977), $4.00.
        number 4 (1977), $4.00.
Volume 9, number 1 (1978), $4.00.
        number 2 (1978), $4.00.
        number 3 (1978), $4.00.
        number 4 (1978), $4.00.
Volume 10, number 1 (1979), $4.00.
        number 2 (1979), out-of-print.
        number 3 (1979), $4.00.
        number 4 (1979), $4.00.
Volume 11, number 1 (1980), $4.00.
        number 2 (1980), $4.00.
        number 3 (1980), $4.00.
        number 4 (1980), $4.00.
Volume 12, number 1 (1981), $4.00.
        number 2 (1981), $4.00.
        number 3 (1981), $4.00.
        number 4 (1981), $4.00.
Volume 13, number 1 (1982), $4.00.
        number 2 (1982), $4.00.
        number 3 (1982), $4.00.
        number 4 (1982), out-of-print.
Volume 14, number 1 (1983), out-of-print.
        number 2 (1983), $4.00.
        number 3 (1983), $4.00.
        number 4 (1983), $4.00.
Volume 15, number 1 (1984), $4.00.
        number 2 (1984), $4.00.
        number 3 (1984), $4.00.
        number 4 (1984), $4.00.
Volume 16, number 1 (1985), $4.00.
        number 2 (1985), out-of-print.
        number 3 (1985), $4.00.
        number 4 (1985), $4.00.
Volume 17, number 1 (1986), $4.00.
        number 2 (1986), $4.00.
        number 3 (1986), $4.00.
        number 4 (1986), $4.00.
Volume 17, number 1 (1986), $4.00.
        number 2 (1986), $4.00.
        number 3 (1986), $4.00.
        number 4 (1986), $4.00.
Index to Geographic Distribution Records for Volumes 1-17 (1967-1986), 44 pages, $5.00.
Volume 18, number 1 (1987), $4.00.
        number 2 (1987), $4.00.
        number 3 (1987), $4.00.
        number 4 (1987), $4.00.
Volume 19, number 1 (1988), $4.00.
        number 2 (1988), $4.00.
        number 3 (1988), $4.00.
        number 4 (1988), $4.00.

CATALOGUE OF AMERICAN AMPHIBIANS AND REPTILES

Loose-leaf accounts of taxa prepared by specialists, including synonymy, definition, description, distribution map and comprehensive list of literature for each taxon. Covers amphibians and reptiles of the entire Western Hemisphere. Issued by subscription, although individual accounts or groups of accounts may be purchased.

CATALOGUE ACCOUNTS: Numbers 1-25, 26-50, 51-75, 76-100, 101-125, 126-150, 151-175, 176-190, 191-205, 206-220, 221-245, 246-270, 271-295, 296-320, 321-345, 346-370, 371-390, 391-410, 411-435, and 436-460. Price for each group of numbered accounts is $20.00. NOTE: Discount of 20% if 3-9 years of accounts are purchased at same time, 30% discount if 10 years or more. Contact the Publications Secretary for details.

INDIVIDUAL ACCOUNTS: $0.50 each (please order by account number). In a number of instances only xerox copies of *individual* accounts can be supplied. However, complete sets and the groups of numbered accounts listed above are all originals.

INDEX TO ACCOUNTS 1-400: Cross-referenced, 64 pages, $5.00.

IMPRINTED POST BINDER: $20.00.

SYSTEMATIC TABS (Ten tabs to fit binder: "Class Amphibia," "Order Caudata," etc.), $4.00.

HERPETOLOGICAL CIRCULARS

Miscellaneous publications of general interest to the herpetological community. Issued irregularly and by subscription. All numbers paperbound, as issued.

No. 1. *A Guide to Preservation Techniques for Amphibians and Reptiles* by George R. Pisani. 1973. 22 p., illus. $3.00.

No. 2. *Guía de Técnicas de Preservación de Anfibios y Reptiles* por George R. Pisani y Jaime Villa. 1974. 28 p., illus. Out-of-print.

No. 3. *Collections of Preserved Amphibians and Reptiles in the United States* compiled by David B. Wake (chair) and the Committee on Resources in Herpetology. 1975. 22 p. Out-of-print.

No. 4. *A Brief Outline of Suggested Treatments for Diseases of Captive Reptiles* by James B. Murphy. 1975. 13 p. $3.00.

No. 5. *Endangered and Threatened Amphibians and Reptiles in the United States* compiled by Ray E.

SOCIETY PUBLICATIONS

Ashton, Jr. (chair) and the 1973-74 SSAR Regional Herpetological Societies Liaison Committee. 1976. 65 p. Out-of-print.

No. 6. *Longevity of Reptiles and Amphibians in North American Collections* by J. Kevin Bowler. 1977. 32 p. $3.00.

No. 7. *Standard Common and Current Scientific Names for North American Amphibians and Reptiles* (1st ed.) by Joseph T. Collins, James E. Huheey, James L. Knight and Hobart M. Smith. 1978. 36 p. $3.00.

No. 8. *A Brief History of Herpetology in North America Before 1900* by Kraig Adler. 1979. 40 p., 24 photographs, 1 map. $3.00.

No. 9. *A Review of Marking Techniques for Amphibians and Reptiles* by John W. Ferner. 1979. 42 p., illus. $3.00.

No. 10. *Vernacular Names of South American Turtles* by Russell A. Mittermeier, Federico Medem and Anders G. J. Rhodin. 1980. 44 p. $3.00.

No. 11. *Recent Instances of Albinism in North American Amphibians and Reptiles* by Stanley Dyrkacz. 1981. 36 p. $3.00.

No. 12. *Standard Common and Current Scientific Names for North American Amphibians and Reptiles* (2nd ed.) by Joseph T. Collins, Roger Conant, James E. Huheey, James L. Knight, Eric M. Rundquist, and Hobart M. Smith. 1982. 32 p. $3.00.

No. 13. *Silver Anniversary Membership Directory*, including addresses of all SSAR members, addresses and publications of the herpetological societies of the world, and a brief history of the Society. 1983. 56 p., 4 photographs. $3.00.

No. 14. *Checklist of the Turtles of the World with English Common Names* by John Iverson. 1985. 14 p. $3.00.

No. 15. *Cannibalism in Reptiles: A World-wide Review* by Joseph C. Mitchell. 1986. 37 p. $4.00.

No. 16. *Herpetological Collecting and Collections Management* by John E. Simmons. 1987. 72 p., 6 photographs. $6.00.

No. 17. *An Annotated List and Guide to the Amphibians and Reptiles of Monteverde, Costa Rica* by Marc P. Hayes, J. Alan Pounds, and Walter W. Timmerman. 1989. 45 p., 32 photographs. $5.00.

RECENT HERPETOLOGICAL LITERATURE

A comprehensive listing of titles and citations, with authors' addresses. Prior to 1983 these lists were published in *Herpetological Review*. Issued irregularly as part of Society membership. Paperbound, as issued. Series now suspended.

Number 1 (1983), 66 p., $3.00.
      2 (1984), 44 p., $3.00.

JOURNAL OF HERPETOLOGY

The Society's official scientific journal, international in scope. Issued quarterly as part of Society membership. All numbers paperbound as issued. Tables of Contents are supplied *gratis* with orders for complete volumes.

Volume 1, numbers 1-4 (1968), out-of-print.
Volume 2, numbers 1-2 (1968), $6.00.
         numbers 3-4 (1968), out-of-print.
Volume 3, numbers 1-2 (1969), out-of-print.
         numbers 3-4 (1969), $6.00.
Volume 4, numbers 1-2 (1970), out-of-print.
         numbers 3-4 (1970), out-of-print.
Volume 5, numbers 1-2 (1971), out-of-print.
         numbers 3-4 (1971), out-of-print.
Volume 6, number 1 (1972), out-of-print.
         number 2 (1972), out-of-print.
         numbers 3-4 (1972), out-of-print.
Volume 7, number 1 (1973), out-of-print.
         number 2 (1973), $6.00.
         number 3 (1973), $6.00.
         number 4 (1973), out-of-print.
Volume 8, number 1 (1974), out-of-print.
         number 2 (1974), $6.00.
         number 3 (1974), $6.00.
         number 4 (1974), $6.00.
Volume 9, number 1 (1975), out-of-print.
         number 2 (1975), out-of-print.
         number 3 (1975), $6.00.
         number 4 (1975), out-of-print.
Volume 10, number 1 (1976), $6.00.
         number 2 (1976), $6.00.
         number 3 (1976), $6.00.
         number 4 (1976), $6.00.
Cumulative Index for Volumes 1-10 (1968-1976), $7.00.
Volume 11, number 1 (1977), $6.00.
         number 2 (1977), $6.00.
         number 3 (1977), $6.00.
         number 4 (1977), out-of-print.
Volume 12, number 1 (1978), out-of-print.
         number 2 (1978), out-of-print.
         number 3 (1978), $6.00.
         number 4 (1978), $6.00.
Volume 13, number 1 (1979), $6.00.
         number 2 (1979), $6.00.
         number 3 (1979), $6.00.
         number 4 (1979), $6.00.
Volume 14, number 1 (1980), $6.00.
         number 2 (1980), $6.00.
         number 3 (1980), $6.00.
         number 4 (1980), $6.00.
Volume 15, number 1 (1981), $6.00.
         number 2 (1981), $6.00.
         number 3 (1981), $6.00.
         number 4 (1981), $6.00.
Volume 16, number 1 (1982), $6.00.
         number 2 (1982), $6.00.
         number 3 (1982), $6.00.

SOCIETY PUBLICATIONS

number 4 (1982), $6.00.
Volume 17, number 1 (1983), $6.00.
number 2 (1983), $6.00.
number 3 (1983), $6.00.
number 4 (1983), $6.00.
Volume 18, number 1 (1984), $6.00.
number 2 (1984), $6.00.
number 3 (1984), $6.00.
number 4 (1984), $6.00.
Volume 19, number 1 (1985), $6.00.
number 2 (1985), $6.00.
number 3 (1985), $6.00.
number 4 (1985), $6.00.
Volume 20, number 1 (1986), $6.00.
number 2 (1986), $6.00.
number 3 (1986), $6.00.
number 4 (1986), $6.00.
Volume 21, number 1 (1987), $6.00.
number 2 (1987), $6.00.
number 3 (1987), $6.00.
number 4 (1987), $6.00.
Volume 22, number 1 (1988), $6.00.
number 2 (1988), $6.00.
number 3 (1988), $6.00.
number 4 (1988), $6.00.

PUBLICATIONS OF THE OHIO HERPETOLOGICAL SOCIETY

OHS was the predecessor to the Society for the Study of Amphibians and Reptiles. All publications international in scope. Paperbound as issued.

Volume 1, numbers 1-4, plus Special Publications 1-2 (all 1958), facsimile reprint, out-of-print.
Volume 2, number 1 (1959), $2.00.
number 2 (1959), $2.00.
number 3 (1960), out-of-print.
number 4 (1960), out-of-print.
Volume 3, number 1 (1961), out-of-print.
number 2 (1961), $1.00.
number 3 (1962), out-of-print.
number 4 (1962), $1.00.
Volume 4, numbers 1-2 (1963), $4.00.
number 3 (1964), out-of-print.
number 4 (1964), out-of-print.
Volume 5, number 1 (1965), $2.00.
number 2 (1965), $2.00.
number 3 (1966), $2.00.
number 4 (1966), $2.00.
Special Publication 3 (1961), out-of-print.
Special Publication 4 (1962), $2.00.

OTHER MATERIALS AVAILABLE FROM THE SOCIETY

Color Prints

FIJI IGUANAS. Two full-color prints (7 x 10 inches) of *Brachylophus fasciatus* and *B. vitiensis*, as published in *Journal of Herpetology*, vol. 15, no. 3, July 1981. Special printing limited to 50 sets, individually signed by the artist David M. Dennis. Set of two, $15.00.

SILVER ANNIVERSARY COMMEMORATIVE PRINT. A full-color print (11 1/2 x 15 1/4 inches) of a Gila Monster (*Heloderma suspectum*) on natural background, from a watercolor by David M. Dennis. Issued as part of Society's 25th Anniversary in 1982. Edition limited to 1000. $8.00 each or $5.00 in quantities of 10 or more.

WORLD CONGRESS COMMEMORATIVE PRINT. A full-color print (11 1/2 x 15 inches) of an Eastern Box Turtle (*Terrapene carolina*) in a natural setting, from a watercolor by David M. Dennis. Issued as part of SSAR's salute to I World Congress of Herpetology, held at Canterbury, United Kingdom, in 1989. Edition limited to 1500. $8.00 each or $5.00 in quantities of 10 or more.

Brochures

HERPETOLOGY AS A CAREER, by Henri C. Seibert, Ralph W. Axtell, Neil B. Ford, and Martin J. Rosenberg. 1985. 4 p. Developed as an aid to students and counselors. Single copies free of charge; additional copies available for $0.25 each.

GUIDELINES FOR THE USE OF LIVE AMPHIBIANS AND REPTILES IN FIELD RESEARCH, by George R. Pisani, Stephen D. Busack, Herbert C. Dessauer, and Victor H. Hutchison, representing a joint committee of ASIH, HL, and SSAR. 1987. 16 p. Covers animal care, regulations, collecting, restraint and handling, marking, housing and maintenance in field, and final disposition of specimens. $4.00 ($3.00 each on orders of five or more copies).

PRODUCTION SPECIFICATIONS

Typesetting: Cornell University, Ithaca, New York, U.S.A. (Dawn Potter, Carol Abarbanell). Type is set in Times face.

Printing and Binding: Thomson-Shore Inc., Dexter, Michigan, U.S.A. (Ned Thomson). The text is printed on 60-pound Warren's Patina stock which is acid- and groundwood-free.

Color Frontispiece: Glundal Color Inc., East Syracuse, New York, U.S.A. (Douglass B. Selover). The frontispiece was laser scanned from a color transparency and is printed on 80-pound acid-free Lustro Offset Enamel dull text stock.

Date and Place of Publication: 11 September 1989, Oxford, Ohio, U.S.A.

Number of Copies: 2000.